Innovative Economic Symposium (IES2021)

New Trends in Business and Corporate Finance in COVID-19 Era

SHS Web of Conferences Volume 132 (2022)

Seoul, South Korea
11 November 2021

Editors:

Jakub Horak
Taehee Lee
Eva Dolejsova

ISBN: 978-1-7138-4210-1

Printed from e-media with permission by:

Curran Associates, Inc.
57 Morehouse Lane
Red Hook, NY 12571

Some format issues inherent in the e-media version may also appear in this print version.

For additional information, please contact EDP Sciences – Web of Conferences
at the address below.

EDP Sciences – Web of Conferences
17, Avenue du Hoggar
Parc d'Activité de Courtabœuf
BP 112
F-91944 Les Ulis Cedex A
France

Phone: +33 (0) 1 69 18 75 75
Fax: +33 (0) 1 69 28 84 91

contact-edps@webofconferences.org

Additional copies of this publication are available from:

Curran Associates, Inc.
57 Morehouse Lane
Red Hook, NY 12571 USA
Phone: 845-758-0400
Fax: 845-758-2633
Email: curran@proceedings.com
Web: www.proceedings.com

TABLE OF CONTENTS

Innovative Economic Symposium 2021

New Trends in Business and Corporate Finance in COVID-19 Era

Innovative Economic Symposium 2021

New Trends in Business and Corporate Finance in COVID-19 Era

http://ies.vstecb.cz/

Editors: Jakub Horák, Taehee Lee, Eva Dolejšová

About the conference IES 2021

Date of the conference: 2021-11-11

The conference wa be held under the auspices of the Rector of The Institute of Technology and Business in České Budějovice, Vojtěch Stehel, MSc, MBA, PhD.

Motto

"Coming together is a beginning; keeping together is progress; working together is success." Henry Ford

Target

The IES 2021 conference brings together scientists and practitioners in economy, international relations, finance and marketing from Asia and Europe. It aims to collect innovative ideas about new trends in business and corporate finance in the current situation of the world economy.

Thematic Areas

- Digital Innovation Development
- Shocks in investment flows; investment decision-making
- Healthcare industry development in relation to the Coronavirus pandemics
- Capital market shocks
- Paradigm shifts in Business Management
- Adaptation of companies to process and robotic automation
- Monitoring of the financial health of businesses
- Changes in international trade
- Progress of macroeconomical variables
- Prediction of the future development of business and corporate finance

About organizers

Kookmin University

Kookmin University, the first private university after Korea's independence, was established by Mr. Shin, Ik Hee and other main members of the Shanghai interim government of Korea. Deeply rooted in the four founding principles, humanism, nationalism, culturalism, and industrialism that Mr. Kim, Sung Gon had emphasized, Kookmin University has grown into a fixture of the academia of Korea with 24,000 students enrolled. Now universities of a knowledge based society have entered into a competition that is more severe than ever before. KMU is trying to solidify its educational foundation to maintain its role as a leading player by employing operative goals. The three operative goals are attracting outside funding, establishing the standards of reasonable distribution of resources, and building an efficient structural system. More information about Kookmin University at https://english.kookmin.ac.kr/.

Institute of Technology and Business in České Budějovice (VŠTE)

The primary focus of VŠTE activities is the implementation of full-time and part-time studies for professional bachelor's and master's degree programmes. The professional study programmes are supported by cooperation with regional entrepreneurs. VŠTE´s other priorities include applied research, the implementation of innovative technologies and knowledge transfer into areas that are current and crucial with regards to the regions needs. More information about VŠTE at www.vstecb.cz.

North China University of Technology (NCUT)

A former active partner North China University of Technology (NCUT), located on the western side of Beijing, is a multi-disciplinary university that combines the natural sciences and engineering with liberal arts, economics, management and law. More information about NCUT at www.ncut.ciss.org.cn.

Samara State University of Economics (SSEU)

The SSEU is the leading university in the Volga region. A practical-oriented approach in teaching is implemented in the University. The cutting-edge educational technologies, including distance learning technology and e-learning are widely used in SSEU. More information about Samara State University of Economics at http://www.sseu.ru/en.

Henan University

Developing for nearly 110 years, HENU is a comprehensive university with 12 academic fields of study covering Humanities, History, Philosophy, Economy, Management, Law, Science, Engineering, Medicine, Agriculture, Education and Arts. More information about Henan University at http://iao.henu.edu.cn/yw/Home.htm.

Conference report

In the light of the ongoing difficult global situation caused by the COVID-19 pandemics, this year, the conference took place online again using the communication platform ZOOM. The whole symposium was greatly organized by the team of organizers from Kookmin University, South Korea and from the Institute of Technology and Business, the Czech Republic.

On 11th November 2021, academics from Russia, South Korea, and the Czech Republic joined the virtual meeting to present their research and exchange expert knowledge.

This year, the thematic focus was New Trends in Business and Corporate Finance in COVID-19 Era which allowed describing the changes that occurred during the whole year in everybody's life.

The conference started with opening speech of prof. Ing. Marek Vochozka, MBA, Ph.D., dr. h.c., Rector emeritus of the Institute of Technology and Business in České Budějovice (VŠTE) and Prof. Hong Jae Yim, Ph.D., president of the Kookmin University in Seoul, South Korea, the main host of this year's conference. The conference programme consisted of two parallel plenary sessions. All presented contributions attracted the attention of wide audience. Each participant received The Book of Abstracts containing brief descriptions of the presented contributions.

Expert guarantor

prof. Marek Vochozka
Rector of the Institute of Technology and Business (CZE)

prof. Taehee Lee
Chair of International School, Kookmin University (ROK)

prof. Zhao Jixin
Dean of the School of Economics and Management, North China University of Technology (CHN)

prof. Svetlana I. Ashmarina
Samara State University of Economics (RUS)

prof. Bingtao Song
Dean of School of Economics, Henan University, China

The Organizing Committee

The organizers in the host country (South Korea):

Prof. Najung Kim - Kookmin University, South Korea

Prof. Taehee Lee, Ph.D. - Dean, Kookmin University, South Korea

Prof. Jaekyung Yi, Ph.D. - Kookmin University, South Korea

Prof. Daeheon Choi, Ph.D. - Kookmin University, South Korea

Organizers in the Czech Republic:

Eva Dolejšová, MS, PhD - ITB in Ceske Budejovice (CZE)

Jakub Horák, MSc, MBA - ITB in Ceske Budejovice (CZE)

Simona Hašková, MSc, PhD - ITB in Ceske Budejovice (CZE)

The Scientific Committee

Statement of Peer review

In submitting conference proceedings to *Web of Conferences*, the editors of the proceedings certify to the Publisher that

1. They adhere to its **Policy on Publishing Integrity** in order to safeguard good scientific practice in publishing.
2. All articles have been subjected to peer review administered by the proceedings editors.
3. Reviews have been conducted by expert referees, who have been requested to provide unbiased and constructive comments aimed, whenever possible, at improving the work.
4. Proceedings editors have taken all reasonable steps to ensure the quality of the materials they publish and their decision to accept or reject a paper for publication has been based only on the merits of the work and the relevance to the journal.

Title, date and place of the conference

Innovative Economic Symposium 2021: New Trends in Business and Corporate Finance in COVID-19 Era.
November 11, 2021, (Seoul, South Korea)

Proceedings editor(s):

Jakub Horák, MSc, MBA
prof. Taehee Lee
Eva Dolejšová, MS, PhD

Date and editor's signature

The Knowledge Based View of Embedding Strategy for Platforms Enterprises**

Chen-Guang Li[123], and *Zhen-Jun* Qiu[3*]

[1]Business Management Department, 100144 Shijingshan Distinct, Beijing, China
[2]Enterprise Management Intelligent Decision Research Centre, 100144 Shijingshan Distinct, Beijing, China
[3]Department of Administration Management, School of Economics & Management, North China University of Technology, Beijing, China

Abstract. Dialectically linking with heterogenous suppliers, manufacturers, consumers and resources, emerging e-commerce platforms achieve and maintain competitive advantage. Based on different levels of innovation network content and structure, YEATION and YOUPIN, two typical e-commerce platforms, adopt different strategies involving the promotion of value-seeking. Many decisions concern how companies perceive the innovation environment and how to engage in the innovation network. Knowledge base view provides useful theoretical lens to understand the network embedding approach for electric business platform enterprises. To put forward, enterprises with wide breadth should adopt structural strategy give priority to relational embeddedness, on the other hand, enterprises of which the depth of knowledge base is deep and width of insufficient shall adopt strategy give priority to the embedding of structure. It is also proposed that platform enterprises should not adopt a single embeddedness strategy but should adopt auxiliary embeddedness strategy in a timely manner.
Keywords: network strategies for SMEs, competitiveness, innovation, value creation, stakeholders

1 Introduction

As Chesbrough suggested [1], the pathway products innovation and promotion have been undergoing a reform during the past decades, manufacturers pursuing a expansion following the value chain. The traditional model of enterprises completing development, manufacturing, marketing, supply and service independently becoming more and more difficult, especially for latecomers in a network circumstance. Prior literature has developed multifarious conceptual models to illustrate the pathway, for example, the "flying geese" paradigm, the "imitation to innovation" transformation [2], among which the track from original equipment manufacturing (OEM) to original design manufacturing (ODM) and original brand manufacturing (OBM) has been one of the most interested among scholars [3,

* Corresponding author: chilqzj@163.com
** This paper is funded by the Ministry of Education of Humanities and Social Science project (No. 20YJCZH066).

4, 5]. To progress to OBM, manufacturers need to possess the skill and resource of marketing [6, 7].

Under innovation network background, internet companies adopt quality e-commerce, which can be seen as integrated vertical e-commerce, and achieve building industrial barriers, competitive brands, and embedding in innovation network by cultivating ODMs into OBMs. In the last decade, many quality e- commerces emerged and achieved financial returns, for instance, YANXUAN, YOUPIN, XINXUAN, JINGZAO, and so on. This paper study how quality e-commerces integrate and exploit knowledge base of its OBMs in embedding innovation network.

This article selected YANXUAN and YOUPIN as typical quality e-commerces, aiming at contributing to the debate on the relationships between innovation, industrial dynamics and industry evolution, by analyzing the links between the knowledge-base and embeddedness.

2 Research background

2.1 Knowledge base

The innovation process was defined as a search process of inventions generated from recombination of existing knowledge elements novelty [8]. In this search process, the knowledge couplings a firm can reach, which is merely a small subset of the knowledge arena, is defined as knowledge base [9].

Firms' knowledge bases drew from diverse resources inside and outside of the organization enable them to establish and consolidate a privileged niche position. Focal companies should examine their characteristics of relevant knowledge accumulation to leverages the existing capabilities. The expansion of knowledge base can be divided into 'widening' and 'deepening'. The former is associated with knowledge searching distally across disciplines or fields, while the latter leads a company to cultivate capabilities in fields it specialized in. Existing study revealed that depth and breadth of knowledge base are important for firm performance and innovation [10, 11]. More specifically, empirical Study has proposed that firms' knowledge acquisition and subsequent innovative process were significantly influenced by their knowledge base characteristic [12]. Knowledge base can be both premises and constraint of particular innovation process. Zhou and Li proposed that firms with a broad knowledge base tend to achieve radical innovation through internal knowledge sharing rather than market knowledge acquisition, while firms with great knowledge base depth are more capable of developing in opposite way [13]. Carnabuci and Operti indicated that knowledge of great breadth hinders firm's recombinant reuse and improve its ability to innovate through recombinant creation, knowledge base depth influencing on the opposite way [14]. Prior Study provides evidence regarding the relationship between change in knowledge base characteristics and industry structure in industries like pharmaceutical [15].

2.2 Embeddedness

In a context of innovation network, enterprises benefit from mutually interactions with suppliers, manufacturers and consumers for joint value creation and innovation. Twenty years after the concept of embeddedness introduced into social network, it has developed two main perspectives: structural and relational. Structural embeddedness shows the impact of a firm's relative position in the network has on performance. Structural Perspective emphasizes properties of ego-network (focal firm) or whole network such as structural hole [16], density

[17], centrality [18], size [19]. Relational embeddedness reflects the degree of trust and information exchange between innovators in a network, usually focus on the strong ties and weak ties within network.

Study of knowledge base and properties of embeddedness about what influence they have on performance show a tendency toward contingency. Characteristics of network have complex influence on performance [20, 21]. This study attempts to investigate the embedding path selection of firms under different knowledge bases through comparative case studies.

3 Case study

3.1 Sample and data collection

This paper involves the selection of strategies for external network embedding based on the internal knowledge base of e-commerce platforms, selecting comparative case study as study method. Of all the quality e-commerces in China, we chose typical and relatively mature ones as case firms, YANXUAN and YOUPIN, who both established recognizable brand styles while of distinctly different knowledge base.

The data sources for this paper are mainly from e-commerce platform websites and disclosures, reports and interviews collected from the Internet. The basic information is as follows.

3.2 YANXUAN

As for the upstream of the value chain, YANXUAN adopts a mechanism called ODM buyer, to eliminate intermediate agents and to exert control over raw materials, production, QA and post-sale. The buyers select and contract with excellent ODMs, especially manufacturers serving top-tier international brands including that of Muji, Uniqlo, Samsonite, Hermes, Shiseido, CK, Nine West, Armani, Ralph Luaren, Chloe, Timberland, Fendi, Stoneline, Coach, HUGO BOSS, Under Armour and so on.

In the midstream, YANXUAN utilizes flexible supply chain to cope with expanding categories. Within six mouths from the launch of YANXUAN, its products include nine primary categories, 52 subcategories and 592 SKUs (Stock Keeping Unit), with is already wide coverage compared to general start-up e-commerces. One year later, SKUs climbed up to over 10000 in a surprising speed. Unlike other large e-commerce platforms' "space for time" solution of dense warehousing network, YANXUAN's flexible supply chain subdivides production plans into batches of smaller quantity. The launch of new products firstly commissioned a few batches, additional orders were adjusted sensitively according to the front-line sales volume. In an emerging of hot cake, it can also leverage control over supplier to fully utilize the production line and meet market demand timely.

From the sales side of the platform, YANXUAN cultivates the user group with the brand concept of "good life, not so expensive". As Walsh and Mitchell argued [22], with the introduction of high quality labeled private merchandise, the confusion proneness associated with quality and emotions have significant impact on marketplace. By combining the user group portraits of Netease Open Class and Netease Cloud Music, YANXUAN targets its core user group at people aged 20-35 years old who seek to improve their quality of life but have limited purchasing power and define the brand style as a mid-range brand rich in artistic temperament, pursuing sophistication and advocating lifestyle aesthetics.

3.3 YOUPIN

YOUPIN intervenes in the upstream supply chain through investment and incubation. Backed on Xiaomi's mobile phone and IOT (Internet of Things), which has been the largest smart hardware platform worldwide, YOUPIN incorporates cooperative enterprises into the Xiaomi ecological chain, exerting values and participating in controlling design concepts, product definition, industrial design and other aspects.

YOUPIN differ from other quality e-commerces mainly in three principles of investment and incubation. Firstly, it cultivates ODMs by exporting enterprise management methodology and product development methodology, providing resources like channels, clients and even supply chain and moreover, values. Secondly, it has skin in the game with ecosystem companies through equity participation but not holding in order to maintain the combativity and innovation of startup teams. Finally, manufacturer can compete with products of YOUPIN. As Hoch and Banerji proposed [23], private label goods should complement, not replace, the manufacturer's brand; that the right choice is for retailers to offer high quality and relatively low prices; and that private labels should try to avoid face-to-face competition with manufacturer-branded goods. Typical example is Huami, being the world's leading wearable maker, differentiating its main product AMAZFIT series from Xiaomi bracelets. Most different from YANXUAN, YOUPIN strives to support manufacturers in establishing their own independent brands, and the eco-friendly Huami, Zmi, Yeelight, Smartmi, Aqara, etc are already at the forefront of their respective industries.

At the horizontal platform level, YOUPIN takes the lead in getting manufacturers to support reciprocally. Zmi, after becoming a leader in the mobile power industry, gradually provides battery for other ecological enterprises and help them improve power management solutions, getting stable customers in return.

3.4 Knowledge base

This paper, specific to the study of quality e-commerces, illustrated the knowledge base as primarily the ability to leverage its own brand resources and select manufacturers as well as products. This paper argues that YANXUAN is superior in terms of knowledge base breadth, while YOUPIN advanced in knowledge base depth.

As NetEase, the parent company of YANXUAN, has accumulated experience in grasping user groups, Internet marketing and establishing brand tone through its community-based products such as open classes and music while lacking experience about goods. On the other hand, the ability to select products need to be more proficient in the mastery of the manufacturing field, YOUPIN's knowledge from Xiaomi of the field of technology products has a more advantageous, more competitive than YANXUAN of the household merchandise.

3.5 Embeddedness

YANXUAN adopted an embedding strategy that was first structural then relational, but did not achieve relational embedding successfully. After scaling back operations， YANXUAN still has far more SKUs on display than its competitors. In spite of strong supply chain management capabilities, the relationship between YANXUAN and ODM manufacturers stays in the contract.

The rapid expansion of the product system makes YANXUAN quickly occupy the central position of the ODM network, but the speed of experience accumulated by the category team can not keep up with the speed of SKU expansion, and can not rely on the relationship established with manufacturers to achieve complementary experience, the control of suppliers is insufficient. In addition, the product itself is neglected and the quality control capability is lost. Limit to contractual relationship, YANXUAN cannot control the manufacturer's R&D activities, and ODM only provides homogeneous products according orders, which to some extent constraints the manufacturer's ability to stimulate innovation.

YANXUAN manages suppliers and a large number of SKUs through Tongzhou, an intelligent cloud collaboration platform for suppliers, which enables sharing of raw material stocking plans, production progress, inventory sharing, standardized quotation process, raw material price reference, as well as supplier optimization and assessment based on the above functions to improve the daily management of suppliers' sales. Tongzhou enables YANXUAN get in touch with ODMs as many as it want. Despite the brand's clear style, there is not too strong a definition of the user group. YANXUAN has opened a product recommendation of styles from Japan, France, Thailand, New Zealand, Vietnam, and Korea, and works with a large number of professional designers and design companies at home and abroad.

YOUPIN, on the other hand, chose a relational to structural embeddedness strategy. About three quarter of the suppliers on YOUPIN come from Xiaomi or its ecosystem companies, and YOUPIN has established close ties with them in many ways, including R&D, patents, sales, QA, inventory, logistics, accounting processing, and IOT ecology etc.

Constrained by the brand tone of the Xiaomi brand, it is difficult to expand consumer base and find added value in terms of emotional value. In the face of surging demand, YOUPIN is incapable of expanding its subcategories with guaranteeing quality.

Though YOUPIN's initial brand position of pursuing cost-effective and IOT ecology makes its consumer groups solid but difficult to expand, forcing the ecological chain companies to try untouched high-end market of YOUPIN itself, forming a virtuous cycle. Successful examples include Huami's Amazfit and Zmi's Zmi 10 etc.

Through a comparative study of the above cases, this paper suggests that the path of embedding strategy should take into account the internal knowledge base of the platform-type enterprises in this paper, enterprises of which the breadth of the knowledge base is wider and lack in depth should adopt a structured strategy, focusing on relational embedding strategy timely; on the other hand, enterprises of which knowledge base is with great breadth and insufficient depth should adopt a relational strategy, gradually structured embedding strategy, instead of a single, consistent embedding strategy.

4 Conclusion

This study, based on a comparative case study of YANXUAN and YOUPIN, argues that the selection of embedded strategies for SMEs based on the internal knowledge base of the enterprise has some implications for the selection of network strategies for SMEs.

Various paradigms between OEM, ODM and OBM are emerging, such as many new formats like "new retail". The relationship between enterprise knowledge heterogeneity and its linkages also needs to be further discussed. The pursuit of differentiation among competing firms, the study of their development direction can more effectively reflect the dynamic and contingent characteristics of innovation network research than focusing merely on performance.

We have contemplated whether there is a need to rephrase knowledge base as a major catalyst for information exchange and value creation within and between firms. As e-commerce platform become more open, more suppliers and manufacturers will be attracted into the platform ecosystem and more knowledge will flow through. Future study should focus on system features through coevolutionary process in the context of heterogeneous of agents and networks.

References

1. H. Chesbrough, The Logic of Open Innovation: Managing Intellectual Property. *California Management Review*, **45**(3), 33–58 (2003)
2. F. Wang, K. Chen, Do product imitation and innovation require different patterns of organizational innovation? Evidence from Chinese firms. *Journal of Business Research*, **106**, 60–74 (2020)
3. M. Hobday, East Asian Latecomer Firms: Learning the Technology of Electronics. *World Development*, **23**(7), 1171–1193 (1995)
4. K. Lee, Making a Technological Catch-up: Barriers and opportunities. *Asian Journal of Technology Innovation*, **13**(2), 97–131 (2005)
5. W. Chu, Can Taiwan's second movers upgrade via branding? *Research Policy*, **38**(6), 1054–1065 (2009)
6. J. Weerawardena, The role of marketing capability in innovation-based competitive strategy. *Journal of Strategic Marketing*, **11**(1), 15–35 (2003)
7. A. Kaleka, A. Morgan, How marketing capabilities and current performance drive strategic intentions in international markets. *Industrial Marketing Management,* **78**, 108–121 (2019)
8. S. Yayavaram, R. Chen, Changes in firm knowledge couplings and firm innovation performance: The moderating role of technological complexity. *Strategic Management Journal,* **36**(3), 377–396 (2015)
9. S. Moorthy, E. Polley, Technological knowledge breadth and depth: performance impacts. *Journal Knowledge Manag*ement, ***14***, 359-377 (2010)
10. S. Awate, M. Larsen, R. Mudambi, EMNE catch-up strategies in the wind turbine industry: Is there a trade-off between output and innovation capabilities? *Global Strategy Journal,* **2**(3), 205–223 (2012)
11. M. Subramanian, W. Bo, C. Kah-Hin, The role of knowledge base homogeneity in learning from strategic alliances. *Research Policy,* **47**(1), 158–168 (2018)
12. Z. Zhou, B. Li, How knowledge affects radical innovation: Knowledge base, market knowledge acquisition, and internal knowledge sharing. *Strategic Management Journal,* **33**(9), 1090–1102 (2012)
13. G. Carnabuci, E. Operti, Where do firms' recombinant capabilities come from? Intraorganizational networks, knowledge, and firms' ability to innovate through technological recombination. *Strategic Management Journal*, **34**(13), 1591–1613 (2013)
14. J. Krafft, F. Quatraro, P. Saviotti, Knowledge characteristics and the dynamics of technological alliances in pharmaceuticals: empirical evidence from Europe, US and Japan. *Journal of Evolution Economy,* **24**(3), 587–622 (2014)
15. T. Lou, J. Tang, Mining structural hole spanners through information diffusion in social networks. *Proceedings 22nd International Conference World Wide Web*, 825–836 (2013)
16. A. Schilling, C. Phelps, Interfirm collaboration networks: The impact of large-scale network structure on firm innovation. *Management Science,* **53**(7), 1113–1126 (2007)
17. T. Caner, J. Sun, E. Prescott, When a firm's centrality in R&D alliance network is (not) the answer for invention: The interaction of centrality, inward and outward knowledge transfer. *Journal of Engeneering and Technology Management,* **33**, 193–209 (2014)

18. N. Lahiri, S. Narayanan, Vertical integration, innovation, and alliance portfolio size: Implications for firm performance. *Strategic Management Journal,* **34**(9), 1042–1064 (2013)
19. M. Tortoriello , R. Reagans, B. McEvily, Bridging the Knowledge Gap: The Influence of Strong Ties, Network Cohesion, and Network Range on the Transfer of Knowledge between Organizational Units. *Organization Science,* **23**(4), 1024–1039 (2012)
20. V. Gilsing, B. Nooteboom, Density and strength of ties in innovation networks: an analysis of multimedia and biotechnology. *European Management Review,* **2**(3), 179–197 (2005)
21. J. Zhang, Facilitating exploration alliances in multiple dimensions: the influences of firm technological knowledge breadth. *R&D Management,* **46**(S1), 159–173 (2016)
22. G. Walsh, W. Mitchell, The effect of consumer confusion proneness on word of mouth, trust, and customer satisfaction. *European Journal of Marketing,* **44**(6) (2010)
23. J. Hoch, S. Banerji, When do private labels succeed?. *MIT Sloan Management Review,* **34**(4), 57 (1993)

Varying importance of the work-life balance dimension of career success for Korean accountants: The effects of gender and generation**

Najung Kim[1,*], *Jaekyoung* You[1], and *Eunhyung* Lee[1]

[1]College of Business Administration, Kookmin University, 77 Jungreung-Ro, Seongbuk-Gu, Seoul, 02707, South Korea

Abstract. In this study, we examine the effects of gender, generation, and the interaction between gender and generation on Korean accountants' perception of career success. With the large survey data collected from 1,000 accountants working in South Korea, we found that Korean female accountants have higher perceived importance of work-life balance dimension of career success than male accountants; and younger generations (Millennials or younger) have higher perceived importance for this dimension than older generations (Generation X or older). No interaction effects between gender and generation were found in relation to the perceived importance of work-life balance dimension of career success. Specifically, female accountants had higher perceived importance of work-life balance than their male counterparts regardless of generation; and while the mean was higher for younger generations the gap between the female accountants' means and the male accountants' means of work-life balance dimension has not been reduced. The insignificant interaction effects between gender and generation regarding the perceived level of the work-life balance dimension of career success suggest that, despite the national and organizational efforts in changing the gender discriminating practices, Korean accounting field may still be making a very slow progress in breaking its glass ceiling.
Keywords: Perceived importance of the work-life balance dimension of career success, career, gender differences, generational differences, South Korea, accountants

1 Introduction

With the abundant research on subjective career success as an important determinant of work-related outcomes [1, 2], several attempts have been made to understand detailed dimensions of subjective career success that were perceived as important by individuals [3, 4]. Subjective career success builds on the idea that individuals have their own criteria about the desired

---

* Corresponding author: najungkim@kookmin.ac.kr

** This study is a part of « A Study on the Talent Nurturing and Leadership Development of Female Certified Public Accountants » conducted at the request of The Korean Institute of Certified Public Accountants.

form of career and use these criteria to assess the extent to which the current state of their career is successful [5, 2].

Reflecting how individuals have various goals that they pursue in relation to their careers [6], scholars have proposed multidimensional notion of subjective career success [7]. The multidimensional notion of career success has drawn various scholars' and practitioners' interests because of its predictability of varying individual and organizational level outcomes as well as its generalizability across multiple cultural contexts [8]. Depending on what their values or desired selves are and what values and ideal selves are imposed on them, individuals can assign varying degrees of importance to each dimension of career success. Building on this notion of career success as a multidimensional construct which can be affected by individual values or orientations, we focus on how the importance of the work-life balance dimension of career success could vary depending on gender or generation or the interaction between the two.

2 Theoretical Framework

2.1 Gender differences in valuing work-life balance as important dimension of career success

Grounded in the literature on gender roles and stereotypes, several suspect the relationship between gender and the importance of work-life balance dimension. They argue that females would have higher level of importance assigned to the work-life balance dimension of subjective career success because females have traditionally been expected to have stronger orientation for family than work. Furthermore, national contexts can influence the perception of career success [9] and due to the Confucianism-based and paternalistic culture, the gender differences in the perceived importance of the work-life balance dimension of career success may be found strongly especially in the South Korean context.

Since the 1987 Equal Employment Opportunity Act, women's participation in the workforce has consistently been increasing [10]. Employers in both private and public sectors have also implemented family-friendly policies (e.g., on-site childcare centers). However, despite these efforts, gender inequality and underutilization of female talent is still prominent in South Korea. The WEF Gender Gap Report ranked South Korea 102nd out of 144, which is lower than several male-dominated Asian countries, due to the wide gap in terms of wage, economic participation, and leadership role position [11]. Korean society is heavily grounded in the Confucianism which promotes acceptance of hierarchy, male dominance, collectivism, and a gender-divided family structure [12, 13, 14]; and the burden of childrearing and household chores falls primarily on women. Paired up with the strong emphasis on prioritizing kids' education, Korean women are considered responsible for children's academic achievements [13].

The situation is harsher for working mothers in the gendered workplace where longer work hours and frequent after-work get-togethers that involve heavy drinking are expected of a competent worker [15]. Male-dominant culture constantly corners working mothers to choose between family and work. Thus, faced with the high expectation about competently managing home and work, females in Korea are likely to feel more desperate about balancing work and life in comparison to males in the gendered culture [16].

2.2 Generational differences in valuing work-life balance as important dimension of career success

Individuals' work values could vary depending on which generation they belong [17, 18] and practitioners and academics have noted that the importance of a certain dimension of career success could vary depending on generation [19, 20]. Some studies have shown how younger generations like Millennials or Gen Xs tend to value leisure and personal life [21]. Younger generations including Millennials scored significantly higher than Boomers on "more vacation time," "work-life balance," and "don't want to work hard" [22, 23, 24]. These results indicate that there could be generational differences in valuing work-life balance dimension of career success.

Similar results were found in the studies on Koreans' work values. Korean Millennials considered leisure, personal time, and work flexibility highly important while the older generations did not perceive so [25]. Furthermore, when asked of the importance of work-life balance, younger generations, in comparison to older generations, perceived work-life balance to be more important in their careers [26]. Overall, younger generations in Korea seem to value work-life balance dimension of career success more than what older generations would value.

Furthermore, extrapolating from the research on how gender stereotypes could be less severe in younger generations than in older generations [27], we would expect the gender differences in work values to be less significant for younger generations. With the growing number of policies and practices regarding work-life balance, the expected difference between female Koreans and male Koreans in their perception of work-life balance dimension as an important dimension of career success may be less prominent for Millennials or younger workers who seem to be less affected by gendered culture than the older generations.

3 Hypotheses

Hypothesis 1: There is a significant difference in the perceived importance of the work-life balance dimension of career success between males and females such that females perceive higher importance than do males.

Hypothesis 2: There is a significant difference in the perceived importance of the work-life balance dimension of career success between younger generations and older generations such that younger generations perceive higher importance than do older generations.

Hypothesis 3: There is a significant interaction between gender and generation on the perceived importance of the work-life balance dimension of career success such that the significance difference in the perceived importance of the work-life balance dimension of career success between males and females would be stronger for older generations than for younger generations.

4 Methodology

4.1 Instruments

We used the career success survey developed by 5C (The Collaboration for the Cross-Cultural Study of Contemporary Career, [8]) to measure the importance of the work-life balance dimension of subjective career success. Each item started with the following statement, "thinking about my career success, I consider this career dimension as …," and was measured using a 5-point Likert-type scale (from 1 = not at all important to 5 = very important). The items are "achieving balance between work and non-work activities," ("having time for non-work interests," and "achieving a satisfying balance between work and family life." The Cronbach's alpha was 0.71.

4.2 Study participants

We chose KICPA (The Korean Institute of Certified Public Accountants) database to invite those who are currently working in South Korea as accountants. Because the perception of career success is not only affected by organizational culture but also by occupational or professional culture [2], we limited the focus to a single profession. Among all the certified accountants registered under KICPA (20453= 7,030+13,423), we invited 13,423 accountants who are currently working as certified accountants at an accounting firm to participate in the study. Because the main focus of our study was to investigate gender differences, we kept the online survey link open until we obtained 500 male accountants' and 500 female accountants' responses. We compensated the participants with a 4 USD worth coffee coupon. A majority of them (512 accountants) worked in Big Four accounting firms and the rest worked in medium-sized (the firm size of 100-499), small-sized (the firm size of 2-99), or one-person firm. In terms of their positions at the firm, the sample consisted of 32 associates, 246 seniors, 317 managers, 155 directors, and 250 partners (including the firm owners). In terms of work experience in the field of accounting, 239 had less than 5 years, 312 had 5-10 years of experience, 198 had 10-15 years, 148 with 15-20 years, and 103 with 20+ years of experience in the field. 164 of them were in their 20s, 490 in their 30s, 255 in their 40s, 83 in their 50s, and 8 in their 60s. Among them, 501 had kids and 499 did not.

5 Results

Number of kids were known to affect one's perception of work-life balance issues [28] and the size of organization can affect the quality of flexible work arrangements [29]. Thus, childcare responsibilities (yes or no) and organizational size (big4 or not) were added as covariates. Hence, considering these potential covariates, we ran two-way ANCOVA for the set of hypotheses. We used SPSS to run the two-way ANCOVA to test the hypotheses.

Table 1. Descriptive Statistics

	Importance of the work-life balance dimension of career success					
	Male	Female	Overall	Non-Millennial	Millennial*	Overall
Mean	4.0780	4.2140	4.1460	4.0425	4.2573	4.1460
Standard Deviation	.57939	.56761	.57727	.56997	.56479	.57727
Number	500	500	1,000	518	482	1,000

*Millennials were those born after 1985, reflecting the unique generational structure in Korea
Source: Authors' current study.

Table 2. ANCOVA Results

(Dependent Variable: Importance of the work-life balance dimension of career success)

	Mean Square	**Df**	**F**	**Sig.**
Child	.101	1	.316	.574
Big4	.794	1	2.479	.116
Gender	2.360	1	7.371	.007
MilGen	5.976	1	18.662	.000
Gender*MilGen	.117	1	.366	.545
Error	.320	994		
R Squared = 0.44(Adj. R Squared = 0.39)				

Notes: n=1000; Child = childcare responsibilities (no = 0, yes = 1) ; Big4 = firm size (not-Big4 = 0, Big4 = 1) ; Gender = gender (male = 0, female = 1) ; MilGen = Millennial generation or younger (Non-Millennials or older = 0, Millennials = 1).
Source: Authors' current study.

ANCOVA analysis revealed statistically significant differences between female accountants and male accountants (F (1, 994) = 2.360, p < .05, partial ɳ2 = .007). We found significant differences between the millennial or younger accountants and older accountants (F (1, 994) = 5.976, p < .05, partial ɳ2 = .018). There was no statistically significant interaction between gender and generation on the importance of work-life balance dimension in career, whilst controlling for organizational size and childcare responsibilities, F(1, 994) = 0.117, p = .545, partial ɳ2 < .000.

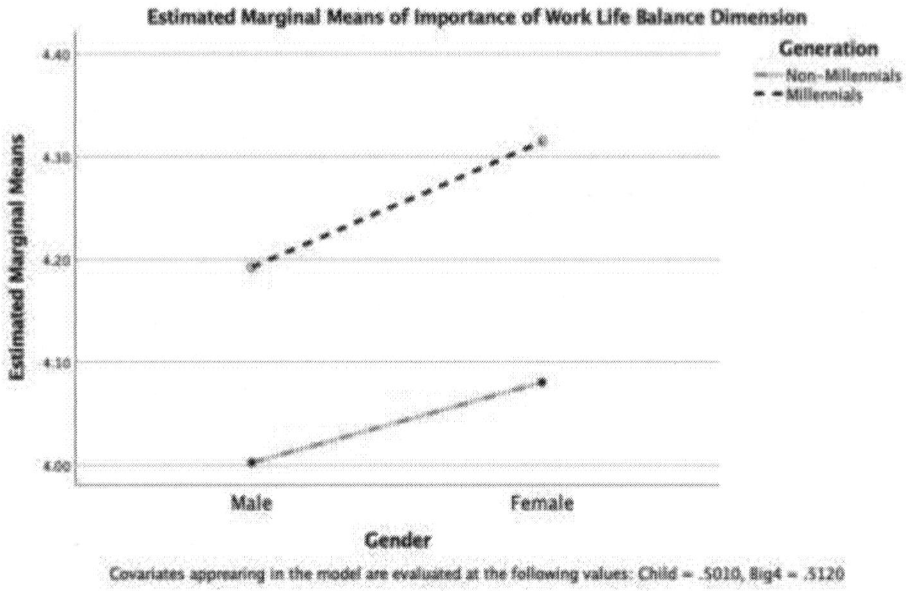

Fig. 1. Estimated marginal means of the importance of work-life balance dimension of career success
Source: Authors' current study.

6 Discussion and Conclusion

The results supported Hypothesis 1 and Hypothesis 2. Female accountants had higher perceived importance of work-life balance dimension of career success than male accountants. Younger generations (Millennials or younger) had higher perceived importance for the work-life balance dimension of career success than older generations (Generation X or older). The results were consistent with the existing studies on the gender differences and the generational differences in work values and subjective career success. However, the interaction effects between gender and generation on the perceived importance of work-life balance dimension were insignificant (Hypothesis 3 not supported). Younger generations had higher perceived importance for the work-life balance dimension but the gap between younger males and younger females was no different from the gender gap found among older accountants.

While studying the situation in Korea, scholars have noted how Korean employees have trouble utilizing the family-friendly practices because of the male-dominated culture of their organizations [30]. The persistent gap between females and males in their perception of work-life balance as an important dimension of career success regardless of their generations shows that, despite the governmental and organizational efforts in developing family-friendly, gender-equality practices, Korean accounting field may still have a long way to break its glass ceiling.

References

1. N. Dries, The meaning of career success: Avoiding reification through a closer inspection of historical, cultural, and ideological contexts. *Career Development International,* **16**(4), 364-384 (2011)
2. P. A. Heslin, Conceptualizing and evaluating career success. *Journal of Organizational Behavior,* **26**(2), 113-136 (2005)
3. J. Pan, W. Zhou, How do employees construe their career success: An improved measure of subjective career success. *International Journal of Selection and Assessment,* **23**(1), 45-58 (2015)
4. K. M. Shockley, H. Ureksoy, O. B. Rodopman, L. F. Poteat, T. R. Dullaghan, Development of a new scale to measure subjective career success: A mixed-methods study. *Journal of Organizational Behavior,* **37**(1), 128-153 (2016)
5. M. B. Arthur, S. N. Khapova, C. P. Wilderom, Career success in a boundaryless career world. *Journal of Organizational Behavior,* **26**(2), 177-202 (2005)
6. B. J. Dik, A. M. Sargent, M. F. Steger, Career development strivings: Assessing goals and motivation in career decision-making and planning. *Journal of Career Development,* **35**(1), 23-41 (2008)
7. N. Dries, R. Pepermans, O. Carlier, Career success: Constructing a multidimensional model. *Journal of Vocational Behavior,* **73**(2), 254-267 (2008)
8. J. P. Briscoe, R. Kaše, N. Dries, A. Dysvik, J. A. Unite, I. Adeleye, ... J. Zikic, Here, there, & everywhere: Development and validation of a cross-culturally representative measure of subjective career success. *Journal of Vocational Behavior,* 103612 (2021)
9. A. Smale, S. Bagdadli, R. Cotton, S. Dello Russo, M. Dickmann, A. Dysvik, J. Unite. Proactive career behaviors and subjective career success: The moderating role of national culture. *Journal of Organizational Behavior*, **40**(1), 105-122 (2019)
10. Statistics Korea [online], Available at: http://kostat.go.kr/portal/eng/index.action (2021)
11. World Economic Forum [online], Available at: http://www3.weforum.org/docs/WEF_GGGR_2021.pdf (2021)
12. S. Chaudhuri, S. Park, S. Kim, The changing landscape of women's leadership in India and Korea from cultural and generational perspectives. *Human Resource Development Review,* **18**(1), 16-46 (2019)
13. Y. Cho, G. McLean, I. Amornpipat, W. Chang, G. Hewapathirana, M. Horimoto, M. Lee, J. Li, N. Manikoth, J. Othman, S. Hamzah, Asian women in top management: eight country cases. *Human Resource Development International,* **18**(4), 407-428 (2015)
14. N. Kim, C. Rowley, The changing face of Korean women managers. In Rowley, C. & Paik, Y. (Eds.). *In the changing face of Korean management* (pp. 184-209), Routledge, London (2009)
15. Y. Cho, J. Park, S. Han, B. Ju, J. You, A. Ju, C. Park, H. Park, How do South Korean female executives' definitions of career success differ from those of male executives? *European Journal of Training and Development,* **41**(6), 490-507 (2017)
16. F. Cheung, D. Halpern, Women at the top: Powerful leaders define success as work+family in a culture of gender. *American Psychologist,* **65**(3), 182-193 (2010)
17. J. I. C. Hansen, M. E. Leuty, Work values across generations. *Journal of Career Assessment,* **20**(1), 34-52 (2012)

18. J. M. Twenge, A review of the empirical evidence on generational differences in work attitudes. *Journal of Business and Psychology*, **25**(2), 201-210 (2010)
19. K. Chudzikowski, Career transitions and career success in the 'new' career era. *Journal of Vocational Behavior,* **81**(2), 298-306 (2012)
20. N. Dries, R. Pepermans, E. De Kerpel, Exploring four generations' beliefs about career: Is "satisfied" the new "successful"? *Journal of Managerial Psychology,* **23**(8), 907-928 (2008)
21. J. M. Twenge, S. M. Campbell, B. J. Hoffman, C. E. Lance, Generational differences in work values: Leisure and extrinsic values increasing, social and intrinsic values decreasing. *Journal of Management,* **36**(5), 1117-1142 (2010)
22. K. Smola, C. Sutton, Generational differences:Revisiting generational work values for the new millennium. *Journal of Organizational Behavior*, **23**, 363–382 (2002)
23. J. M. Twenge, S. M. Campbell, Generational differences in psychological traits and their impact on the workplace. *Journal of Managerial Psychology*, **23**, 862–877 (2008)
24. M. Wong, E. Gardiner, W. Lang, L. Coulon, Generational differences in personality and motivation: Do they exist and what are the implications for the workplace? *Journal of Managerial Psychology*, **23**, 878–890 (2008)
25. H. Lee, G. Yu, Work-life balance of Generation Y: The meaning and the role in view of generational work value, *Quarterly Journal of Labor Policy,* **13**(4) 1-31 (2013)
26. H. Lee, Y. Kim, Generational differences in values of life and factors influencing turnover intention among Korean nurses, *Journal of Digital Convergence,* **19**(1) 217-228 (2021)
27. K. E. Cichy, E. S. Lefkowitz, K. L. Fingerman, Generational differences in gender attitudes between parents and grown offspring. *Sex Roles*, **57**(11-12), 825-836 (2007)
28. M. Hilbrecht, S. M. Shaw, L. C. Johnson, J. Andrey, 'I'm home for the kids': contradictory implications for work–life balance of teleworking mothers. *Gender, Work & Organization,* **15**(5), 454-476 (2008)
29. B. Kotey, I. Koomson, Firm size differences in financial returns from flexible work arrangements (FWAs). *Small Business Economics*, **56**(1), 65-81 (2021)
30. S. Kim, Women and family-friendly policies in the Korean government. *International Review of Administrative Sciences,* **74**(3), 463-476 (2008)

Relationships between mental workload, job burnout, and organizational commitment

Altanchimeg Zanabazar[1*], and *Sarantuya* Jigjiddorj[1]

[1] Management Department, Business School, National University of Mongolia, Ulaanbaatar, Mongolia

Abstract. The business environment has becoming exceptionally unsteady and competitive recently. Information technology advancements and ever changing customer preferences has played a substantial role in an increase of the significance of highly productive, committed and loyal employees. These changes bring both advantages and challenges need to be addressed. The more challenges accelerate, the more issues come up to solve in timely manner. For any organizations, not only taking care of the customers but also employees of the organization should be treated equally taking a better care about their psychological and physical health as well as creating a healthy work environment. The objective of the current research is to study correlations of workload, job burnout and organizational commitment in the case of nurses working in the Health Center of Selenge aimag, Mongolia. According to the results, an increase of mental workload nurses by one unit results in job burnout increases by 0.578 (beta) unit and leads to the decrease of the organizational commitment by 0.437 (beta) unit. Moreover, one-unit increase of job burnout leads to decrease of the organizational commitment by 0.301 (beta) unit.
Keywords: Mental workload, Job burnout, Organizational commitment, nurses

1 Introduction

Employee loyalty ensures the organization competitive advantages [1]. Recently technology based businesses like banks and medical institutions are getting more focused on the upgrading employee loyalty, rather than years in service for the organization, bearing in mind their further contribution [2]. Therefore, for maintaining highly productive, loyal and highly committed employees the organizations are more than ever required to address employee physical and psychological health, balanced workload and healthy work environment.

An excessive mental workload, work stress and exhaustion bring negative impacts on employee loyalty as well as organizational outcomes. As a poor performance and productivity due to the job related issues often contributes to the severe damage in employee wellbeing, the scholars and business practitioners pay more attention on the mental workload [3-4].

Specially, evaluation of efficiency and mental workload is essential in a man-machine interacted work environment [5]. The significance of human factor is vital for maintaining high performance in sustainable manner [6]. To reduce unforeseen risks associated with the human activities, it is crucial to evaluate the mental workload of employees in a certain interval of time. The time pressure, multi-tasks, and hastiness not only result in high mental workload but also cause depression and job burnout which negatively impact employee's life.

* Corresponding author: altanchimegz@num.edu.mn

Researchers provided different studies that identified the relationship between numerous conditions such as intense workload, the job burnout and the employee loyalty to the organization [7-8], mental workload and organizational loyalty [9], mental workload and poor performance [10], decrease of employee loyalty to the organization [11-14], decrease of job satisfaction [12] and demonstrated how these conditions might have impacted on employee wellbeing and organizational outcomes.

2 Literature review

2.1 Workload

Workload issues have been studied by various researchers, and they approached differently explaining the source of workload, mechanism and measurement. Gopher and Donchin (1986) described the work load as the interaction between mental needs of employee to perform his assigned tasks. Work quality related workload is the physical and mental challenges the employee experience while performing his work tasks. Workload is identified as the quantitative and qualitative variables [15], or mental and physical variables [16].

Mental Workload: It is important to distinguish between mental and physical workloads [16] as the causes and effects can be different in terms of theory and practice. The concept of mental workload is originated from cognitive load theory that concerns with the relations between human mind and its limitations [17]. By this theory, if a human is over or low loaded, learning new things and performing tasks are influenced negatively [18]. Taking in account this perspective, Hart described the mental workload as the adaptation and interaction between work requirements and employee's perception, talents, abilities and emotions [19]. For instance, two people performing the same tasks may end up with different outcomes. To reveal the reasons serving behind differences the mental workload may be considered and analyzed carefully [20]. Moreover, the amount of information obtained during work tasks is related to mental workload as the ability to memorize the information can affect in balancing between task demand and cognitive resources [21]. Work demand with time pressure, complex in nature increases the workload perception [22].

Vidulich and Tsang considered that mental workload is consisted of exogenous task demands and endogenous supplies of resources. Clearly, task priority, task difficulty, situational contingencies are classified as exogenous task demand while updating, perceiving memory, decision-making, planning and processing are considered as endogenous ones [20]. Mental workload is related with the characteristics of the tasks, employee's skills and abilities and environmental features of the workplace and the organization [22].

Cognitive demands for performing tasks prompt mental workload perception of employees. In this sense, mental workload is considered with the complexity of the tasks that are required to perform for daily activities [22]. Vidulich and Tsang described workload as an interaction between employee's capacity and task requirements [20]. Deviations between structure of the required tasks and human capacities affect the occurrence of the workload. It means imbalance between task demands and employee capacity, mental comes to exist.

Mental workload get attention for being a significant occupational risk factor in the workplace that leads to loss of performance, productivity, welfare and well-being [3, 4, 23]. Performance failure may increase the perception of the workload [24] and consequently, overload and fatigue speed up making mistakes and failures in the job [25].

On the other hand, when mental workload increases, control variability and errors increase and an individual performance tend to decrease. The workload related with the volume and quantity is considered in relation to the volume of the employee job tasks [26]. The workload is one of the risks in the job that may threaten employee commitment, loyalty, productivity, satisfaction and wellbeing [3-4].

2.2 Job Burnout

Job burnout is a concept that is discussed in the organizational behavior literature by Freudenberger and Maslach. Burnout described by these researchers as an overwhelming exhaustion, feelings of frustration, anger, cynicism and a sense of ineffectiveness and failure [8]. Job burnout is a negative phenomenon that is originated from the chronic job stress and causes harms regardless the nature of the occupation [27]. Burnout is an emotional and mental exhaustion that make people stranded [28].

Work overload, conflict with other people in contact often reveal the sign of burnout and it is often visibly demonstrated by loss of trust and negative attitudes towards others. When time passes feeling unhappiness, dissatisfaction about the job accomplishment lead to inability coping with their job and the suffered one evaluates himself, gets aware of the burnout. The main dimensions of job burnout considered as emotional exhaustion, depersonalization and personal accomplishment [29].

Demerouti and colleagues (2001) considered that burnout may be found in any occupational context and seen both within and outside the human services [30]. While in the beginning the theoretical concept development of burnout, it principally discussed work associated factors, however recent researchers consider that environmental and individual factors also have an effect on burn out syndrome [29-32].

Burnout occurs as a result of complicated interaction between individual characteristics and factors in the work environment [10]. According to some studies causes of burnout with a focus on the work environment indicating role stressors, organizational structures, and job characteristics. Particularly, what is stressful for one may not be stressful for another [29].

Mainly, burnout develops after continues, repeated and unsuccessful confrontation of undesirable working conditions [24]. As Karabay, Tezergil and Köse (2014) considered the reason for the burnout among people work related stress factors [33]. The factors cause the burnout situation at work and create negative circumstances. Moreover, these factors influence the performance of employees and reduce the productivity. According to Maslach (2003) burnout is a response to chronic work related stress that attempts to adapt and protect oneself from it [34].

Emotional exhaustion: The current component is considered as the key symptom of burnout, and it is strongly related with other components. Emotional exhaustion refers to the feelings of being emotionally over extended and exhausted by one's work. It is regarded as a lack of energy and feeling that one's emotional resources are used up. As Maslach (1982) described it as it as a response for the overwhelming emotional demands of other people. Individuals who are emotionally exhausted lack enough energy to face another day, and often report that they are filled with dread at the prospect of returning to work for another day.

Depersonalization: The next component of burnout is depersonalization. Researcher consider it as defensive coping response to deal with the effects of emotional exhaustion when alternative coping resources are not in a place. When the one suffer from burnout often developed a detached, callous or dehumanized responses to the people they should to care for. It also involves the development of negative attitudes and feelings towards persons for whom work is done. It may be expressed as unprofessional comments towards coworkers, blaming others for their problems. Generally, it is a distant attitude towards work.

Diminished personal accomplishment: It is the tendency to negatively evaluate the worth of one's work, feeling insufficient to perform his job, and a generalized poor professional self-esteem. Individuals with burnout syndrome show some symptoms like frustration, anger, fear and anxiety as well as some physical discomfort including insomnia, headache, paroxysm and even digestive track discomfort. Diminished personal accomplishment is a tendency where professional value their own capacity to carry out tasks and to interact with coworkers and feeling dissatisfied with achievements. Often job burnout results from an unpleasant work environment, frequent incident of unfavorable conditions and a long time progressing stress [10] that negatively affects on employee productivity, job satisfaction and encourage employee turnover intention.

2.3 Organizational commitment

Organizational commitment can be perceived as the trust built up among the employees. Committed employees possess a sense of obligation to provide a loyal service to the organization in the future as the one feel confident within the organizational climate [35].

Loyal and committed employees are a valuable asset for the organization as they are capable generating significant amount of profit. The employee emotional attachment and connection to the organizational decides if the employee will be committed or not. Employee loyalty and job satisfaction are crucial in the organization's achievements and only committed and satisfied with their job employees are key for maintaining an effective cooperation with the business's main stakeholders. As an employee communication and job satisfaction are contributing to the increase of the commitment, giving a special focus on these issues enable organizations increasing their employee commitment [36].

Meyer and Allen (1991) identified 'three component model of commitment', namely affective commitment, continuance commitment and normative commitment.

Affective Commitment: Affective commitment refers to a strong psychological engagement that make the ones to be loyal to the organizational goals, values and coworkers. If the he or she considers that the organizational goals and values are consistent with personal values, aspirations they tend to become more committed and loyal to the organizations. Numerous factors such as work tasks challenge the employees, clarity of job responsibilities, clarity of goals, management attitude towards to the employees, team unity, equality, respects for individuality, feedback from managers, participation and dependence on others also impact to the employee commitment.

Continuance Commitment: It relates to the degree employee's sense if they need to stay at their organization based on their reflection of costs and benefits of staying or moving to another organization. When an employee feels the need continuing with their present employer because of their wage and fringe benefits would not improve if they move to another organization cause this type of commitment and make them stay with the present employer. Employees with multiple years of experience of working at the same organization or seniority often contribute enjoying multiple benefits and make them stay committed to the organization, however it is not always because of enthusiasm.

Normative Commitment: Normative commitment refers to employee's feeling of obligation to the organization and the belief that continuing their work at the organization is the right thing. Regardless the benefits and job satisfaction the employees remain committed to the organization. High employee commitment can be seen by their devotion to fulfill the organization's goals, improvement of productivity and efficiency and provision of a quality service to the customers.

Employees contribution to the achievement of the organizational goals and objectives are valuable. Only maintaining a favorable work environment, work life balance as well as their wellbeing enable employees to become committed to the organization. Ensuring these factors in sustainable manner require HR managers to understand and research the state of employees' mental workload and job burnout to better recognize how their employees are committed to the organization.

Numerous studies are conducted to realize the relationships between these factors, including the relationships between job burnout and organizational commitment by Brown and his associates (2019), the relationships between psychological condition, organizational commitment and job burnout by Peng and researchers (2013)and correlations between organizational commitment, job burnout, psychological condition and anxiety by Zhou and researchers (2018). Moreover, Lin and researchers (2013) conducted the study on the impact of stress, job burnout, employee satisfaction to the organizational commitment. Also Abbas, Kanwal, & Iqbal (2014) and Usman & Raja (2013) conducted similar studies on the relationships between above said factors.

2.4 Research design

Upon reviewing the research designs and results of the above researches conducted earlier we proposed the following research design.

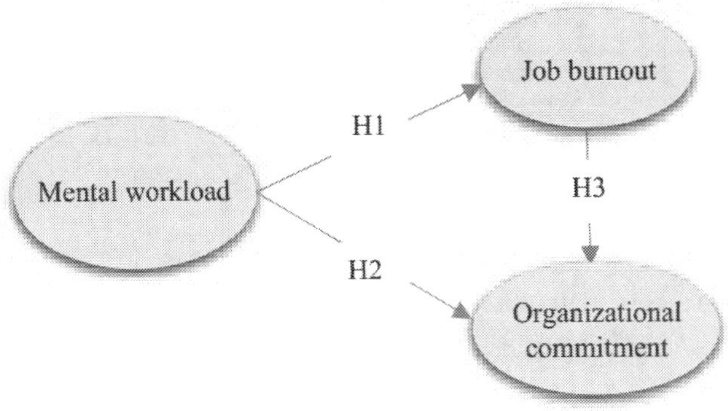

Figure 1. Research design
Source: Own processing.

2.5 Research hypothesis

We proposed the following hypothesis for the study:
H1: Mental workload exposure causes job burnout.
H2: Mental workload influences to the organizational commitment.
H3: Job burnout influences to the organizational commitment.

3 Research methodology

3.1 Research method

The current survey is conducted to identify if mental workload and job burnout impact on the organizational commitment in the case of nurses working in the Health Center of Selenge aimag. The survey data is processed using SPSS 23.0 and Smart PLS 3.0. The results are produced based on the results of factor analysis, reliability test, correlation and regression analysis and PLS analysis.

Sample*:* The population of nurses working in the above mentioned medical facility is 175 and selection of 120 nurses for the survey could present the population with 95 % confidence level of probability. The data of 124 questionnaires is considered for the further analysis.

Data collection: Online survey is conducted between June 28 and July 19, 2021 and 124 nurses participated in the survey. We used various questionnaires such as CarMen-Q workload survey developed by Rubio-Valdehita & et al (2017), MBI-HSS survey on job burnout by Maslach and Jackson (1981) and OCQ survey on organizational commitment by Meyer and Allen (1997).

Demographic characteristics of the participants are summarized in the below.

Table 1. Demographic characteristics of the respondents

Age	No	%	Sex	Number	%	Marital status	No	%
20-24	4	3.2	Male	19	15.3	Married	104	83.9
25-29	4	3.2	Female	105	84.7	Single	20	16.1
30-34	40	32.3	**Work experience**	**Too**	**%**	Other	-	-
35-39	16	12.9	up to 1 year	8	6.5	**Amount of salary (MNT)**	**No**	**%**
40-44	34	27.4	1-4 years	4	3.2	Up to 600,000	4	3.2
45-49	15	12.1	5-9 years	52	41.9	600,001-750,000	4	3.2
50-54	7	5.6	10-14 years	35	28.2	750,001-900,000	51	41.1
55 and above	4	3.2	More than 15 years	25	20.2	900,001-1,100,000	35	28.2
If you have children	**No**	**%**	**Type of work**	**Too**	**%**	1,100,001-1,300,000	15	12.1
Yes	104	83.9	Regular day	108	87.1	1,300,001-1,500,000	15	12.1
No	20	16.1	By shift	16	12.9	1,500,001 and more	-	-

Source: Own processing.

Only 15.3% of the participants of the survey were male and it can be concluded that the majority of the professionals are female. By marital status, majority of nurses are married (83.9%) and only minority (16.1%) are single ones. By age, 51.6% of the respondents were young people who are below 40 years old, 39.5% of them were at their early 50s. Majority of the married professionals had children and only unmarried people tended to have no children.

By work experience, 6.5% of the respondents have up to one-year experience, 3.2 % worked 1-4 years of experience, 41.9% have 5-9 years of work experience, 28.2% of them worked 10-14 years and 20.2% of respondents worked more than 15 years. By type of work, 87.1% engaged in regular day work and 12.9 % of them worked by shifts. By wage size, 47.5% earn up to 900.000 MNT, 28.2 % earn between 900,001-1,100,000 MNT, 12.1% earn 1,100,001-1,300,000 MNT, and 12.1% of them earn 1,300,001-1,500,000 MNT. It shows that the salary level of the medical professionals, particularly nurse's salary is very low.

It is said that 67.7 % of the respondents had a sickly leave within a year due to Covid-19 (33.9%), heart and vascular disease (27.4%) and gastroenterological, joint and muscle diseases (3.2%). It can be concluded that nurse's job is one of the riskiest job among the medical professions.

4 Result and Discussions

4.1 Reliability test

For verification of the reliability of the questionnaire Cronbach's Alpha value has been checked as it must reach 0.7 [37] and in our case Cronbach's Alpha value for all variables fully reliable for measuring the proposed variables. Moreover, composite reliability coefficient all presented more than 0.5 [38] which means that all composites included in the questionnaire are capable to measure. Calculation of an Average Variance Extracted showed 0.5 [38] which means the reliability of the model is acceptable. As composite reliability coefficient estimation presented 0.5 as compatible [39] and the composites with coefficient less that 0.5 excluded [40]. In the analysis we used only composites with 0.5 values [41]. The results of the test are summarized in the table 2.

Table 2. The results of reliability test

Factors	Number of questions	Factor loading	Cronbach's α	AVE	CR
Mental workload-MW	25	0.628-0.885	.970	.964	.825
Job burnout -JB	15	0.611-0.883	.881	.896	.673
Organizational commitment -OC	15	0.609-0.889	.830	.851	.591

Source: Own processing.

4.2 Correlation analysis

Results of correlation analysis is shown in Table 3.

Table 3. Results of correlation analysis

Correlations				
		MW	JB	OC
MW	Pearson Correlation	1		
	Sig. (2-tailed)			
JB	Pearson Correlation	.455**	1	
	Sig. (2-tailed)	.000		
OC	Pearson Correlation	-.370**	-.575**	1
	Sig. (2-tailed)	.000	.000	

** Correlation is significant at the 0.01 level (2-tailed).
* Correlation is significant at the 0.05 level (2-tailed).

Source: Own processing.

The results of the analysis show that the correlation between mental workload of nurses and organizational commitment have a weak correlation [$r = -.370$, $p< 0.01$] while job burnout and organizational commitment presented a visible correlation [$r =-.575$, $p< 0.01$].

4.3 Structural Equation Modeling

To ensure the reliability of data and evaluation of the model we estimate R-Squared to better predict the structural equation modeling. Through an estimate of relations between movements of a dependent variables based on an independent variables movement we found that the data and predictions unbiased or significance ($p <0.05$). The results are shown in the following Table 4:

Table 4. Research Results of PLS-SEM

Hypothesis	Path	Standardize Beta	t value	P value	Remarks
H1	MW→JB	0.578***	5.641	.000	Supported
H2	MW→OC	-0.437***	9.763	.000	Supported
H3	JB→OC	-0.301***	10.896	.000	Supported

Source: Own processing.

As the results show that mental workload and job burnout [$\beta=0.578$; $p<0.01$] notably affects to the organizational commitment [$\beta=0.437$; $p<0.01$], while job burnout has a weak impact to the organizational commitment [$\beta=0.301$; $p<0.01$].

Determination coefficient (R squared) of 0.434 can be explained that these factors explain the relationships by 43.4% and remaining are explained by other factors that are not inclusive in the equation.

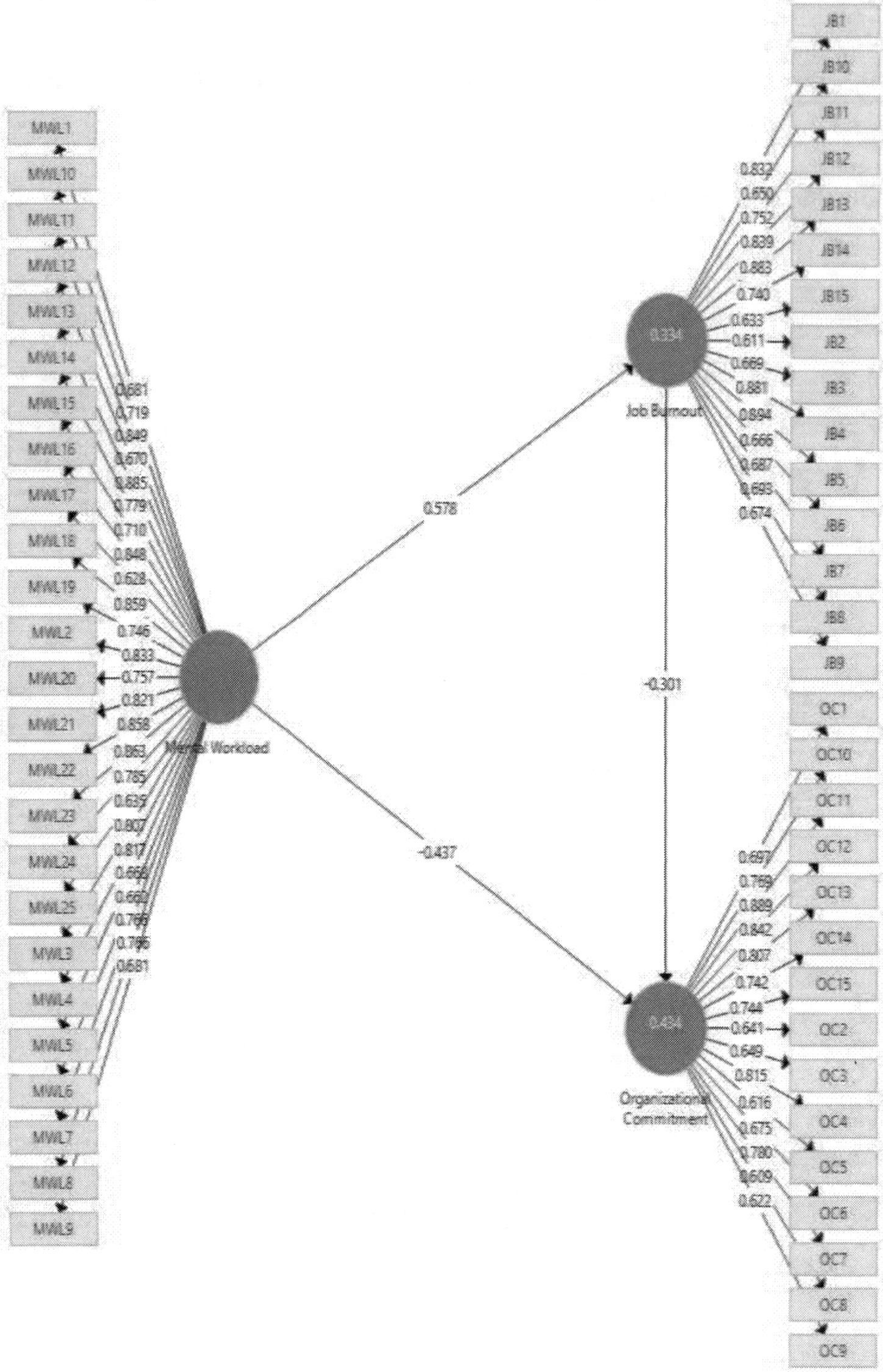

Figure 2. Path Model and PLS-SEM Estimate
Source: Own processing.

We conclude that all proposed hypothesizes are supported as the beta coefficient of all variables are positive and statistically significant ($p < 0.05$).

5 Conclusion

Employee commitment ensures the long term organizational sustainability maintaining the prestige, values and organizational culture that is an asset reaching the competitive advantage over competitors. Therefore, for keeping the employee efficiency, commitment, their active participation in organizational life the employers need to give a significance on mental and physical health of employees, balancing the workload and creation of healthy work environment.

The current research is examined the relationships between mental workload, job burnout and organizational commitment. Upon reviewing related theoretical framework and researches on mental workload, job burn syndrome, organizational commitment we proposed hypothesis on the relationships between the above said variables. We developed a questionnaire with Likert 5 scale measurement based on the frameworks and instruments developed earlier such as CarMen-Q developed by Rubio-Valdehita & et al (2017), MBI-HSS instrument by Maslach and Jackson (1981) and OCQ survey Meyer and Allen (1997). We survey is conducted among 124 nurses working for the Health Center of Selenge aimag.

According to the results, the case of the nurses of The Health Center of Selenge aimag show that an increase of mental workload nurses by one unit results in job burnout increases by 0.578 (Beta) unit and leads to the decrease of the organizational commitment by 0.437 (Beta) unit. Moreover, one-unit increase of job burnout leads to decrease of the organizational commitment by 0.301 (Beta) unit. Based on the results of the study we make the following conclusions:

1. Time pressure, multitask handling, urgency and heavy workload often lead to the stress and job burnout which negatively affect one's life [7].
2. We observed that the greater mental workload, the job burnout increases simultaneously [$\beta=0.578$; $p<0.00$]. In other words, the more people feel more time pressure, stress and cognitive needs they feel exhaustion and job burnout. The results of study by Keser & Yılmaz (2014), Rubio-Valdehita & et al (2017) and Liu & Lo (2018) all evidenced the above said.
3. The more the employees feel the mental workload, the organizational commitment decreases accordingly [$\beta=-0.437$; $p<0.00$]. Clearly, the more people feel more time pressure, stress and cognitive needs they feel exhaustion and job burnout on their jobs they tend to under deploy their knowledge and skills to perform the job as well lose interests to continue the work. The results of study demonstrated that mental workload negatively affect the organizational commitment is fully consistent with the results of the study conducted by Ziaei & et al (2015), Kanchanapa & Karnsomdee (2019), Ahuja & et al (2002), Rodríguez-Fernández, Herrera, & de las Heras-Rosas (2021).
4. The study results show that job burnout contributes to the decrease of the organizational commitment [$\beta=-0.301$; $p<0.00$]. Job burn syndrome causes various negative impacts towards to the job such as decrease of productivity, depersonalization, absenteeism, decreased level of job satisfaction, abuse of tobacco, alcohol and even coffee, loss of self-confidence and often contributes to the organizational performance which can be observed by decline of productivity, an increase of employee turnover, increase of training & development expenditure and decrease of customer satisfaction level. The current finding is also consistent with the results of the studies conducted by Peng & et al (2013), Lin, Lin, & Cheng (2013),

Abbas, Kanwal, & Iqbal (2018), Zhou & et al (2018) and Brown, Walters, & Jones (2019).

We should note that the current study has some limitations in terms of scale and scope of the case of study as it conducted among the employee of only one local medical institution. Further, it needs to expand the coverage of the institutions of the current sector as well as the number of professionals to be involved in the study to improve its' practical benefits for the human resource management of the medical institutions operating throughout the country to enable them to better understand the job burnout syndrome that prevailing in the institutions and better address the issues preventing such negative phenomena in the job.

References

1. C. Foster, P. Whysall, L. Harris, Employee Loyalty: An Exploration of Staff Commitment Levels towards Retailing, the Retailer and the Store. *International Review of Retail, Distribution & Consumer Research*, **18** (4), 423-435 (2008)
2. S. K. Aityan, T. P. Gupta, Challenges of Employee Loyalty in Corporate America. *Business and Economics Journal,* BEG-55 (2012)
3. A. Keser, G. Yılmaz, Workload, Burnout, and Job Satisfaction Among Call Center Employees. *Sosyal Siyaset Konferansları Dergisi*, (66–67), 1–13 (2014)
4. H. L. Liu, V. Lo, An integrated model of workload, autonomy, burnout, job satisfaction, and turnover intention among Taiwanese reporters. *Asian Journal of Communication,* **28**(2), 153–169 (2018)
5. M. Ortiz-Solis, et al., Human Error Identification and Mental Workload Index Setting During a Computer Power Source Change Task: A Case Study. Advances in Manufacturing. *Production Management and Process Control,* 427-438 (2019)
6. K. C. Soto, et al., Effects of Human Factors in Planning and Production Control Activities in Remanufacturing Companies. *Advances in Ergonomics of Manufacturing: Managing the Enterprise of the Future,* 465-474 (2016)
7. M. Ziaei, et al., Level of Workload and Its Relationship with Job Burnout among Administrative Staff. *International Journal of Occupational Hygiene*,7, 53-60 (2015)
8. S. Kanchanapa, P. Karnsomdee, The influence of perceived Workload and Job burnout on Organizational commitment of employee of Thailand Post Co., Ltd. Sakon Nakhon Post Office, Thailand. *International Academic Research Conference*, 286-292 (2019)
9. M. Rodríguez-Fernández, J. Herrera, C. de las Heras-Rosas, Model of Organizational Commitment Applied to Health Management Systems. *International Journal of Environmental Research and Public Health,* **18**(9), 4496 (2021)
10. C. M. Kokkinos, Job stressors, personality and burnout in primary school teachers. *The British Journal of Educational Psychology,* **77**(1), 229–243 (2007)
11. J. Peng, et al., The Impact of Psychological Capital on Job Burnout of Chinese Nurses: The Mediator Role of Organizational Commitment. *PLoS ONE,* **8**(12), e84193 (2013)
12. J. Lin, C. Lin, C. Cheng, A Study of the Relationship between Job Stress, Job Burnout, Job Satisfaction and Organizational Commitment among Medical Radiologists in Taiwan. *Journal of Information and Optimization Sciences,* **34** (2-03), 149-167 (2013)
13. Z. Abbas, A. Kanwal, J. Iqbal, Impact of Job Burnout on Organizational Commitment of Universities in Lahore, Pakistan. *Minhaj Journal of Economics and Organization Science – MJEOS,* **1**(1), 84-99 (2018)
14. J. Zhou, et al., Serial multiple mediation of organizational commitment and job burnout in the relationship between psychological capital and anxiety in Chinese female nurses: A cross-sectional questionnaire survey. *International Journal of Nursing Studies,* **83,** 75-82 (2018)

15. L. E. Tetrick, et al., A comparison of the stress–strain process for business owners and nonowners: Differences in job demands, emotional exhaustion, satisfaction, and social support. *Occupational Health Psychology,* **5**(4), 464–476 (2000)
16. D. J. Dwyer, D. C. Ganster, The effects of job demands and control on employee attendance and satisfaction. *Organizational Behavior* **12**(7), 595-608 (1991)
17. C. D. Wickens, Multiple Resources and Mental Workload. *Human Factors The Journal of the Human Factors and Ergonomics Society,* **50**(3), 449-455 (2008)
18. L. Longo, Experienced mental workload, perception of usability, their interaction and impact on task performance. *PLoS ONE,* **13**(8), e0199661 (2018)
19. S. G. Hart, NASA-Task Load Index (NASA-TLX); 20 years later. *Proceedings of the Human Factors and Ergonomics Society Annual Meeting,* **50**, 904-908 (2006)
20. M. A. Vidulich, P. S. Tsang, Mental workload and situation awareness. In G. Salvendy, *Handbook of human factors and ergonomics (4th Eds),* (pp. 243–273). Hoboken, NJ: Wiley (2012)
21. R. Parasuraman, T. B. Sheridan, C. D. Wickens, Situation Awareness, Mental Workload, and Trust in Automation: Viable, Empirically Supported Cognitive Engineering Constructs. *Journal of Cognitive Engineering and Decision Making,* **2**(2), 140-160 (2008)
22. M. S. Young, et al., State of science: Mental workload in ergonomics. *Ergonomics,* **58**(1), 1-17 (2015)
23. S. Rubio-Valdehita, et al., Development of the CarMen-Q Questionnaire for mental workload assessment. *Psicothema,* **29**(4), 570-576 (2017)
24. E. Demerouti, et al., The job demands-resources model of burnout. *Journal of Applied Psychology,* **86**(3), 499–512 (2001)
25. K. A. Brookhuis, D. de Waard, Monitoring drivers' mental workload in driving simulators using physiological measures. *Accident Analysis & Prevention*, **42**(3), 898-903 (2010)
26. X. X. Che, et al., Stressors beget stressors: The effect of passive leadership on employee health through workload and work–family conflict. Work and Stress, **31**(4), 338-354 (2017)
27. H. J. Freudenberger, *Burn-out: the high cost of high achievement.* New York: Anchor Press, (pp. 211) (1980)
28. M. Westman, D. Eden, Effects of a respite from Work on burnout: Vacation relief and fade-out. *Applied Psychology,* **82**(4), 516-27 (1997)
29. P. S. Lau, et al., Do Demographic Characteristics Make a Difference to Burnout Among Hong Kong Secondary School Teachers? *Social Indicators Research,* **71**(1-3), 491-516. https://doi.org/10.1007/s11205-004-8033-z (2006)
30. H. J. Kim, K. H. Shin, W. T. Umbreit, Hotel job burnout: The role of personality characteristics. *International Journal of Hospitality Management*, **26**(2), 421-434 (2007)
31. E. Demerouti, et al., Job Crafting Interventions: Do They Work and Why? *Positive Psychological Intervention Design and Protocols for Multi-Cultural Contexts*, 103–125 (2019)

Fuzzy-based non-bankruptcy trend forecast

Simona Hašková[1*]

[1]The Institute of Technology and Business, School of Expertness and Valuation, Okruzni 517/10, 37001 Ceské Budejovice, Czech Republic

Abstract. Many companies face an economic downturn due to the Covid-19 pandemic outbreak, which makes their future uncertain. The practical aim of the paper is to establish a procedure for an effective prediction of a business tendency to bankrupt in the short-term period. The tool is a three-stage fuzzy model formulated in the theoretical-methodological part and applied on the real data of an examined company. The model input parameters are objective and subjective measured data between 2008-2020 of a nature affecting the output. The output is an interval of subjectively expected values determining the non-Bankruptcy trend (non-B) of a company. The paper shows advantages of the interval fuzzy approach for bankruptcy prediction and identifies the measure of business safety.
Keywords: fuzzy approach, non-bankruptcy index, fuzzy interval prognosis

1 Introduction

The Covid-19 pandemic is a health crisis, that has stimulated a serious economic crisis. In the affected economies the basic economic indicators are expected to be lower compared to 2018 until a functional medical treatment, that permanently stabilizes the situation, is found.

In this paper we notice the relationship between the trend of a company to bankrupt/non-bankrupt and the influencing variables of microeconomic and macroeconomic nature that have, to a certain extent, a direct relationship to a bankruptcy threat.

The level of bankruptcy threat results from the crisis future development. A clearer picture is expected in the short time period of 1-3 years and is given by real situation. A current development of basic economic indicators within the first half of 2021 shows an incomplete recovery in most of countries [3]. It is clear that crisis impact, and the way the individual EU states get over, is going to be very uneven. How well and quickly the countries will pull together relies upon external and internal factors that individual economies face ([12] [8]). The pandemic spread and the strictness of containment measures affects the future situation to a certain extend; nevertheless, a great role belongs to economic exposures, pre-existing conditions and the policy responses [11].

Regarding the status of the Czech economy the economists and politicians have often discussed the strong dependence on the automotive industry and a strong export focus on Germany. This dependence has always been manifested very clearly during economic crises.

* Corresponding author: haskovas@post.cz

The last "coronavirus" crises brought a severe decline in the production and sales of cars, automotive components and technologies. For example, Škoda Auto factory, that is understood as the heart of the Czech industry, creates on average for about five percent of GDP and accounts approximately 9 % of exported goods ([1], [14]). The pandemic situation forced the Škoda Auto factory management to stop all lines for several month in 2020.

The paper's aim is to show sensible and trustworthy procedure of the short-term prediction for 2022-2023 of Škoda Auto factory trend to go bankrupt/non-bankrupt. The suitable tool is the fuzzy procedure, that enables to incorporate the human knowledge and expertise. This allows the experts to formulate the inference rules and incorporate specifics of the task into the fuzzy model. To take into account the extreme condition in which the economies currently appear the input variables are expressed as the interval values of occurrence of the historically measured data. The interval output value signals possible trend of the firm to bankrupt/not to bankrupt. We explain why the left limit value performs the important indicator of business safety. This approach fully takes into account the uncertainty of a turbulent environment and offers the external users a more realistic prediction compared to a one-point estimate, which statistical methods standardly offer.

2 Methodology: The Introduction to Fuzzy Approach

In the tasks of expert analysis of the vaguely defined systems, the vague terms are usually presented by linguistic variables and their linguistic values (the terms), where the word „linguistic" is used in the sense of „*communicated in natural language* ". The relationship between combinations of terms of the input linguistic variables and the output linguistic variable is given by a set of rules [16].

The tool to solve the tasks of vague nature is the fuzzy logic on which the fuzzy approach is based (more in [15]). The fuzzy approach purposefully modifies the assumptions and thus conclusions of the „*principle of extension*" of fuzzy logic. The individual fuzzy approach phases are demonstrated in Fig. 1 and are concisely described as follows (for detailed view see [9]):

- The fuzzification phase means that the numerical vector $x = (x_1,\dots,x_n)$ of real input variables is transformed to internal dimensionless input values $u = (u_1,\dots,u_n)$ (mostly expressed as percentage) and as such enters the fuzzy system of the block K (see [7]). In the block K the vector u selects inferential rules relevant to its multi-criteria evaluation. These are the rules of the type $((A_{1i},\dots,A_{nj}), B_k) \in P$, in which terms' formalization of the left side of the rule $A_{1i},\dots,A_{nj}$ all nonzero numbers $\mu_{\underline{A}1i}(u_1),\dots, \mu_{\underline{A}nj}(u_n)$ are assigned.
- Moving from the symbolic notation $((A_{1i},\dots,A_{nj}), B_k)$ of the selected inference rule to its logic notation dependent on u in the form $(\min\{\mu_{\underline{A}1i}(u_1),\dots, \mu_{\underline{A}nj}(u_n)\}, \mu_{\underline{B}k}) = (\mu_{\underline{A}^*}(u), \mu_{\underline{B}k})$ we obtain a number $\mu_{\underline{A}^*}(u)$ indicating not only the degree of truthfulness that the situation characterized by the vector u actually occurred, but also the degree of significance of the term B_k in the multi-criteria evaluation of the vector u.
- $\mu_{\underline{B}k}$: $Y \rightarrow \langle 0,1\rangle$ is the standard membership function of the elements $y \in Y$ to the fuzzy set $\underline{B}_k$. On the other hand, $\mu_{\underline{B}k}^*$: $Y \rightarrow \langle 0,1\rangle$ in the form $\mu_{\underline{B}k}^*(y) = \min\{\mu_{\underline{A}^*}(u), \mu_{\underline{B}k}(y)\}$ limits from above the function $\mu_{\underline{B}k}$ by means of the number $\mu_{\underline{A}^*}(x)$.
- The components of the fuzzy system of the block K will then be the elements of the set $P_u = \{B_k\text{: } \exists y\text{:}(\min\{\mu_{\underline{A}^*}(u),\mu_{\underline{B}k}(y)\} > 0), ((A_{1i},\dots,A_{nj}), B_k) \in P, y \in Y\}$. These are the partial intermediate results by aggregation (logical sum) of which we obtain the formalized term $\underline{B}_{agg} = \{(y, \mu_{agg}(y))\text{: } y \in Y\}$ on Y, for which it applies (1)

$$\mu_{agg}(y) = max\left\{min\{\mu_{\underline{A}^*}(u), \mu_{\underline{B}k}(y)\}\text{: }\left((A_{1i},\dots,A_{nj}), B_k\right) \in P, y \in Y\right\} \quad (1)$$

– The defuzzification phase is the last step to obtain the output value y_x of the multicriteria model of an alternative $(a_1,\ldots, a_c) \in A$ as the mean value of the elements $y \in \underline{B}_{agg}$ weighted by the values $\mu_{agg}(y)$ of their significance – see (2). Therefore

$$y_x = \int y \cdot \mu_{agg}(y)dy \,/ \int \mu_{agg}(y)dy \tag{2}$$

where ∫ is the symbol of a definite integral over the universe Y – more in detail in [6] and [2].

The process of the multi-criteria assessment of a short time prediction of the trend of the Czech company Škoda Auto factory not to go bankrupt/not to go bunkrupt is demonstrated in other sections.

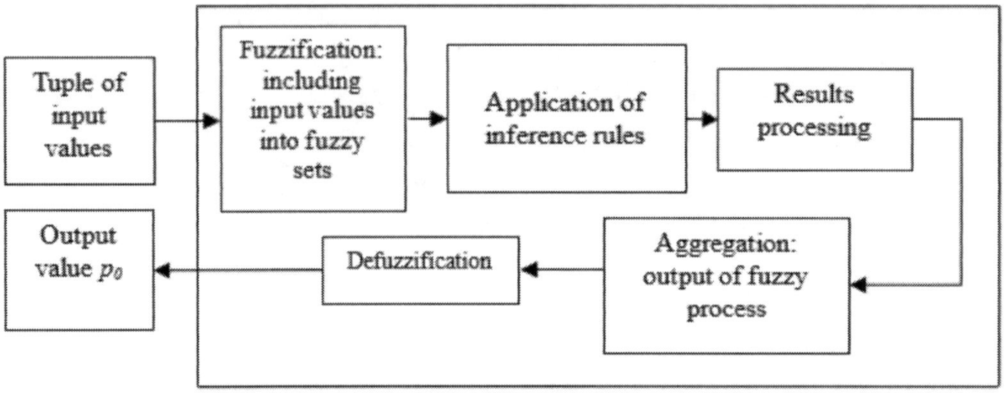

Fig. 1. Schematic representation of the multicriteria problem solving process within the fuzzy approach
Source: Author´s own processing.

3 Application – Introduction to the Problem

The strong dependence of the Czech economy on the automotive industry with a strong export focus on Germany have often been critically discussed by economists and politicians. This dependence had clearly manifesed itself during economic recession as the data of car production volume in Tab. 1 show.

Table 1 The passenger cars´ production in the Czech Republic

Year	2008	2009	2010	2011	2012	2013	2014	2015	2016	2017	2018	2019	2020
Passenger car production (ths. units)	934	976	1070	1192	1172	1128	1247	1298	1344	1414	1437	1428	1005

Source: [10], [14] / author´s own processing.

The data in Tab. 1 show that in the year 2020 the Czech (and also German) automotive sector returned to the values of 2009 crisis year. The coronavirus outspread at the beginning of 2020 brought a rapid decline in the production and sales of cars, automotive components, and technologies. The reasons are obvious: the whole economy decline accompanied by uncertainty on the demand side, the cessation of investment and renewal of vehicle fleets, and last but not least the closure of car showrooms and car bazaars.

At this point we should remind ourselves that the automotive sector had already been decreasing for some years before Covid-19 outbreak. It was mainly caused by the decline in demand in China; the markets of North America and Europe were roughly stagnant. The drop in production and sale was also caused due to the transition to the electric car production as some factories were modified to produce electric cars and their capacity decreased. This transition goes hand in hand with the additional costs that increase the price of cars ([10], [13].

3.1 Assumptions and Data of the Fuzzy Predictive Model

The formulation of the fuzzy model of prediction of the non-Bankrupt trend (non-B) has the following specifics:

- We can specify indicators affecting the target variable (non-B) – these are historically measured data of influencing input variables.
- The trend of influencing input variables can be used for the one-point prediction purposes only if an economy is not at the state of an extreme, which deviates to one or the other side (extreme growth/decline) from the "normal" (expected) framework (in terms of quantification of the decline, growth or stagnation). Its prediction is justified by the development of the historical part of influencing input values on which a target output vaguely depends – under these circumstances the fuzzy procedure was succesfully applied e.g. in [8].
- Under extreme conditions and/or the high uncertainty of input data the prognosis relies on the possible values of input variables specified by intervals; from this it is obvious that the prediction outcome is presented in the form of the interval of possible values.

Tab. 2 shows the developement of the input variables between 2008 – 2020 affecting the target variable non-B, which are $\Delta GDP_G\%$ (a yearly change in Germany's gross domestic product), $\Delta AP_{CZ}\%$ (a yearly change in the Czech cars´ production) and Z-score (a bankruptcy criterium Z-score based on a modified version of the Taffler´s model). From the viewpoint of external and internal interest groups these entry data perform results, for which the highest possible value is desirable.

Table 2 Basic macroeconomic input data for fuzzy model of non-B prediction

Variable/Year	2008	2009	2010	2011	2012	2013	2014	2015	2016	2017	2018	2019	2020
$GDP_G\%$	0,8	-5,7	4,2	3,9	0,4	0,4	2,2	1,7	2,2	2,5	1,5	0,6	-5,8
$AP_{CZ}\%$	1,4	4,5	9,5	11,5	-1,7	-3,7	10,5	4,2	3,5	5,2	1,7	-0,7	-29,6
Z-score	1,1	0,92	1,06	1,1	1	1,1	0,9	1,2	1,5	1,6	1,3	1,1	0,9

Source: [4], [14] / author´s own processing.

From Tab. 2 we see two significant slowdowns (2008/2009 and 2020). The 2008/2009 downturn is primarily caused by the financial crisis 2008 followed by the economic crises; the second slowdown is due to the Covid-19 pandemic outbreak.

The Czech economy was not affected by the global 2008 financial crisis; it was rather hit by the global economic crisis, which followed and which was mainly a consequence of a sharp decline in foreign demand (a GDP yearly change in the Czech republic in 2009 was 4,7 % / the yearly drop of GDP of eurozone in 2009 was -4,5 %). The current Covid-19 crisis has been a major economic recession, which significantly affected almost all economies in the world (the 2020 yearly drop in GDP of the Czech republic was 5,8 %, in eurozone countries it was -6,5 %).

Needless to say that any one-point economic prognosis does not have much informative value as enormous uncertainty is still presented [5]. The coronavirus spread and precautionary measures have not disappeared. On the other hand, we cannot ignore that the population immunization is growing. Furthermore, the development of new drugs and advanced vaccines indicate a real possibility of a positive reversal. The economy release phases followed by restrictions´ alleviation were accompanied by growing demand for goods and services. Moreover, the abnormal rise in the amount of private savings made during the strict pandemic measures (although the level of savings is very uneven in the population) makes many economists believe that economies embark on a path of growth in demand and supply, and thus in the growth of GDP in the forcomming period.

3.2 Procedures and Results

A direct prediction of a non-bankrupcy trend from the historical series of values GDP_G, AP_{CZ} and Z-score is not possible. At this stage, the state of affairs described above and the opinion of experts is a major specificity of the fuzzy model formulation, i.e., determination of the model structure and formulation of inference rules.

Regarding the uncertainty and the aforementioned arguments for a positive medical and economic development we will consider the influencing variables within the interval limits, in which the "worst" data are non-negative lowest values, or respectively the "best" data are the highest values of the period 2008-2020 ($GDP_G\% \in \langle 0{,}4; 4{,}2\rangle$, $AP_{CZ}\% \in \langle 1{,}4; 11{,}5\rangle$, Z-score $\in \langle 0{,}9, 1{,}6\rangle$) – see Tab. 2.

The model works with the dimensionless values (%) $u_{Z\text{-score}}$, u_{AP_CZ}, u_{GDP_G} of input universes U_{Zscore}, U_{AP_CZ}, U_{GDP_G}, or respectively with dimensionless value (%) y of output universe Y_{BT}. The dimensionless input values are obtained by converting the given basal input interval values $x_{Z\text{-score}} \in \langle 0{,}9; 1{,}6\rangle$, $x_{AP\_CZ} \in \langle 1{,}4; 11{,}5\rangle$, $x_{GDP\_G} \in \langle 0{,}4; 4{,}2\rangle$ of the respective input linguistic variables into interval $\langle 0, 100\rangle$, where the 100 % signals zero threat to bankrupcy; analogically, the left extreme indicates a definite bankrupt.

The fuzzy system in block *K* includes numerical vectors in % (GDP_G, AP_{CZ}, *Z-score*), $GDP_G \in \langle 10, 100\rangle$, $AP_{CZ} \in \langle 12, 100\rangle$, $Z\text{-}score \in \langle 56, 100\rangle$ of the triple of input linguistic variables GDP_G, AP_{CZ} and Z-score with terms $\underline{L}_i$, $\underline{M}_i$ and $\underline{H}_i$, i = gdp_g, ap_{cz}, z_{score} ($\underline{L}$ – low, $\underline{M}$ – middle, $\underline{H}$ – high).

Let us suppose that fuzzification tables for linguistic input variables are symmetric and identical except for the subscripts (i.e., for all i = gdp_g, ap_{cz}, z_{score} it applies: $a_i = 20$, $b_i = 40$, $c_i = 60$, $d_i = 80$), which applies for the output linguistic variable non-B as well, see Table 3.

Table 3 Fuzzification table valid for i = gdp_g, ap_{cz}, z_{score}

Interval	$u_i < 20$	$20 \leq u_i < 40$	$40 \leq u_i < 60$	$60 \leq u_i < 80$	$u_i \geq 80$
$\underline{L}_i$	1	$(40 - u_i) / 20$	0	0	0
$\underline{M}_i$	0	$(u_i - 20) / 20$	1	$(80 - u_i) / 20$	0
$\underline{H}_i$	0	0	0	$(u_i - 60) / 20$	1

Source: authors' own processing.

The following fuzzy procedure in block *K* enables us to calculate a number $y \in \langle 0, 100\rangle$ of the output linguistic variable of a non-Bankrupt trend (non-B) taking into account the setting of the terms $\underline{L}$, $\underline{M}$ and $\underline{H}$. The set of inference rules of the type $((\underline{T}_{gdp\_g}, \underline{T}_{ap\_cz}, \underline{T}_{Z\text{-score}}), \underline{T})$, $\underline{T}_i \in \{\underline{L}_i, \underline{M}_i, \underline{H}_i\}$, $\underline{T} \in \{\underline{L}, \underline{M}, \underline{H}\}$ has 27 elements formed by the strategies of the predominant element. It assigns to the given left side the very term $\underline{T}$, which predominates on the left side. If there is no such term, the term $\underline{M}$ is assigned.

For the recalculated vector (u_1, u_2, u_3) it holds: $u_1 = gdp_g$, $u_2 = ap_{cz}$ and $u_3 = z_{-score}$. To obtain the outputs y_{min} and y_{max}, the interval values combination (u_{1min}, u_{2min}, u_{3min}) for y_{min} and (u_{1max}, u_{2max}, u_{3max}) for y_{max} is processed as follows:

- $h(u_{1min}) = h(10) = \{\underline{L}_{gdp\_g}\}$, $h(u_{2min}) = h(12) = \{\underline{L}_{ap\_cz}\}$, $h(u_{3min}) = h(56) = \{\underline{M}_{Z\text{-}score}\}$, $H = \{(\underline{L}_{gdp\_g}, \underline{L}_{ap\_cz}, \underline{M}_{Z\text{-}score})\}$. The triad ($\underline{L}_{gdp\_g}$, $\underline{L}_{ap\_cz}$, $\underline{M}_{Z\text{-}score}$) of the input fuzzy sets of the projection F assigns the output fuzzy set $\underline{H}$ according to the already mentioned strategy of the predominant element. Therefore, $M_{\underline{M}} = M_{\underline{H}} = 0$, $M_{\underline{L}} = \mu_{AGG}(y) = \mu_{\underline{L}}(y)$.
  For the value $\boldsymbol{y_{min}}$ it holds: $\boldsymbol{y_{min}} = \int y \cdot \mu_{\underline{agg}}(y)dy \,/ \int \mu_{\underline{agg}}(y)dy = 30 / 2 =$ **15 %**.
- $h(u_{1max}) = h(100) = \{\underline{H}_{gdp\_g}\}$, $h(u_{2max}) = h(100) = \{\underline{H}_{ap\_cz}\}$, $h(u_{3max}) = h(100) = \{\underline{H}_{Z\text{-}score}\}$, $H = \{(\underline{H}_{gdp\_g}, \underline{H}_{ap\_cz}, \underline{H}_{Z\text{-}score})\}$, $M_{\underline{L}} = M_{\underline{M}} = 0$, $M_{\underline{H}} = \max\{\{\min\{1, 1, 1\}\}\} = 1$.

  For the value $\boldsymbol{y_{max}}$ it holds: $\boldsymbol{y_{max}} = \int y \cdot \mu_{\underline{agg}}(y)dy \,/ \int \mu_{\underline{agg}}(y)dy = 2\,533.33 / 30 =$ **84 %**.
- The average interval fuzzy value $y_u = (15 + 84) / 2 =$ **49,5 %**.

4 Results Discussion and Practical Usage

The fuzzy procedure of deffuzification step (see Fig. 1) provided us with the output of the fuzzy interval y (%) = ⟨15; 49,5; 84⟩ expressing the tendency of the company not to get bankrupt in the short term period 2022-2023. According to the model results, the non-B = y_u = 49,5 %, i. e., the average susceptibility of Škoda Auto factoty not to bankrupt is medium.

At this point it is necessary to take into account the facts regarding the positive trend in retreat of Covid-19 pandemy, the positive trend in indicators´ development within the first half of the year 2021 (GDP_G% = 3,6, AP_{CZ} % = 24,7) and the values of Z-score > 0,3 within the period 2008-2020 (the Taffler´s model has excluded an imminent threat of bankruptcy of Škoda Auto factory across the reference period). This entitles us to argue that the threat of bankruptcy in the next two years is below average; the fuzzy values of non-bankrupcy trend are therefore y (%) = ⟨49,5; 84⟩ with the non-B = y_u = 67 % as the average indicator of non-bankrupcy.

Given the fact that the left extreme y_{min} = 49,5 is the bottom limit of possible (expected) values, which will not be likely lower (with the 70 % probability) in the next two years´ period, the value y_{min} can be interpreted as a measure of business safety from the viewpoint of the bankruptcy evaluators.

5 Conclusion

The aim of the paper was to present a general fuzzy algorithm formulated within the Zadeh's fuzzy approach for solving an economic short-term prediction under conditions of external uncertainty (relating to the data entering the model) and internal uncertainty (related to the algorithmic model). Such a procedure offers a short-time period prediction more than equivalent to an alternative statistical prognosis.

The presented fuzzy algorithm builds its core procedures on one of the important concepts of fuzzy logic, which is the "extension principle", at the level of linguistic variables and their terms within which the problem is solved. The algorithm was utilized to estimate the non-bankrupcy trend (non-B) of the Škoda Auto factory within an interval of possible values.

Under „normal", i.e., an expected behaviour of economy, the entry data are partial components of the historical development of the indicators that directly and, to a certain extent expectidely, influence the behaviour of a target output and its prediction (the influencing indicators here are a change in German gross domestic product, a change in the

Czech cars´ production and a bankruptcy criterium Z-score based on a modified version of the Taffler´s model, the target indicator is a non-bakrupcy index (non-B) of the Škoda Auto factory). Due to the Covid-19 outbreak such a direct procedure is excluded.

Instead, the possible development of influencing input variables is taken into account in the form of interval values capturing the extreme data of the downturn and growth within the monitored period 2008-2020. The input data are used in the fuzzy model formulation phase, within which an expert, through the interventions in the structure of the model and the formulation of inference rules, takes into account one´s own knowledge and experience. The own process of solving the problem is mechanical and independent of the human factor.

The fuzzy process consists of five steps, which are *fuzzification*, application of *inference rules*, further *processing of intermediate results*, *aggregation* and *defuzzification*. The here formulated fuzzification table makes the solution more transparent and easier. Regarding the inference rules in the function of tools for partial transformations of inputs to an output, only one of the total number of 27 was used: (($\underline{L}$, $\underline{L}$, $\underline{M}$), $\underline{L}$). The right-hand side of the rule corresponds with the predominant element strategy in formulation of the system of inferential rules.

The result of the solution at the level of terms of the linguistic variable non-B % was an interval of possible values y (%) = ⟨15; 49,5; 84⟩ expressing the tendency of the company not to get bankrupt in the short term period 2022-2023. Other objective facts had to be taken into account when evaluating the results and interpreting them: the positive trend in retreat of Covid-19 pandemy, the positive trend in GDP_G and AP_{CZ} indicator development within the first half of the year 2021 and the values of Z-score > 0,3 within the period 2008-2020 excluding an imminent threat of bankruptcy of Škoda Auto factory. This entitled us to argue that the threat of bankruptcy of Škoda Auto factory in 2022-2023 is below average: it can be expected in the interval of possible values y (%) = ⟨49,5; 84⟩ with the non-B = y_u = 67 % as the average indicator of non-bankrupcy. The results can be interpreted as a measure of business safety from the viewpoint of the bankruptcy evaluators, which is an indicator that can be widely utilized in practice.

References

1. F. Aliu, D. Pavelkova, B. Dehning, Portfolio risk-return analysis: The case of the automotive industry in the Czech Republic. *Journal of International Studies* (2017)
2. L. Běhounek, P. Cintula, From fuzzy logic to fuzzy mathematics: A methodological manifesto. *Fuzzy Sets and Systems*, ***157***(5), 642-646 (2006)
3. Eurostat, *Your key to European statistics – euro indicators*.[online], Available at: https://ec.europa.eu/eurostat/news/euro-indicators (2021)
4. Eurostat, *Germany: Real GDP growth rate – volume*. [online], Available at: https://ec.europa.eu/eurostat/databrowser/view/tec00115/default/table?lang=enStatista (2020)
5. A. Evgenidis, S. Papadamou, C. Siriopoulos, The yield spread's ability to forecast economic activity: What have we learned after 30 years of studies?. *Journal of Business Research*, ***106***, 221-232 (2020)
6. M. Grabisch, The application of fuzzy integrals in multicriteria decision making. *European Journal of Operational Research*, ***89***(3), 445-456 (1996)
7. S. Hašková, Estimation of Long-Term Projects in Terms of Uncertainty of Inputs. *38TH INTERNATIONAL CONFERENCE ON MATHEMATICAL METHODS IN*

ECONOMICS (MME 2020). (Eds.). Kapouek, S. & Vranova, H., (pp. 154-159), Brno, CZECH REPUBLIC (2020)

8. S. Hašková, S, Fuzzy approach to short-term youth unemployment forecast. In *SHS Web of Conferences*, 73, 01007. EDP Sciences (2020)
9. S. Hašková, P. Fiala, A fuzzy approach for the estimation of foreign investment risk based on values of rating indices. *Risk Management*, ***21***(3), 183-199 (2019)
10. International Organization of Motor Vehicle Manufacturers [online], Available at: https://www.oica.net/category/production-statistics/2019-statistics/ (2021)
11. R. M. del Rio-Chanona, P. Mealy, A. Pichler, F. Lafond, D. Farmer, Supply and demand shocks in the COVID-19 pandemic: An industry and occupation perspective. *ArXiv preprint arXiv:2004.06759* (2020)
12. T. A. Skvortsova, T. A. Pasikova, L. N. Vereshchagina, A. N. Pozdnishov, A. V. Sukhovenko, The problem of bankruptcy of business entities as a consequence of the COVID-19 pandemic (2020)
13. M. Srivastava, L. Tyll, The effect of industry-specific networking behaviour on the internationalization performance of Czech SMEs. *European Business Review* (2020)
14. Škoda, Výroční zprávy [online], Available at: https://www.skoda-storyboard.com/cs/vyrocni-zpravy/ (2021)
15. L. A. Zadeh, Fuzzy logic and approximate reasoning. *Synthese*, ***30***(3-4), 407-428 (1975)
16. L. A. Zadeh, K. S. Fu, K. Tanaka. (Eds.). *Fuzzy sets and their applications to cognitive and decision processes: Proceedings of the us–japan seminar on fuzzy sets and their applications, held at the university of california, berkeley, california, july 1-4, 1974.* Academic press (2014)

Do companies in the Czech Republic feel a decline in ICT graduates?

Jiří Kučera[1*], *Yaroslava Kostiuk*[1], and *Daniel Kortiš*[2]

[1]University of Žilina, The Faculty of Operation and Economics of Transport and Communications, Department of Economics, Univerzitná 8215/1, 010 26 Žilina, Slovakia
[2]Institute of Technology and Business in České Budějovice, School of Expertness and Valuation, Okružní 517/10, 370 01 České Budějovice, Czech Republic

Abstract. The aim of this paper is to determine the possible cause of lagging Czech companies in the field of HR transformation. The basic source of data is data from the Czech Statistical Office. The paper uses the method of classification analysis of graduates in the field of information and communication technologies. The paper is divided into two parts, where the first part deals with the evaluation of tabular data and the second with the testing of the established hypothesis (H0). The number of graduates in the field of information and communication technologies in the Czech Republic has been steadily declining since 2015, although the results achieved so far do not indicate a significant change, which could be the main cause of Czech companies lagging behind in HR transformation. The low involvement of graduates in this field is also caused by older and backward employees in companies, who do not like to change established systems.
Keywords: graduates, digital transformation, information and communication technologies, human resource management
JEL codes: A22, A23, O15, O31

1 Introduction

So far, the digital transformation in HR (Human Resources) is rather reminiscent of backward or outdated practices. Everyone is waiting for the change to come on its own, but on the contrary, there is nothing to wait for [1]. Nothing will come by itself; you need to start working on it. ICT (Information and Communication Technologies) are already so advanced that this cannot be overlooked even in the field of HR. The next generation of workers has been coming to work for many years with a mobile phone and with certain expectations, but the request for days off work is still written out on paper or sent by e-mail. Managers always fill in the same Excel forms sent back and forth. The only thing that hinders the digital transformation in HR is the rigid thinking of HR and IT (Information Technology) departments [2].

Anyone who monitors technological development, if only a little, will admit that it has exceeded expectations, for better or worse. Today, it is not a question of whether or not something can be technologically solved [3]. It is a question of the extent to which employees

* Author's contact information: kuceraj@mail.vstecb.cz

are informed about technological innovations and whether they are willing to participate in the digital transformation themselves [4]. The digital transformation cannot be imagined without disturbing established work orders. Of course, the digital transformation in the corporate sphere follows different rules than in the private sphere. They differ in terms of the goal to be achieved and also in motivation [5].

The digital transformation in the company environment in the field of HR is specific, in many companies it is delayed or has not even started at all. Even so, recording information in many complicated Excel spreadsheets, searching hundreds or thousands of e-mail messages, or constantly redrawing the organizational structure in PowerPoint cannot provide 21st-century productivity.

The digital transformation in the HR area is often delayed not only in smaller, but also in a number of medium-sized, and especially large companies, where the effects on productivity are evident [6]. Despite the visible development of ICT technologies, business practice in the HR field remains stuck on models 20 to 30 years old. It is not difficult to find and eliminate the causes under one assumption if there is someone in the company who recognizes the opportunity in the digital transformation of HR technologies [7].

Today, ICT technologies in the Czech Republic are somewhat backwards compared to other countries, in my opinion mainly because young people devoted to ICT do not get many opportunities in their jobs, because companies do not like to change established and used procedures. Another reason can also be considered the lack of graduates in ICT fields, which would mean that companies do not have the opportunity to keep up with the times in the field of HR.

The aim of this paper is to determine the possible cause of Czech companies lagging behind in the field of HR transformation.

For this purpose, the following research question (RQ) was formulated:

- RQ1: How many students graduated in the field of ICT?

Furthermore, the null hypothesis (H0) will be tested:

- H0: The composition of the total number of graduates did not change during the monitored years.

Otherwise, the alternative hypothesis (H1) will be valid.

- H1: Companies in the Czech Republic will not be able to keep up with other countries, due to significant changes in the level of education of graduates of ICT fields.

2 Literature research

Because automation and structural transformation mark the new digital reality of the 21st century, today's organizations require a highly resilient and committed workforce to maintain a competitive advantage [8]. At the turn of the century and millennium, humanity faces global challenges, which it seeks to understand and to which it seeks theoretical and practical answers [9]. The spread of digital technologies increases the density of the communication network between all parties, devices, people and institutions involved [10]. Through digital transformation, small service businesses seek to increase their competitive advantage, improve business performance, and achieve business growth [11].

The development of artificial intelligence and digital transformation has brought with it the challenge of harnessing the potential of transformation of human resources (HR) processes, systems and organizations through new digital platforms, applications and ways of providing services [12]. The aftermath of the Covid-19 pandemic has caused a deep global economic downturn and a major rethink of enterprise technology [13]. Digital transformation processes have intensified, which is reflected in the development of global supply chains [14].

An increasing number of companies are moving forward in the digital transformation, which is the creation of new systems that connect people, things and companies with digital technologies [15]. Strohmeier [16] focuses on conceptual clarification of digital human resource management and related concepts of digitization of human resource management, referring to the general literature on digital organizations to develop terminology and typology, which together clarify the concept of digital human resource management and related concepts.

Hajnic and Boshkoska [12] introduce a method implemented in a decision support platform for employee transfer developed to address three constraints. The proposed platform is designed to process all available information and human-defined rules to support HR managers in planning and organizing their strategic activities in a timely manner, speed up the candidate selection process in the organization for more than 82%, provide some of the best alternatives for employee utilization, reduce operating costs by more than 87% and reduce the administrative burden of human resource management.

Dahlbom et al. [17] discuss the implications of role transformation, HR-induced human resource functions. These are qualitative interviews in nine leading Finnish companies. The results show that both technical and human barriers, working with very basic HR processes and traditional information systems and poor data quality and hindering the adoption of advanced HR, further point to the need for increased collaboration with data analysts and HR professionals in providing and understanding data related to HR for business decision making.

Caputo et al. [18] explore the relationship between soft skills, information technology and Big Data in order to create a possible bridge capable of connecting human and technological dimensions to increase company performance. Using the Business-Inventory of Personality, the working personality of 4,758 human resources involved in 72 high-tech European companies was analyzed and its relationships with companies' investments in Big Data and the economic performance of companies were tested using structural equations. The results show the existence of strong relationships between some elements of the human resources personality, such as work motivation and social competences and the economic performance of companies. At the same time, they also clarify the mediated effect of investing in Big Data in the relationship between the organizational behavior of human resources and the economic performance of companies.

Intelligent manufacturing is a general concept that is constantly evolving [19]. Del Giudice et al. [20] analyze the impact of circular economy practices on company performance in the circular supply chain and examine the moderating role that the big data supply chain plays in these relationships. They use data collected through an online survey distributed to managers of 378 Italian companies that have adopted the principles of the circular economy. The results provide a number of original contributions to the practices of circular economy in a supply chain based on big data and provide useful insights, as well as taking on the challenge of capturing digital transformation trends and expanding sustainability research in supply chain management, further enhancing the literature by examining relationships between three different types of circular economy supply chain practices and company performance and also clarify the moderating role of big data in decision making and implementing circular supply chain solutions to achieve better environmental, social and economic benefits.

Choain et al. [21] claim that a professional services company has implemented Richard Florida's "Florida class" concept from the urban to the corporate environment for organizational change. Talent, technology and tolerance make up a high-value triptych driving the growth and attractiveness of the city - HR can appropriate it to bring about profound changes in the company's management and culture. Therefore, the return on experience from the project, which was implemented in the course of one year in the company

Mazars, is presented here. The results offer a model for identifying members of the "creative class" in the corporate environment and a tried and tested approach to integrating the "creative class" into the exclusive and institutional exercise of the determination strategy.

The assessment of psychosocial risks at work aims to carry out interventions in the field of health protection at work based on the assessment of working conditions. However, it is expected that the ongoing industrial digitization will be accompanied by a fundamental transformation of these working conditions. The results of qualitative interviews are critically reflected and build on existing research [22].

3 Data and methods

As mentioned in the introduction, the biggest weakness of ICT technology may be the inaction of the employer to involve young graduates in individual positions or research. The second proclaimed assumption is the lack of graduates of the ICT field.

Data from the Czech Statistical Office will be used and analyzed to find answers to the set research questions. Here, a classification analysis of students' involvement in the fields of ICT technology will be performed. The analysis itself will be divided into two parts, the first of which will be the evaluation of tabular data, in which we have data that divide our graduates from several perspectives. The first aspect divides graduates according to whether they have foreign or Czech citizenship. The second aspect will be the level of education completed within the education system (Bc., Ing. / Mgr., Ph.D.). This will find the answer to VO1.

The second part of the analysis will consist of a percentage expression from the above points of view. Furthermore, a statistical test of the set of analyzed that will be used. Specifically, the chi square test will be used to confirm or refute the established null hypothesis (H0).

4 Results

The table below presents data obtained from the Czech Statistical Office. Data is from 2010–2019, more current data for 2020 is not yet available.

Table 1. Graduates of ICT fields from universities in the Czech Republic

	2010	**2011**	**2012**	**2013**	**2014**	**2015**	**2016**	**2017**	**2018**	**2019**
Total	3 787	3 928	4 225	4 762	4 468	4 599	4 361	3 916	3 802	3 638
Citizenship										
Czech	3 259	3 352	3 646	4 024	3 822	3 831	3 550	3 131	2 921	2 668
Foreign	528	576	579	738	646	768	811	785	881	970
Study program										
Associate degree	-	-	-	258	214	249	-	-	-	-
Bachelor's	2 503	2 529	2 543	2 596	2 479	2 554	2 291	2 080	2 055	1 989
Master's	1 205	1 326	1 599	1 817	1 696	1 703	1 986	1 750	1 667	1 580
Doctoral	79	73	83	92	79	93	84	86	80	69

Source: Czech Statistical Office, 2021, processed by authors.

From the above table, it is clear at first glance that the number of ICT graduates in the Czech Republic has been declining since 2015. Over the last four years, there has been a decrease from 4,599 graduates to 3,638 graduates. It is interesting that the number of foreign graduates, on the other hand, is constantly increasing. And as for the last evaluation, according to the studied program, most graduates completed a bachelor's degree program, while the overall shares in individual levels of education do not differ much, as shown in Table 2. Data in 2010-2012 and 2016-2019 did not contain information on the number of graduates Associate degree.

Table 2. Graduates of ICT fields from universities in the Czech Republic in percentage

Study program	2010	2011	2012	2013	2014	2015	2016	2017	2018	2019
Associate degree	-	-	-	5,41	4,78	5,41	-	-	-	-
Bachelor's	66,09	64,38	60,18	54,51	55,48	55,53	52,53	53,11	54,05	54,67
Master's	31,81	33,75	37,84	38,15	37,95	37,02	45,54	44,68	43,84	43,43
Doctoral	2,08	1,85	1,96	1,93	1,76	2,02	1,92	2,19	2,1	1,89
Total	100	100	100	100	100	100	100	100	100	100

Source: Czech Statistical Office, 2021, processed by authors.

Based on the above tables, it can be stated that the proportions of graduates of individual levels of education do not differ significantly, despite the declining total number of graduates of ICT disciplines.

However, this statement was tested using the chi square test. A significance level of p 5% was chosen. According to the statistical test, it can be stated that in the case of variability in the number of bachelor's and master's degree graduates, there were no significant changes in the monitored years ($p_{bachelor's\ degree} = 5.37015^{-41}$, $p_{master's\ degree} = 6.75981^{-56}$). However, in the case of a doctoral degree, H0 cannot be refuted ($p_{doctoral} = 0.71173$).

5 Discussion

In this part, the research questions and hypotheses will be answered. RQ1: How many students graduated in the field of ICT?

The total numbers of ICT graduates varied considerably from 2010 to 2019. Most graduates were in the ICT field in 2013-2015, when the field of ICT was a promising field for many young people. In these years, 4,762 (2013), 4,468 (2014) and 4,599 (2015) graduates were recorded. After 2015, however, this interest dropped significantly. In the last monitored year (2019) there were only 3,638 graduates in the ICT field. This can be considered a problem especially for companies, which are forced due to constant technological development to expand the number of their employees by experts in the field of ICT.

Based on the performed chi-square test at the selected level, the significance of p 5% was refuted in the case of the bachelor's degree and the master's degree of study H0 and can therefore be considered a valid alternative hypothesis H1. However, in the case of the number of graduates of the doctoral study program, H0 was confirmed. However, in the case of digital transformation in HR in companies, it can be argued that this high level of education is not necessary for companies and the digital transformation of the HR system can also be carried out by bachelor's or master's degree graduates.

At the same time, it can be assumed that this can reduce the wage costs for these workers than in the case of employment of graduates with doctoral degrees.

6 Conclusion

The aim of this paper was to determine the possible cause of lagging Czech companies in the field of HR transformation.

To achieve the set goal, one RQ was determined and the above hypothesis (H0) was further tested.

Based on the achieved results, it was found that the number of graduates in ICT fields has been declining recently, which could indicate that Czech companies do not have much opportunity to improve their position in the field of digitization in HR compared to other countries. However, the above hypothesis H0 was tested for clarification. Based on the performed statistical test, it was found that H0 cannot be refuted and digital transformation of the HR system can also be performed by graduates of bachelor's or master's studies.

However, based on the achieved results with a gradually increasing trend of demands on ICT professionals, it can be assumed that the need for graduates of ICT disciplines will continue to increase. In the following research, an analysis of the entire downward trend in the number of graduates could be performed with the identification of a specific moment when a critical shortage of graduates of ICT fields will be achieved. Last but not least, the creation of recommendations or methodologies to prevent this tragic scenario does not occur at all.

References

1. R. Kubík, Quantity and quality of human capital: Does the average years of schooling increase the quality of student? *Littera Scripta*, **6**(1), 61-69 (2013)
2. C. Jane Hollowell, Z. Rowland, T. Kliestik, J. Kliestiková, V. V. Dengov, Customer loyalty in the sharing economy platforms: How digital personal reputation and feedback systems facilitate interaction and trust between strangers. *Journal of Self-Governance and Management Economics*, **7**(1), 13-18 (2019)
3. M. Connolly-Barker, T. Kliestik, P. Šuleř, K. Zvaříková, Real-time decision-making in the information technology-driven economy: Planning, managing, and operating smart sustainable cities. *Geopolitics, History, and International Relations*, **12**(1), 73-79 (2020)
4. E. Nica, V. Konečný, M. Poliak, T. Kliestik, Big data management of smart sustainable cities: Networked digital technologies and automated algorithmic decision-making processes. *Management Research and Practice*, **12**(2), 48-57 (2020)
5. F. Míka, *SystemOnline* [online], Available at: https://m.systemonline.cz/hrm-personalistika/digitalni-transformace-hr-zatim-pripomina-cekani-na-godota-1.htm (2017)
6. H. Kearney, T. Kliestik, M. Kováčová, M. Vochozka, The embedding of smart digital technologies within urban infrastructures: Governance networks, real-time data sustainability, and the cognitive internet of things. *Geopolitics, History, and International Relations*, **11**(1), 98-103 (2019)
7. P. Malik, P. Garg, Learning organization and work engagement: the mediating role of employee resilience. *The International Journal of Human Resource Management*, **31**(8), 1071-1094 (2020)
8. I. Krajnik, R. Demeter, The impact of digital technologies on accounting in light of the challenges of Society 5.0. *Civil Szemle*, **17**(3), 115+ (2020)
9. S. Grumbach, O. Hamant, Digital revolution or anthropocenic feedback? *The Anthropocene Review*, **5**(1), 87-96 (2018)

10. C. L. Chen, Y. C. Lin, W. H. Chen, C. F. Chao, H. Pandia, Role of government to enhance digital transformation in small service business. *Sustainability*, **13**(3) (2021)
11. R. Maňák, Approach to human resources performance measurement. *Littera Scripta,* **6**(1), 82-94 (2013)
12. M. Hajnic, B. M. Boshkoska, A disruptive decision support platform for reengineering the strategic transfer of employees. *IEEE Access*, **9**, 29921-29928 (2021)
13. Dias R., N. Teixeira, V. Machová, P. Pardal, J. Horák, M. Vochozka, Random walks and market efficiency tests: Evidence on US, Chinese and European capital markets within the context of the global Covid-19 pandemic. *Oeconomia Copernicana*, **11**(4), 585-608 (2020).
14. O. Lomakina, V. Kookueva, A. Makarenko, Redistribution of economic resources in the digital society. *Business and Society Review,* **126**(1), 25-35 (2021)
15. Y. Kunii, T. Hasegawa, Fujitsu's activities to support digital transformation. *Fujitsu Scientific & Technical Journal*, **55**(1), 3-8 (2019)
16. S. Strohmeier, Digital human resource management: A conceptual clarification. *German Journal of Human Resource Management: Zeitschrift für Personalforschung,* **34**(3), 345-365 (2020)
17. P. Dahlbom, N. Siikanen, P. Sajasalo, M. Jarvenpää, Big data and HR analytics in the digital era. *Baltic Journal of Management,* **15**(1), 120-138 (2019)
18. F. Caputo, V. Cillo, E. Candelo, Y. P. Liu, Innovating through digital revolution. *Management Decision*, **57**(8), 2032-2051 (2019)
19. J. Zhou, P. Li, Y. H. Zhou, B. C. Wang, J. Y. Zang, L. Meng, Toward new-generation intelligent manufacturing. *Engineering*, **4**(1), 11-20 (2018)
20. M. Del Giudice, R. Chierici, A. Mazzucchelli, F. Fiano, Supply chain management in the era of circular economy: the moderating effect of big data. *The International Journal of Logistics Management,* **32**(2) (2020)
21. L. Choain, T. Malzy, Leading change through your creative class. *Journal of Organizational Change Management*, **32**(3), 377-384 (2019)
22. M. Diebig, F. Jungmann, A. Müller, I. C. Wulf, Content and process-related requirements for the risk assessment of psychological stress in the context of Industry 4.0. *Zeitschrift für Arbeits-und Organisationspsychologie A&O*, **62**(2), 53-67 (2018)
23. Czech Statistical Office. *Digital economy in numbers* [online], Available at: https://www.czso.cz/csu/czso/digitalni-ekonomika-v-cislech?fbclid=IwAR1ZerEKXHQdKalIdMAl6zQpPO7DSj9LCSaYPWt0FD3ym1ikcj5pusdTTmE (2020)

Prediction of supply and demand for jobs in the Czech Republic

Veronika Machová[1*], *Veronika* Šanderová, *Petr* Šuleř[1], and *Anna* Hodinová[1]

[1]Institute of Technology and Business in České Budějovice, School of Expertness and Valuation, Okružní 517/10, 370 01 České Budějovice, Czech Republic

Abstract. The topic of supply and demand for jobs is a very current and important topic, as their prediction contributes to the future operation of an organization. The base source of data is information and data obtained from online databases of the Ministry of Labor and Social Affairs of the Czech Republic. The aim of this work is to predict the supply and demand for jobs up to 2022, using the method of exponential alignment of time series. With the existence of natural unemployment, not all job positions will be filled. In such a situation, organizations will have to demand labor from abroad. In the event that organizations are unable to secure labor from abroad, they will have to reduce their activities and the associated volume of production they produce. Last but not least, the organization can start going bankrupt. In the forecast, we do not address the structure of unemployment by the highest attained education or by profession. We see the application benefits of the work in companies that manage human resources. At present, the prediction of future development is mostly at the national level and in the Czech Republic it is at the beginning of its journey. The developed methodology will enable organizations to monitor the current situation on the labor market and respond to changes in the short and long term.
Keywords: human resources, supply, demand, job position prediction
JEL codes: J22, J23

1 Introduction

The most important area in an organization is human resources and the organization should pay more attention to them. Members of the Human Resources Management Department provide the knowledge, tools, training, administrative services, coaching, legal and management consulting, and talent management oversight that the rest of the organization needs to function successfully [1, 2]. Over the past thirty years, a number of empirical studies have argued that the content of human resource management (HRM) has a positive effect on an organization's performance [3].

Labor demand is an important part of any organization for its future operation. According to Šikýř [4], human resource planning serves to realize the strategic goals of the organization, allows to determine the need of employees, estimate the coverage of these needs from both

[*] Author's contact information: machova@mail.vstecb.cz

internal and external sources, prevents shortage or surplus of employees and also serves to meet the needs of employees and allows planning their development and career.

However, the impact of HRM content on organizational performance across different settings still remains unanswered [5]. These different settings may represent boundary conditions for the effectiveness and efficiency of human resource management content for individual and organizational performance [6].

The organization anticipates how many and which employees it will need over the course of one year or over the next few years in order to know how much work it must provide and thus achieve the expected goals. In other words, the organization estimates the demand for labor [4]. Too many studies, especially in the area of strategic human resource management, ignore the impact of context or, at best, attempt at "context control", leading to a "(mostly) closed system". The irony is that the neglect of human and social outcomes over the past thirty years has made HRM a success [7,8,9].

When planning human resources, it is effective to start from the planned volume of production of the organization, on the basis of which the total need for employees is estimated and the total needs of employees are covered from internal resources [4]. Although concepts such as human resource management and engagement help us understand how employees respond to HRM, several authors still raise concerns that HRM research remains too management-focused and exclusively addresses managers' initiatives to explain employee perceptions and responses to HRM.

The point of view can also be extended to interface regarding interactions between the labor market and the state or, more precisely, the government. The labor market is highly influenced by the government, especially from the position of a major employer. The public sector, which consists of, for example, the police, the army, but also health institutions and education, offers many job positions, which are valued by set salaries [10].

Although evidence suggests that the effectiveness and efficiency of human resource management content depends on unforeseen events, such as product or industry, there are still unanswered questions about the effectiveness and efficiency of human resource management in recession situations or situations where the economy moves dynamically around its equilibrium conditions [5].

The aim of this work is to map the labor market in the Czech Republic with a focus on predicting supply and demand up to 2022.

To achieve this goal, we set the following research questions:

RQ1: How will supply develop up until the year 2022?

RQ2: How will demand develop up until the year 2022?

2 Literature research

On the labor market, both supply and demand are key factors in determining the appropriate salary for both jobseekers and employers. However, it is difficult to discover supply and demand in any labor market. For each labor market, the supply and demand of the labor market is constructed using job offers and the response of candidates. Equilibrium wage and equilibrium workload are calculated on the basis of supply and demand [11]. Predicting the difference between supply and demand is a type of traffic flow prediction that aims to predict the difference between supply and demand for the next period of time when historical traffic data is given [12]. Many researchers create various algorithms to obtain information about occupations and specific skills from online job advertisement databases. These sources provide a wealth of information on the labor market. However, this data is generally unstructured. Online job vacancy databases generally have unstructured text boxes that contain information about the occupation and skills required. However, such fields are manually filled in by company representatives and this requires data preparation procedures and algorithmic techniques to be implemented in order to obtain relevant information in a standardized form [13]. Boon et al. [14] present an argument regarding skills that is supported by the results of Green and Henseke [15] and Harvey and Turnbull [16]. These authors emphasized the growing importance of specific skills for labor market outcomes, such as wages. Green and Henseke [15] emphasizes the fact that employees with a combination of social and mathematical skills receive higher rewards than those with other skills, but Harvey and Turnbull [16] show that workers with certain skills (i.e. higher levels of self-organization) can receive higher rewards in digitally demanding industries compared to less digitally demanding industries. However, only the study of changes in employment shares can neglect important aspects of real skills demand, as shown by Katou [17] and Green and Henseke [15]. On the one hand, the job serves as a representative for skills requirements and experience requirements at the same time. This is because employment is a balanced outcome of labor demand and supply, which is assumed to be inflexible in the short term. In addition, due to the growing technological capacity to replace human labor, it is essential to gain a better understanding of changing skills and skill combinations. Ternikov and Aleksandrova [18] point to this, increasing productivity has positive effects, thanks to automation, which under certain conditions leads to growing demand for labor, skills shortages can have significant negative effects with far-reaching consequences for inequality. Way et al. [19] stated that "improved technology and capabilities contribute indirectly to economic growth through extensive scientific and technical training and learning by practice". The first goal is to model labor markets using the theory of supply and demand. In this approach, we use job position and application data collected from large job portals.

3 Data and methods

3.1 Data

The base source of data will be data from the Ministry of Labor and Social Affairs of the Czech Republic, where we collected data on labor supply and demand from the last five years from the file Analysis of the demand for labor and supply of labor. The data was collected from December 31, 2015, to March 31, 2021. All obtained data is from the whole of the Czech Republic for a monthly period. We exported the data obtained from the Ministry of Labor and Social Affairs to Microsoft Excel, where we graphically displayed time series of supply and demand using a line graph.

The number of jobs offered has been steadily declining since January 2016 from 467,403 to 215,010 jobs in November 2018. The supply curve will rise sharply in December 2018 to 457,506 jobs. From January 31, 2019, the supply curve will be declining slightly and in December 2019 the number of jobs offered will fall to 215,532. The supply as of January 31, 2020 amounts to 230,022 jobs offered, and since then the supply curve has been declining slightly up until March 2020. From April 31, 2020, the number of jobs offered is growing and at the end of 2020 the number of jobs offered is 291,977 in 2021, the number of positions offered will change slightly in the first three months, but the value will not fall below 300,000.

From December 31, 2015, the demand for jobs increased slightly from 102,564 jobs in demand to 141,024 until September 2016. From September 2016 to the end of 2016, the demand curve slightly decreased to 132,519 jobs in demand. From the beginning of 2017, the number of requested places has been constantly increasing up to November 31, 2018, to 323,570. In December 2018, the number of demanded jobs will fall to 225,972, and then the demand will start to grow again until August 2019, when 350,556 jobs are demanded. From then on, the demand curve fluctuates slightly between 330 thousand and 341 thousand. As of

August 31, 2020, the number of jobs in demand is 340,824 and the value decreases slightly to 310,729 in October 2020. From November 2020 to March 2021, the demand curve rises to 338,864 jobs in demand.

Chart No. 1 shows the number of jobs offered and demanded from December 31, 2015 to March 31, 2021.

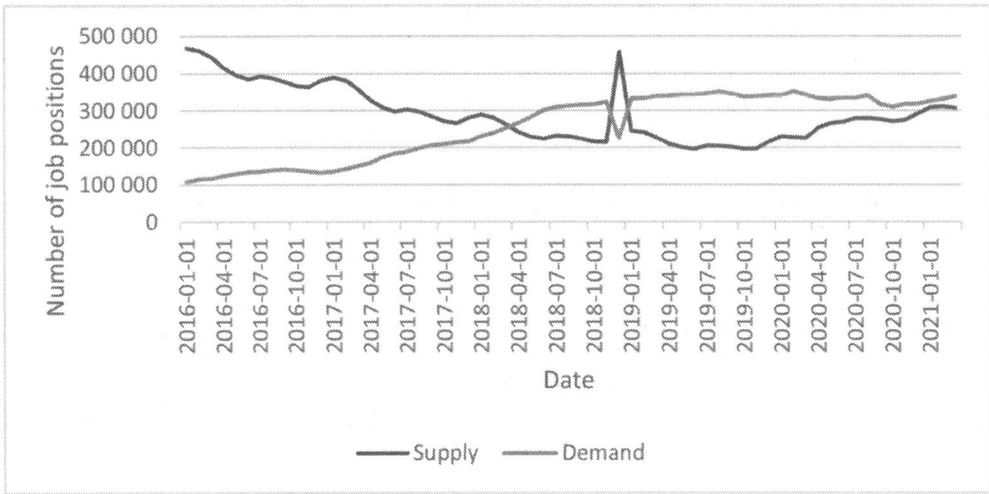

Fig. 1. Actual course of time series
Source: Authors.

The data will be provided by the internet portal https://data.mpsv.cz/web/data/analyza-poptavky-po-pracovni-sile-a-nabidky-pracovni-sily [online]. [cit. 2021-5-23].

3.2 Methods

To predict supply and demand, we choose the method of exponential alignment of time series. We will create a file in Microsoft Excel, where we will record the data obtained from the Ministry of Labor and Social Affairs, which will be divided into columns A, B, C and the number of rows will be 64. The first row will contain the name of the variables Month, Supply, Demand, and the other rows will be variables. We then import the generated file from Microsoft Excel into Statistica software from Tibco in version 14.0.0.15 and click on the Statistics tab. In the Statistics tab, we select advanced models and then time predictions. We will select variable in the time prediction and we will deal with the supply first. Then we select the exponential alignment and prediction. The seasonal component will be twelve, because we will monitor the seasonality monthly. This means dividing the time series into twelve time series. The forecast of the demand is given up until the end of 2022. In 2022 it is twelve months and in 2021 we had data for the first three months. Therefore, we will predict for 21 months. We will not change the division of individual scales and leave it on the default settings. We set the maximum number of iterations to 100; in doing so we should achieve greater accuracy. We will choose a model that works with seasonal fluctuations. We use the Linear Trend and choose the Multiplicative model that works with Alpha, Delta and Gamma components. This Multiplicative model means that the trend will be linear downwards or upwards and deviations will develop from it. We will prepare a demand forecast in the same way. By this point, we will have both time series calculated.

We then insert the processed data into Microsoft Excel, where we create a line graph from the supply and demand prediction data. The next step will be residue detection. In the Excel file, we will create columns A, B, C named Month, Supply - Residues and Demand - Residues. If we want a higher accuracy of the model, we find it using the correlation coefficient. We find the correlation coefficient from the data Supply or Demand and Smoothing of the time series.

4 Results

As already mentioned in the methodology, using exponential alignment of time series, we predicted the supply and demand for jobs from April 30, 2021 to December 31, 2022. The development of job supply was presented, which is shown in Figure No. 2. The graph shows that the supply curve will increase slightly from the second quarter of 2021 and this situation will persist until the end of the forecast period. Compared to 2019, the subsequent demand for value will increase relatively with smaller deviations. The prediction graph is a smoothed time series of supply and demand.

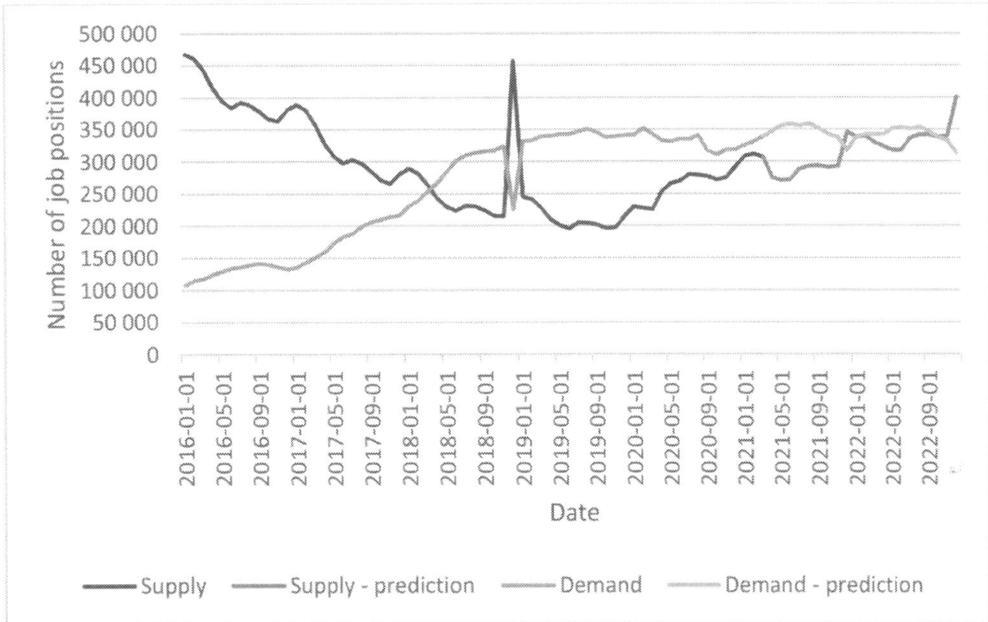

Fig. 2. Actual course of time series with subsequent prediction of supply and demand
Source: Authors.

From Figure No. 2 we can read that the number of predicted offered jobs from April 30, 2021 at 274,642 will start to decrease slightly by June 2021 to 271,449 predicted offered jobs. From July 31, 2021 to November 2021, the number of predicted positions offered will not exceed 300,000. At the end of 2021, the number of predicted jobs offered will increase to 346,130 and from then on the supply curve will begin to decline slightly to 317,984 in June 2022. The number of predicted jobs offered will increase slightly up until September 2022 and then begin to decline slightly until November 2022. At the end of the year 2022, 400,783 offered jobs are predicted.

The demand forecast curve from March 2021, when 338,864 positions in demand are registered, increases to 358,683 positions by August 31, 2021. From August 2021 to December 2021, the number of predicted positions will decrease slightly to 317,769 positions. From the beginning of 2022, the number of predicted demanded jobs will increase slightly until August 2022, when the number of demanded jobs will be 353,574. From then on, the number of predicted demanded jobs will decrease slightly until the end of 2022 to 313,221 predicted demanded jobs.

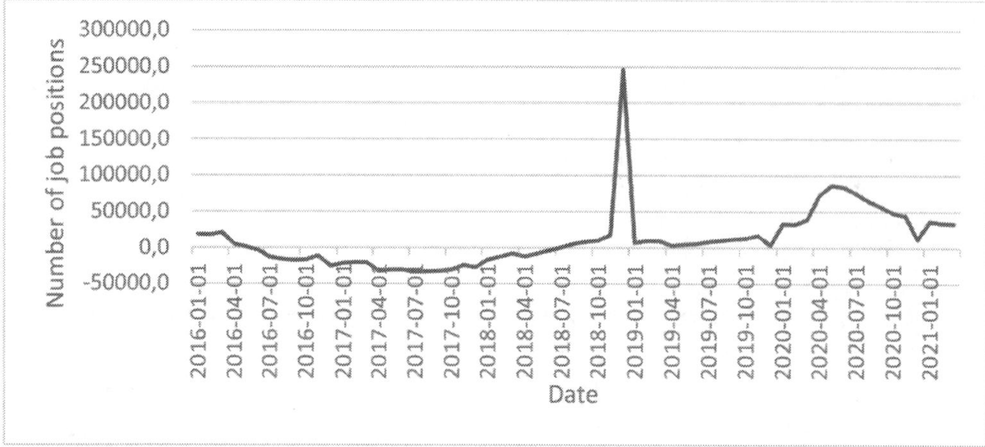

Fig. 3. Supply - residues
Source: Authors.

The chart Supply - Residues and Demand - Residues graphically shows us a line graph created in Microsoft Excel. The obtained data is the difference between the actual and predicted variables. In this case, they show us how the model differs from reality.

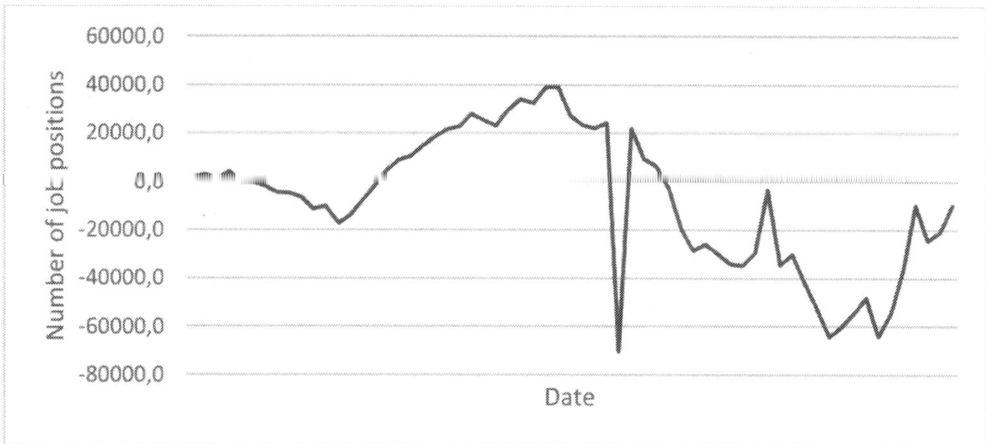

Fig. 4. Demand – residues
Source: Authors.

Figure No. 2 and 3 show sharp fluctuations in values, which were also reflected in the models. The reason for such sharp fluctuations in the supply and demand curve is the fact that the models were unable to detect them in time.

Comparing charts 2 and 3, we can see that the model is relatively accurate. By applying time series alignment, we obtained a model and then applied this model to all data. If we want a higher accuracy of the model, we find it using the correlation coefficient. The closer the correlation coefficient is to 1, the more accurate the model will be. The supply correlation rate is relatively high at 0.86, but the demand correlation rate of 0.96 is even higher. A statistically significant value of the correlation coefficient is 0.7 and more. As a result, this means that the result is relatively valid.

5 Discussion

Based on the obtained results, we are able to answer research questions that were determined:

RQ1: How will supply develop up until the year 2022?

We predicted the supply up to 2022 with a suitable chosen method of exponential alignment of time series and displayed it using a line graph created in Microsoft Excel. We have found that the supply curve will grow from March 31, 202. The supply of job positions will be affected by the demand for jobs, which will be higher than the number of jobs offered. However, at the end of the forecast period on December 31, 2022, there will be a change in which the supply curve will rise sharply. In this case, the demand curve for jobs will start to decline gradually.

RQ2: How will demand develop up until the year 2022?

In the same way as we calculated the development of supply until 2022, we also found out the development of demand until 2022. The demand for jobs will increase relatively from March 31, 2021, with small deviations. The value of demand will decline December 31, 2022 and will subsequently increase again.

In this paper, we also created two charts focusing on supply residues and demand residues. The values in these charts were from the period of December 31, 2015 to March 31, 2021 The duration of the residue is only from the beginning of the period to the end of the period when we had the data obtained from the Ministry of Labor and Social Affairs of the Czech Republic. We then compared the chart of residue supply and residue demand with the graph of actual time series. The aim of these charts was to record how the actual and predicted variables differ.

6 Conclusion

The aim of the paper was achieved. Through exponential alignment of time series, we predicted the supply and demand for jobs in the labor market in the Czech Republic. When creating charts, predicting supply and demand, we chose a model that works with seasonal fluctuations. The presence of a seasonal component was present in all of the time series analyzed. We chose a multiplicative model with a number of iterations of 100 to achieve greater accuracy. For higher accuracy, we also chose to calculate the correlation coefficient.

Predicting supply and demand is not easy because we do not know which direction the labor market will take. In particular, technological change cannot be determined exactly at what rate they will be used or forgotten. In such a case, organizations will have to demand labor from abroad. If organizations are unable to secure labor from abroad, they will have to reduce their activities and the associated volume of production they produce. Last but not least, the organization can start going bankrupt.

We see the application benefits of this work in companies that manage human resources. At present, the prediction of future development is mostly at the national level and in the Czech Republic it is at the beginning of its journey. Predicting future developments in the labor market is one of the most effective tools of active employment policy. Thanks to this, we can address the unemployment factor before it happens. However, the biggest advantage of prediction is the provision of timely information on changes in supply or demand for job positions. This information serves as a basis for organizations in decision-making and setting strategic goals. The strategic goals will make it possible to determine the need for employees and the volume of production that the organization will be able to implement in the coming years. In the forecast, we did not address the structure of unemployment by educational attainment or by profession. The developed methodology will enable organizations to monitor the current situation on the labor market and respond to changes in the short and long term. The output of the prediction was information that corresponds to the needs of the labor market in a given time interval. In this case, up to the end of 2022.

References

1. R. Maňák, Approach to human resources performance measurement. *Littera Scripta,* **6**(1), 82-94 (2013)
2. S. M. Heathfield, What is human resource management? [online], Available at: http://humanresources.about.com/od/glossaryh/f/hr_management.htm (2021)
3. L. Dienesová, M. Šuta, Conference Review. "Entrepreneurial competences and human resource management in the 21st century. *Littera Scripta,* **10**(2), 180-184 (2017)
4. M. Šikýř, *Human resources for managers and human resources*. 2. Eds. Prague: Grada (2016)
5. B. Okay-Somerville, D. Scholarios, A multilevel examination of skills-oriented HRM and perceived skill utilization during recession: implications for the wellbeing of all workers. *Human Resource Management*, **58**(2), 139-154 (2019)
6. Y. Hong, Y. Jiang, H. Liao, M. C. Sturman, High performance work systems for service quality: boundary conditions and influence processes. *Human Resource Management*, **56**(5), 747-767 (2017)
7. C. Jane Hollowell, Z. Rowland, T. Klieštik, J. Kliešťiková, V. V. Dengov, Customer loyalty in the sharing economy platforms: How digital personal reputation and feedback systems facilitate interaction and trust between strangers. *Journal of Self-Governance and Management Economics*, **7**(1), 13-18 (2019)
8. M. J. Donate, I. Pena, J. D. S. de Pablo, HRM practices for human and social capital development: effects on innovation capabilities. International Journal of Human Resource Management, **27**(9), 928-953 (2016)
9. W. Wikhamn, Innovation, sustainable HRM and customer satisfaction. International Journal of Hospitality Management, **76**, 102-110 (2019) (za číslo 7)
10. H. Lipovská, *Modern economics: simply about everything you should know*. 1 Eds. Prague: Grada (2017)
11. H. S. Sugiarto, E. Lim, N. L. Sim, On analysing supply and demand in labor markets: Framework, model and system. *IEEE International Conference on Data Science and Advanced Analytics 6th DSAA*, pp. 511-520 (2019)
12. D. Wang, W. Cao, J. Li, J. Ye, DeepSD: Prediction of supply and demand for online voting services using deep neural networks. *Proceedings IEEE 33. International Conference on Data Engineering. (ICDE)*, pp. 243-254 (2017)
13. R. Boselli, M. Cesarini., F. Mercorio, M. Mezzanzanica, Classifying online job advertisements through machine learning. *Future Generation Computer Systems*, **86**, 319-328 (2018)
14. C. Boon, D. N. Den Hartog, A systematic review of human resource management systems and their measurement. *Journal of Management*, **45**(6), 2498-2537 (2019)
15. F. Green, G. Henseke, Europe's evolving graduate labour markets: supply, demand, underemployment and pay. *Journal for Labour Market Research,* **55**(2) (2021)

16. G, Harvey, P. Turnbull, Ricardo flies Ryanair: Strategic human resource management and competitive advantage in a single European aviation market. *Human Resource Management Journal,* **30**(4), 553-565 (2020)
17. A. A. Katou, Human resources flexibility as a mediating mechanism between high performance work systems and organizational performance: A multilevel quasi-longitudinal study. *EuroMed Journal of Business* (2021)
18. A. A. Ternikov, E. A. Aleksandrova, Demand for skills on the labor market in the IT sector. *Business Informatics*, **14**(2), 64-83 (2020)
19. S. A. Way, P. M. Wright, J. B. Tracey, J. F. Isnard, HR flexibility: precursors and the contingent impact on firm financial performance. *Human Resource Management*, **57**(2), 567-582 (2018)

Evaluation of inflation and unemployment and plotting the Philips curve of the Czech Republic

Tomáš Krulický*, *Veronika* Šanderová, and *Dominik* Dolejš

Institute of Technology and Business in České Budějovice, School of Expertness and Valuation, Okružní 517/10, 37001 České Budějovice, Czech Republic

Abstract. The Phillips curve was supposed to mean an expansion of the doctrine based on the original regulatory ideas of J. M. Keynes. At the time of its inception (1950s), it gave governments theoretical hope, coming from the possibility of choosing a negative correlation between the price level (P) and the product (Y). Her early denial (at least in the short term) by Milton Friedman, on the other hand, has not changed anything about other applications that are still relevant until present time. In the fact, advantage of Phillips curve is her ability based on broad-spectrum use for any type of national economy. The aim of this work is to determine the shape of the Phillips curve for the Czech Republic in the period from 2000 to the present and to compare its shape with the shape of the original Phillips curve. The method of regression analysis is used here, comparison and prediction are performed using time series. In this paper, we find out what the short-term Phillips curve looks like for the Czech Republic, that it does not coincide with the original Phillips curve, and that in the future we can count on a growing correlation between inflation and unemployment.
Keywords: Inflation, Phillips curve, regression analysis, time series, unemployment, prognosis

1 Introduction

The results of a model estimate showed a significant and negative impact of inflation and unemployment on long-term economic growth, suggesting that inflation and unemployment have reduced economic growth in the long run [1].

Inflation is the weakening of the purchasing power of a currency against items that can be bought for that currency. This means that, for example, you will need more currency units each year for the same market basket. By comparing the year-on-year difference in the price of the market basket, we then obtain the percentage rate of year-on-year inflation.

Huber [2] found that his bounded bivariate model is better at prediction than many alternatives, including a version of his model with unlimited unobserved components. The model also provides reasonable estimates of trend inflation, NAIRU, inflation persistence, and the slope of the Phillips curve.

The paper will be processed by comparing data on unemployment and inflation. For this comparison, we decided to make it easier by using the Phillips curve, which describes the

* Corresponding author: krulicky@mail.vstecb.cz

relationship between the unemployment rate and wage growth. Therefore, it should be a very useful source of information.

Powell [3] does not see as likely that the Phillips curve is dead or that it will happen soon. Many factors, including the better conduct of monetary policy over the last few decades, are more likely to have significantly reduced, but not completely eliminated, the effects of tight labor markets on inflation. However, no one fully understands the nature of these changes or the role they play in the current context. Common sense suggests that we should be careful when prognoses predict events rarely before they are observed in the economy.

It will be necessary to find a connection between the Philips curve and the inflation rate. The connections found need to be entered into charts and their accuracy evaluated. In particular, it will be necessary to work with information on the situation in which the Phillips curve remains in its original form.

It is interesting to follow, for example, the development of the Phillips curve in the USA. As average inflation expectations have been anchored in the US inflation target since 2000, the long-term Phillips curve may have skewed downwards. Since the Great Recession of 2008-2009, the US economy has recorded an average 0.22% point below the inflation target in 2008-2017Q4. Sayeed et al. [4] found that the long-term Phillips curve for the United States had flattened since 2008. The study also estimates that the US economy faced 1.6% higher average unemployment than if average inflation were equal to the target. These are significant unemployment costs resulting both from a broader long-term Phillips curve and from exceeding the inflation target.

Furthermore, it will also be important to find out whether the Phillips curve corresponds with the data from the Czech Statistical Office and if it is possible to apply it to the Czech economy, and thus able to aid us. However, it will most likely be necessary to derive data for a given region, in this case the Czech Republic, so that the data is sufficiently relevant for that use. The rate of disinflation and deflation is a topic that will also need to be included in the results. Their cause, the situation where there is a turnaround between them and the correlation with unemployment [5, 6].

Meaningful estimates of the non-accelerating inflation rate of unemployment (NAIRU) within the Phillips curve require an identified trade-off between inflation and unemployment. However, observations of inflation and unemployment are equilibrium points that cause the problem of simultaneity [7]. Kajuth [8] assessed conventional identification assumptions in the literature on the German NAIRU in the general bivariate system of inflation and unemployment equations. To identify the trade-off for Germany, he used a data-based identification method based on shifts in the relative volatility of shocks to unemployment and inflation. The results support models that estimate the current impact of unemployment on inflation and those that model inflation and unemployment together [8].

The aim of this article is to determine the shape of the Phillips curve for the Czech Republic in the period from 2000 to the present and to compare its shape with the shape of the original Phillips curve.

In order to fulfill the aim of the paper, the following research questions were set:
RQ1: What is the shape of the Phillips curve in the Czech Republic from 2000 to the present?
RQ2: How does the shape of the Phillips curve in the Czech Republic differ from the short-term Phillips curve?
RQ3: What development can be expected of the shape of the Phillips curve in the Czech Republic until 2025?

2 Literature research

Onifade et al. [9] used a bound co-integration test approach to test for the presence of relationship levels between variables. Their findings show that openness to trade and domestic investment are significant for Nigeria's unemployment during the study period. Furthermore, the validity of the Phillips curve was confirmed. Zhu, Kavanagh and O'Sullivan [10] chose BoE to compare inflation projections with the actual impact of substituting values using the substitution method. They prove its functionality with their results. Barnichon and Masters [11] used a scientific description, a graphical method, synthesis and analysis, and a logical construction method to calculate lost GDP from youth unemployment in the EU in 2016. They found that youth unemployment was a serious microeconomic and macroeconomic problem. Peneva and Rudd [12] used a vector autoregressive model to investigate whether and to what extent the passage of labor costs to price inflation changed over time. Their findings call into question the explanation for recent inflationary behavior. Research by Busseti, Caivano and Delle Monache [13] has shown that the domestic output gap has a greater effect on the right part of the conditional inflation distribution graph than on the left part of the graph, suggesting a nonlinearity of the Phillips curve. The aim of Kadeřábková, Jasová and Holman [14] was to determine the nature and intensity of the slope of Phillips curves (PC) for workers at risk of unemployment in individual phases of the economic cycle between 2000 and 2016 in the Visegrad Group (V4). They used linear regressions to replace the household unemployment deflator. Also, from Barnichon's and Master's [11] research, we learn that the derivation of the Phillips multiplier is based on a simple instrumental variable regression of cumulative inflation and cumulative unemployment using monetary shocks as tools. Subsequently, I found that Álvarez and Correa-Lopéz [15] used a model with gradual regressions for their prognosis and evaluated their results using meta-regression. In addition, Damane and Sifat [16] used a combination of seemingly unrelated regression (SUR) and marginal regression techniques based on a multidimensional cumulative function to investigate the existence of a common Phillips curve between CMA members. Similarly, Furtula, Durkalic and Simionescu [17] examined the paradigm of the Phillips curve in macroeconomics and examined the relationship between inflation and the unemployment rate, which they analyzed on empirical data using Bayesian linear regression models. Berge [18], who examined the behavior of inflation expectations in the United States, is cited as the latest example of the regression model. His algorithm creates a regression model of expected inflation using a large panel of macroeconomic data as possible variables. However, for example, Daman's and Sifat's [16] model estimates were made using data on the annual time series of specific countries for inflation, unemployment and imports. Tule, Salisu and Chiemeke [19] also propose a Phillips curve based on supply for an oil-dependent (Nigerian) economy. They compare the predictive results of the Phillips curve based on oil data with traditional variants and time series models such as ARIMA and ARFIMA. Stuart [20] encountered the problem of lack of data, and therefore first solves how to proceed in this situation and then also predicts the Phillips curve for the next 50 years using time series and the method of combining forecasts. In a similar time frame, Ali et al. [21] use data from the time series, i.e. from 1961 to 2014, and apply descriptive statistical analysis, the autoregressive distributed delay model (ARDL) and the paired Granger causality test. Annual time series data was also used by Agboola and Bekun [22]. Specifically, from 1981-2014. The data was used to determine the stationarity properties of the considered variables. Tallman and Zaman [23] use a comparison of time series prediction models with the aggregated forecast and standard Phillips curve prediction models to refine their generation of aggregate inflation forecasts. Lopéz-Peréz [24] is the latest example of using time series; here it is used to examine unobservable individual heterogeneity among predictors, although true, in panel data, the Survey of Professional Forecasters dataset.

Correlation analysis, simple regression and time series regression seem to be the most suitable methods for obtaining answers to the research questions.

3 Data and methods

The data will be obtained from the MLSA website of the Ministry of Labor and Social Affairs of the Czech Republic [25]. It will be entered into the table in MS Excel so that in column A there will be time data, in column B the relevant data on unemployment and in column C the relevant data on inflation. This paper will further be based on this table. The values of unemployment and inflation are entered into Chart 1, which shows their development in the period from 01.01.2000 to 28.02.2021.

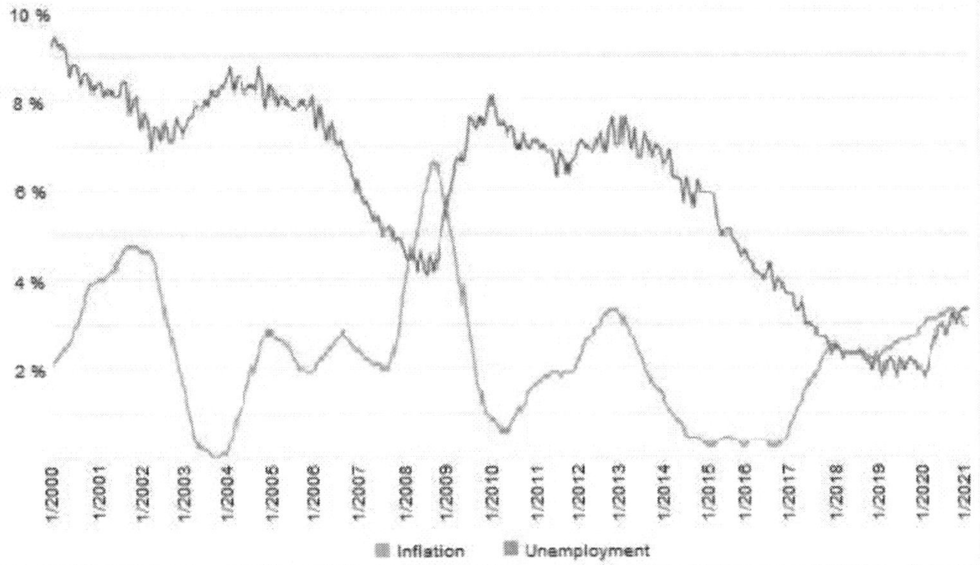

Fig. 1. Unemployment and inflation 2000-2021
Source: MLSA [25].

Next, a scatter plot will be created using regression analysis, and the most accurate trend line will be determined, according to the value of the reliability coefficient (R2). Specifically, the linear trend line with the value R2 = 0.69 work best here. And this line is a Phillips curve for the Czech Republic. This is then compared with a short-term Phillips curve. This will provide answers to the first two research questions. We decided to answer the third research question using time series. Thus, in the program Statistica in version 13.0 from the company TIBCO, through advanced methods, time series, unemployment is selected as a variable and time series alignment is processed. We insert the resulting table into MS Excel again. We will do the same with the variable inflation. We assign the result to the correct date, in our own columns for prediction. We then insert all values from 2000 to 2025 into a line chart, which will be extended by prediction up to 2025.

Next, we create a chart depicting the development of the Phillips curve in the future up to 2025. The values on which the graph is based are rounded to two decimal places. The graph will be made by the same method as the Phillips curve based on already known values, namely regression analysis and trend line.

4 Results

Using regression analysis, we found that the shape of the Phillips curve for the Czech Republic looks like this in the given time period (Figure 2).

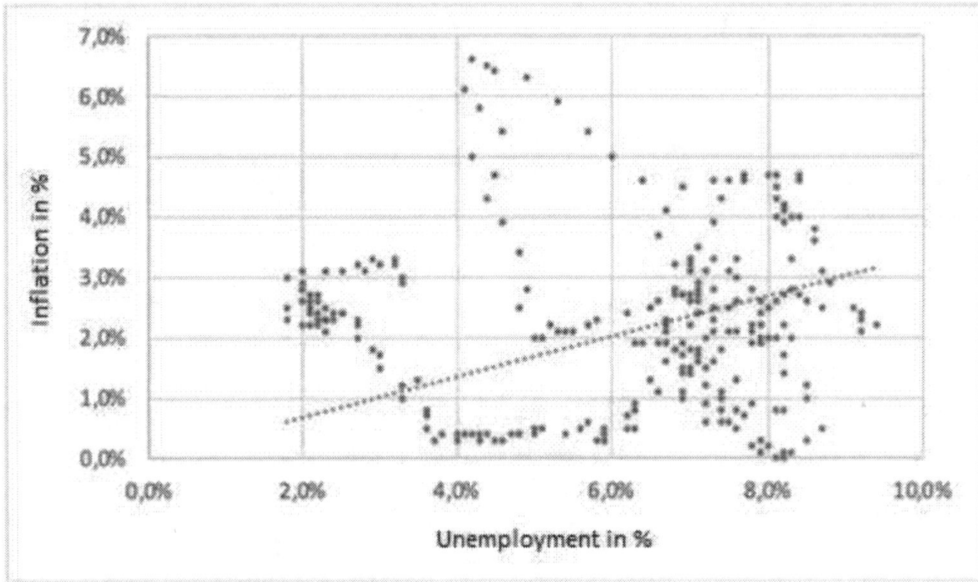

Fig. 2. Philips curve of the Czech Republic
Source: Own processing.

The reliability coefficient for this curve corresponds to R2 = 0.6359, y = 3355x.
To compare the two curves, we use the already created graph and enter the Phillips curve (Figure 3).

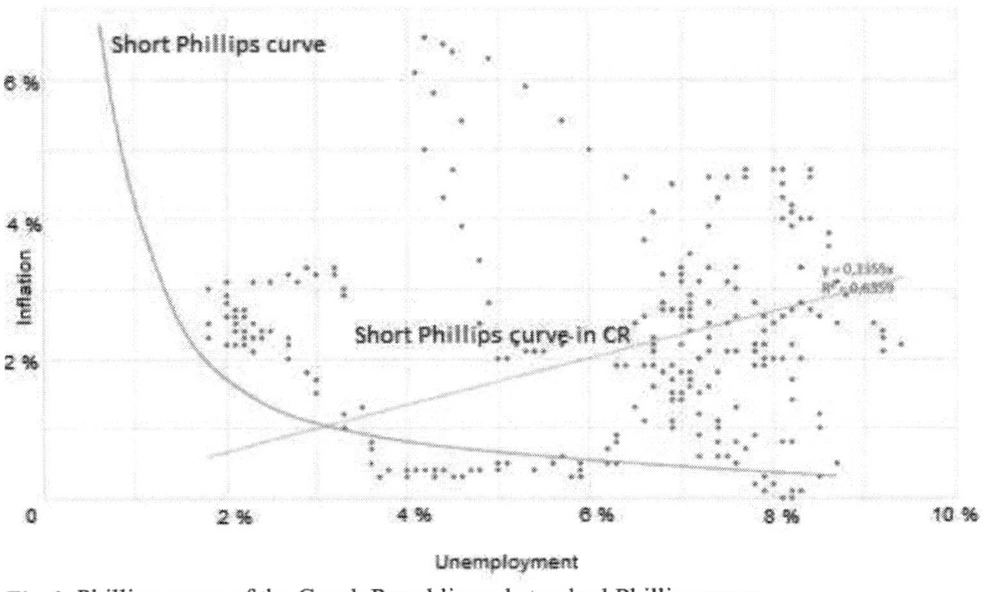

Fig. 3. Phillips curve of the Czech Republic and standard Phillips curve
Source: Own processing.

Therefore, I found out that these two curves do not match, the Phillips curve for the Czech Republic has a completely different shape and slope than the traditional Phillips curve.

The aforementioned time series are used to predict the Phillips curve. We plotted the results in Figure 4.

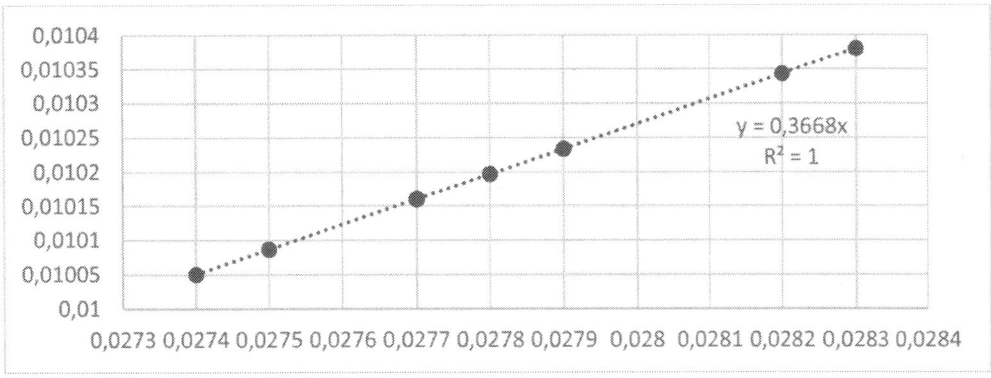

Fig. 4. Phillips curve prediction
Source: Own processing.

The result is practically an extension of the original Phillips curve. Thus, the Phillips curve for the Czech Republic will continue to move away from the short-term Phillips curve. Its slope and position in the chart will move further from the prediction of the short-term Phillips curve. Furthermore, the reliability coefficient for this curve is higher than for the curve based on historical data. We also used the line chart in Figure 5, which was created by combining already known data and predictions up to 2025. According to the examined prediction, it is assumed that inflation and unemployment will stabilize in the next five years.

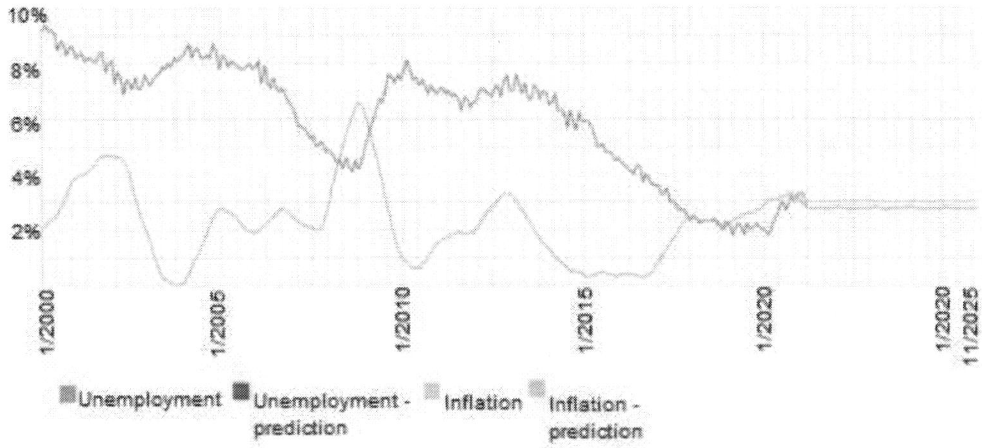

Fig. 5. Prediction of unemployment and inflation in the Czech Republic
Source: Own processing.

5 Discussion

Based on the results, we are able to answer the research questions which we set out.

So, what is the shape of the Phillips curve from 2000 to the present? The short-term Phillips curve appears to have a decreasing tendency reminiscent of an exponential decline. However, such a tendency did not appear in the Phillips curve for the Czech Republic. The Phillips curve for the Czech Republic has an upward trend, and moreover, the typical shape of the curve has not been shown here. For the Czech Republic, it came out almost practically flat. This may be due to the fact that no methods were used for our work that would show a more accurate result, however, the difference in the original and in my curve is so drastic that even with a more accurate method we would come to the same conclusion that the curves do not match. The credibility of the method of regression analysis is also evidenced by its use in the work of Barnichon and Masters [11], or Álvarez and Correa-Lopéz [15].

How does the shape of the Phillips curve in the Czech Republic differ from the short-term Phillips curve? As we have already mentioned, they really are very different. There may even be doubts about the validity of the original Phillips curve, but this has already been confirmed by several studies. For example, Onifade, Asongu and Bekun [9] add in their research that the Phillips curve was valid for their use. It should therefore be about the differences in our and foreign economies that make such a drastic difference.

What development of the Phillips curve can be expected in the Czech Republic up to 2025? Based on the results, the answer is that a further rise of the Phillips curve can be expected in the future. This will continue to move away from the original curve. This conclusion can be drawn from Figure 4, but also from Figure 5 where it can be seen how both quantities approach each other more and more over time and therefore their correlation increases and therefore it is logical that the curve will have an upward trend. The validity of the results obtained by this method proves the use and accuracy of the results in the works of Lopéz-Peréz [24], or Agboola and Bekun [22].

5 Conclusion

The aim of the paper was to determine the shape of the Phillips curve for the Czech Republic in the period from 2000 to the present and to compare its shape with the shape of the original Phillips curve.

The shape of the Phillips curve was determined by verified methods. We came to the surprising conclusion that the original short-term Phillips curve is unusable in the case of the Czech Republic because the curve based on domestic data is different and has an increasing tendency, and from the prediction it can be concluded that it will have this tendency in the future. That is, the relationship between unemployment and inflation will strengthen. Furthermore, the effectiveness of the methods was confirmed to some extent, but a more efficient method could be chosen for the prediction. While regression analysis was evaluated as a sufficiently accurate method.

It would therefore be appropriate to ask how economic development differs in the Czech Republic and in other countries of the world. In the same way, other questions arise, which unfortunately cannot be answered from the given results. How is it possible that the investigated curves are so different? Is it possible for national economies to have such different unemployment and inflation rates? It is therefore appropriate to propose follow-up research examining the relationship between unemployment and inflation in other countries of the European Union, or to include in the research even more distant countries for greater relevance of the results. We can also recommend the use of a more accurate method for predicting inflation, if such a prediction needs to be made. This is how research limits can be defined.

References

1. M. Mohseni, F. Jouzaryan, Examining the effects of inflation and unemployment on economic growth in Iran (1996-2012). *Procedia Economics and Finance*, **95**, 381-389 (2016)
2. F. Huber, D. Kaufmann, Trend fundamentals and exchange rate dynamics. *Economica*, **87**(348), 1016-1036 (2015)
3. J. H. Powell, Monetary policy and risk management at a time of low inflation and low unemployment. *Business Economics*, **53**(4), 173-183 (2018)

4. J. Sayeed, M. D. Islam, S. Yasmin, Does the US economy face a long run trade off between inflation and unemployment? *International Journal of Monetary Economics and Finance*, **12**(2), 118-132 (2019)
5. S. Hašková, P. Volf, V. Machová, Economic convergence of Czech regions in terms of GDP and unemployment rate in response to FDI flows: Do businesses and regions flourish? *Ad Alta: Journal of Interdisciplinary Research*, **9**(1), 326-329 (2019)
6. I. Kmecová, J. Stuchlý, J., L. Polanecký, M. Šuta, Analysing structure of employed and unemployed population of Czech Republic as part of human capital on labour market. *Littera Scripta*, **12**(1), 121-141 (2019)
7. D. Stroukal, The short-term relationship between unemployment and home ownership. *Littera Scripta*, **9**(1), 140-153 (2016)
8. F. Kajuth, NAIRU Estimates for Germany: New evidence on the inflation–unemployment tradeoff. *German Economic Review*, **17**(1), 104-125 (2016)
9. S. T. Onifade, A. Ay. S. Asongu, F. V. Bekun, Revisiting the trade and unemployment Nexus: Empirical evidence from the Nigerian economy. *Journal of Public Affairs*, **20**(3), (2020)
10. S. Zhu, E. Kavanagh, N. O'Sullivan, Inflation targeting and financial conditions: UK monetary policy during the great moderation and financial crisis. *Journal of Financial Stability*, **53**, (2021)
11. R. Barnichon, G. Mesters, The Phillips multiplier. *Journal of Monetary Economics*, **117**, 689-705 (2021)
12. E. V. Peneva, J. B. Rudd, The passthrough of labor costs to price inflation. *Journal of Money, Credit and Banking*, **49**(8), 1777-1802 (2017)
13. F. Busetti, M. Caivano, D. Delle Monache, Domestic and global determinants of inflation: Evidence from expectile regression*. *Oxford Bulletin of Economics and Statistics*, **83**(4), 982-1001 (2021)
14. B. Kadeřábková, E. Jasová, R. Holman, Analysis of substitution changes in the Phillips curve in V4 countries over the course of economic cycles. *International Journal of Economic Sciences*, **9**(2), 39-54 (2020)
15. L. J. Álvarez, M. Correa-López, Inflation expectations in euro area Phillips curves. *Economics Letters*, **195**, (2020)
16. M. Damane, I. Sifat, Testing for a common Phillips curve in common monetary area of Southern Africa. *Journal of Economic Studies*, **47**(6), 1401-1436 (2020)
17. S. Furtula, D. Durkalić, M. Simionescu, Testing Phillips curve for Serbian and Romanian economy. *Romanian Statistical Review*, **3**, 40-56 (2018)
18. T. J. Berge, Understanding survey-based inflation expectations. *International Journal of Forecasting*, **34**(4), 788-801 (2018)
19. M. Tule, A. Salisu, Ch. Chiemeke, Improving Nigeria's inflation forecast with oil price: The role of estimators. *Journal of Quantitative Economics*, **18**(1), 191-229 (2020)
20. R. Stuart, A quarterly Phillips curve for Switzerland using interpolated data, 1963-2016. *Economic Modelling*, **70**, 78-86 (2018)
21. S. Ali, Y. Liu, T. Shah, A. Tariq, A. A. Chandio, I. Ali, Analysis of the nexus of CO2 emissions, economic growth, land under cereal crops and agriculture value-added in Pakistan using an ARDL approach. *Energies*, **12**(23), (2019)
22. M. O. Agboola, F. V. Bekun, Does agricultural value added induce environmental degradation? Empirical evidence from an agrarian country. *Environmental Science and Pollution Research*, **26**(27), 27660-27676 (2019)

23. E. W. Tallman, S. Zaman, Forecasting inflation: Phillips curve effects on services price measures. *International Journal of Forecasting*, **33**(2), 442-457 (2017)
24. V. López-Pérez, Do professional forecasters behave as if they believed in the New Keynesian Phillips curve for the euro area? *Empirica*, **44**(1), 147-174 (2017)
25. Ministry of Labor and Social Affairs. MPSV [online]. Available at: https://www.mpsv.cz/ (2021)

Optimization of the capital structure of an agricultural company in the Czech Republic

Jaromír Vrbka[1], *Eva* Kalinová[2*], and *Zuzana* Dvořáková[1]

[1]Institute of Technology and Business in České Budějovice, School of Expertness and Valuation, Okružní 517/10, 370 01 České Budějovice, Czech Republic
[2]University of Žilina, The Faculty of Operation and Economics of Transport and Communications, Department of Economics, Univerzitná 8215/1, 010 26 Žilina, Slovakia

Abstract. The topic of optimizing capital structure is very important for a company to work efficiently and reliably. It is important for every company to optimize everything so that they have the highest possible efficiency. Entrepreneurs also try to make this optimization last them as long as possible. Therefore, the aim of this paper is to determine the optimal capital structure of an agricultural company operating in the Czech Republic. The base source of data is the closing data of individual agricultural companies from the Albertina database of Bisnode. More than 9,000 agricultural enterprises operating in the given sector of the national economy in the Czech Republic are recorded in this data set. The calculation of the WACC method and the level of debt of individual agricultural companies are used. Subsequently, the equation for calculating the cost of capital is determined using power regression. Here we also obtain the value of reliability, which in this particular case is not ideal, but still reasonable. A line chart is used to determine the optimal interval for the agricultural company. The optimal debt interval comes out to 20 to 25%, at which the cost of capital is declared to be from 22 to 24%. If agricultural companies had higher or lower debt ratios, it would be inefficient for the enterprise.

Keywords: Optimal capital structure, debt, power regression, cost of capital, agriculture, optimization
JEL codes: D24, Q14

1 Introduction

The decision regarding the capital structure of a company is very important, because a bad decision can affect the profitability of the company, which leads to a reduction in shareholder value [1, 2]. Good financial decisions can increase the market value of equity and bad financial decisions can reduce it [3]. An optimal capital structure is a key prerequisite for a business, although the determining of an optimal capital structure is difficult [4]. The available studies represent many different and contradictory factors that need to be taken into account in managers' strategic financial decisions. Their importance and intensity differ for individual business entities [5].

* Corresponding author: kalinova@mail.vstecb.cz

When making the correct adjustment of the capital structure, more precisely the ratio of equity and debt, it is necessary to take into account a number of factors affecting the capital structure, such as various sources of financing [3, 6]. Using factor analysis, three key external factors that entrepreneurs came up with were derived: the economic and political development of the country, the market environment of the country and the rates of taxes and interest rates [7]. European agricultural activity makes a significant contribution to European and global food supply, as well as to the preservation of the landscape and livelihoods. The main objective of this paper is to analyze the sustainability of agriculture in the 28 Member States of the European Union in order to promote sustainable agriculture, food security and rural development in the European region [8].

The measurement is then valid for all companies in the industry. Based on the results of the empirical study, the relationship of one of the factors in the Czech dairy sector is not constant. This limits the recommendation to use a common model for all companies in one sector. When making financial decisions, it is necessary to analyze the individual ability of a company to use foreign capital after quantifying the factors of the capital structure [9]. "Powerful managers" have a capital structure that differs significantly from "inefficient management" cooperatives [10]. The aim of the paper is to determine the optimization of the capital structure of companies focused on agriculture. The following research questions were set out to achieve the goal of the paper:

RQ1: What is the relationship between the price of capital and a company's debt ratio?

RQ2: What is the optimal capital structure?

2 Literature research

Woltz and Pencáková [11] found that in the Czech Republic, agricultural production is characterized by a dualistic structure, i.e. private farmers on the one hand and corporate farms on the other. According to Hudečková and Lošťák [12], the dual concept of social capital makes it possible to explain some circumstances that can be presented in a relatively simplified manner in the case of a one-way orientation of understanding social capital. In 2007, Ortman and King [13] introduced cooperative theories and a new theory of institutional economics (including transaction cost economics, agency theory, and property rights theory) and its applicability. They are also presented in the organizational form of a cooperative, as well as the inherent problems of conventional cooperatives, namely free-rider, horizon, portfolio, monitoring and influencing cost issues caused by vaguely defined property rights. According to Stekla and Grycová [14], companies can use debt or equity to finance their assets. The best option is to mix debt and equity. The effect of capital structure decisions on a company's profitability and value is that it increases value through the present value of tax savings from the use of debt. Byer and Hinke [15] described the resulting differences in profitability broken down using the additive decomposition method, which addresses the quantified impact of several influencing factors. More profitable companies were less dependent on debt financing, while fast-growing companies had limited access to the financial market, said Fenyves et al. [16].

Keishinick and Moussawi [17] interpret their evidence by suggesting that, over time, managers allow their risk preferences to dominate their firm decisions about capital structure when protected from being disciplined. Yen, Miranda and Katchova [18] found that long-term use and size of debt have a positive impact on cooperative growth, which is contrary to Gibrat's law, which assumes that the size and growth of a company are independent. Cash flow, unallocated capital and long-term debt financing in particular make a decisive contribution to the growth of the assets of small and medium-sized cooperatives.

Wysokiński et al. [19] state that modern agriculture depends on industrial energy sources and on how socio-economic developments are changing into increasingly energy-intensive production technologies. Dimitrova [20] is convinced that society can focus on compliance with certain values of the indicator – these are the values that, according to the financial management of the company, if achieved, will mean that there will be no serious problems in repaying current liabilities. Kalusova and Bedura [21] used the methods of multiple regression and correlation analysis as the main research methods. The results showed that, in addition to internal business factors, external macro-environmental factors were also statistically highly significant for certain groups of agricultural enterprises. Wang [22] states that when the model is solved numerically on a numerical illustration, in equilibrium, an optimal capital structure is derived for an agricultural cooperative. The effects of changes in personal income tax rates and discount rates are also examined. The calculation of economic income is a reasonable way to calculate the requirements for economic viability [23]. Zeshan [24] argues that the WACC variable is selected as a dependent variable, while the Quality of Corporate Governance (QCG) variable was used as an independent variable. The endogeneity of the QCG variable was investigated using the 2SLS model, and the results show that the QCG variable significantly and negatively affects the WACC variable. In addition, the variables of leverage, ROA, revenue growth and stock volatility have significant positive links with WACC, while company size does have a significant negative relationship with WACC.

The increase in production compared to the farm's equity plays a crucial role in increasing profitability for all groups of farms but is especially important for smaller entities. In addition, agricultural company managers should monitor the level of debt, as the debt-to-assets ratio is a highly significant negative determinant of farm profitability in most groups. An increase in the subsidy rate is usually reflected in a higher ROA, but this variable has a negative impact on the largest enterprises [25]. Belás, Gaurová and Tóth [5] used ordinal logistic regression and estimated five models. These models were used to predict the SME attitude towards the capital structure. The analysis was performed in R Software. The main finding is that the size of the company measured by the number of employees affects its attitude to the volume of foreign and equity capital used to finance its activities.

A regression analysis is selected for this work to determine the optimal capital structure for agricultural enterprises in the Czech Republic. The debt ratio of selected agricultural enterprises in the Czech Republic will be assessed. The detected data is transferred to a scatter plot.

3 Data and methods

The source for finding important information that will help us calculate the optimal capital structure will be a file that contains the closing results of individual agricultural holdings. The WACC method is used to determine the optimization of individual companies and the regression analysis method is used to learn the optimal capital structure from all companies. The file contains a table listing information on 9,814 agricultural companies that had final accounts at the end of the accounting year (material, inventory, capital, etc.), but also contains personal data (name, ID number, place of business etc.) on individual companies. The columns needed to calculate the optimal capital structure will be added to the table. The data is downloaded from Bisnode's Albertina database.

First, the total equity, total debt and total capital of individual companies will be added up. We also need to add up the individual costs for debt and equity. Once we have this completed, we will calculate the optimization for all companies. We calculate it using the WACC method. The formula for this method is:

$$WACC = E/C * re + D/C * rd * (1 - t) \quad (1)$$

where:

$WACC$ - weighted average cost of capital,
E - equity,
D - debt,
C - total value of capital (equity plus debt),
r_e - cost of equity,
r_d - cost of debt,
t - tax rate.

Another value will be needed to represent individual companies, namely the ratio of debt This will be calculated using the formula:

$$debt\ ratio = D / C \quad (2)$$

Once all the unknowns for each farm have been identified, they will be able to be plotted under the coordinates [x; y]. The variable x (independent) will show the debt ratio and the variable y (dependent) will show the cost of capital. The individual points will be scattered over the entire area of the chart, so we will have to determine the position of the line. The general power regression formula is used: $y = b_1 * x^{b2}$, but this equation is nonlinear. Therefore, we adjust this equation to the form:

$$ln(y) = ln(b1) + b2 * ln(x) \quad (3)$$

where:

y - represents the cost of capital,
x - debt ratio,
b_1 - first parameter,
b_2 - second parameter.

Using power regression, we calculate the parameters $ln(b_1)$ and b_2. We will use the LinkRegression function in Excel. Once we know the parameters b_2 and $ln(b_1)$, we will have to obtain the parameter b_1, de-logarithmize parameter $ln(b_1)$. To determine the optimal capital structure for an agricultural enterprise, we use differentiation to calculate the minimum. The general formula for power regression will be derived.

4 Results

To display the points in Figure 1, a few companies had to be excluded due to the lack of information in the data file or for the clarity of the chart. 5,077 agricultural companies are shown here. The unknown x-axis shows the debt ratio of individual companies and the y-axis their cost of capital. These unknowns are listed in percentages.

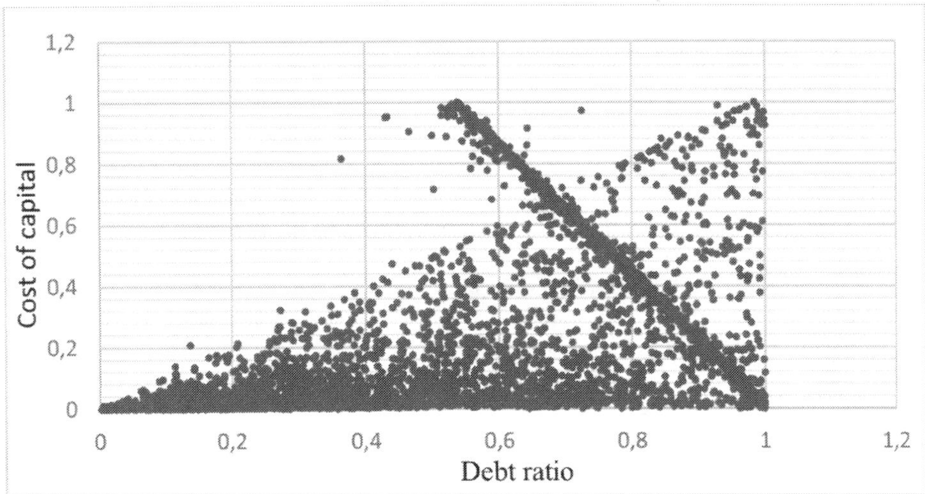

Fig. 1. Relationship between the cost of capital and the debt ratio in a scatter plot
Source: Authors.

Using the trendline function in Excel, a power line is displayed. As we can see in Figure 2. This line has a reliability value of R^2 = 32.02 %. The value of reliability determines how two values are dependent on each other. For this example of a pair of values, the debt ratio and the cost of capital. This percentage is not large, so the relationship between the debt ratio and the cost of capital is not reliable in this chart, but is still fair. Next, the equation of the power line: $y = 0.2957 * x^{1.3871}$. According to the general formula $y = b_1 * x^{b2}$ we derive unknown parameters b_1 and b_2. We are not certain that these parameters are already linear for the line or are still nonlinear for the line. So to be sure, we will have to logarithmize the unknown x and use the LinkRegression function, which gives us the value $ln(b_2)$ = 1.3871 and the value $ln(b_1)$ = -0.5292, after de-logarithmizing the value $ln(b_1)$ we get the value b_1 =0.2957. For the value of $ln(b_2)$, it comes out to $ln(b_2) = b_2$. Therefore, we will no longer work with this parameter. As we can see here, the equation shown to us by the chart is the same as the one we calculated using the function in Excel.

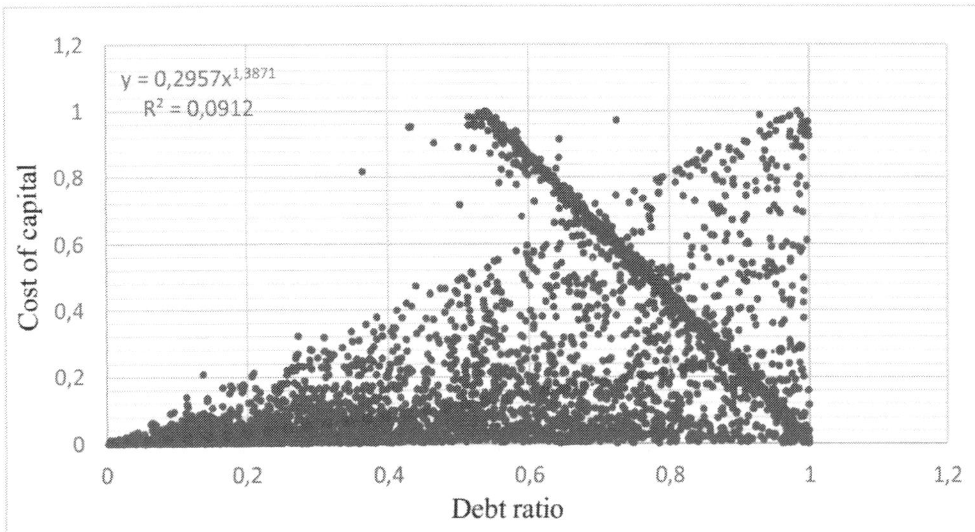

Fig. 2. Power regression in a scatter plot
Source: Authors.

If we already know the parameters b_1 and b_2 for a linear line, we substitute them into the general power equation. So we get the equation: $y = 0.2957*x^{1.3871}$. And we differentiate this equation so that we can then calculate the minimum cost of capital. We get the equation:

$$y' = 0.41016 * x\, 0.3871 \quad (4)$$

Using this equation, we can calculate the minimum cost of capital from certain percentages of the debt ratio of individual companies. For this reason, I decided to create a table with the debt ratio from 5-150% in increments of 5%. It is shown in Table 1. In this equation, we can substitute any percentage of the debt ratio of a particular agricultural enterprise to calculate the minimum cost of capital.

Table 1. Minimum cost of capital from a single percentage of debt

Debt ratio (%)	Cost of capital (%)
5	12.86
10	16.82
15	19.68
20	22.00
25	23.98
30	25.74
35	27.32
40	28.77
45	30.11
50	31.36
55	32.54
60	33.66
65	34.72
70	35.73
75	36.69
80	37.62
85	38.52
90	39.38
95	40.21
100	41.02
105	41.80
110	42.56
115	43.30
120	44.02
125	44.72
130	45.40
135	46.07
140	46.72
145	47.74
150	48.00

Source: Authors.

To better compare the debt ratio and the cost of capital, a line chart was compiled. Here we can see how the line of minimum cost of capital increases with higher debt ratio. It shows the point where the line rose. That point shows the debt ratio for the optimal capital structure for the farm, and that is the debt ratio of 23%.

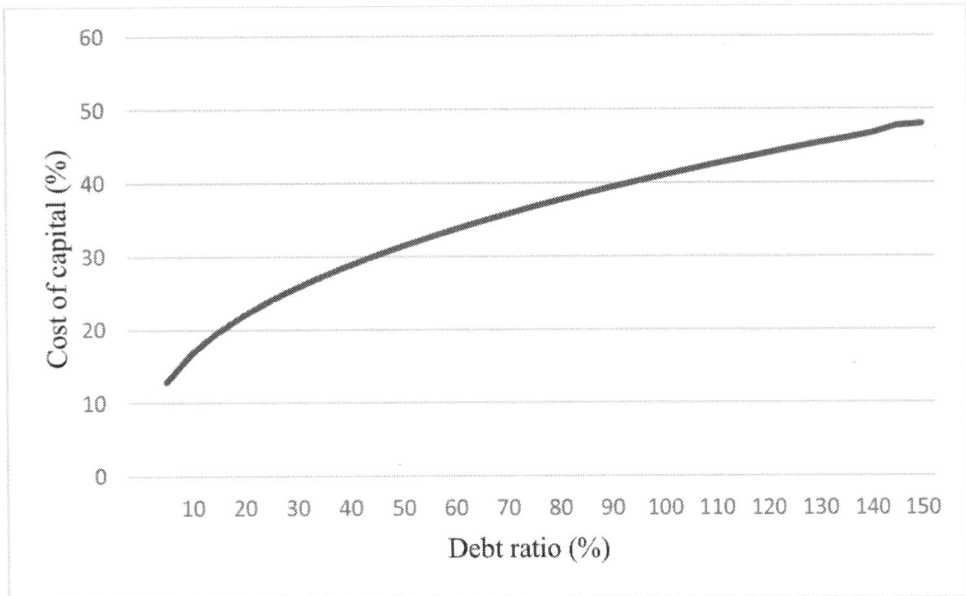

Fig. 3. Relationship between the debt ratio and cost of capital in a line chart
Source: Authors.

5 Discussion

Based on the obtained results, we can answer the research questions which we have set out earlier:

RQ1: What is the relationship between the price of capital and a company's debt ratio?

The results of Stekla and Grycová [14] showed that all three hypotheses must be rejected because there is a relationship between short-term debt and total assets and profitability, between long-term debt to total assets and profitability, and between total debt to total assets and profitability. Kalusova and Badura [21] found that the negative correlation between total indebtedness and the size of the company, i.e. an increase in the size category of the company is accompanied by a decrease in the company's indebtedness. Only in terms of credit indebtedness, our research shows a positive correlation – so an increase in the size category also leads to an increase in the company's interest rate debt.

RQ2: What is the optimal capital structure?

From Chart 3: Looking at the relationship between the debt ratio and the cost of capital in the line graph, we know that for the optimal capital structure according to the constructed table (see Table 1), the debt ratio is between 20-25%. For this value, the cost of capital shows us 22-24%. In doing so, the company does not use any sales profits or other cash flows.

6 Conclusion

The aim of this research was to determine the optimal capital structure for agricultural enterprises in the Czech Republic. Using the WACC method and the debt ratio of individual companies, this goal was achieved. In the results chapter, the individual debt ratios and the cost of capital for individual companies were calculated. Subsequently, they were transferred to a scatter plot (see Figure 1), for which a power line was drawn (see Figure 2). To assess the unknowns x and y, the debt ratio for the unknown x and the cost of capital for the unknown y were used. Using a function in Excel, the power equation and the reliability indicator were displayed. The reliability indicator showed the reliability of the given equation. To find out if the equation is for a linear or for a nonlinear line, an operation had to be performed, in which the unknown x and y had to be logarithmized. Furthermore, the parameters for the linear line had to be calculated with the results of the operation.

After substituting the correct parameters into the general power equation, the equation had to be recalculated using derivation in order to calculate the minimum cost of capital. Furthermore, a table was compiled in the results (see Table 1), where a differentiated equation was used to calculate the minimum cost of capital for various debt ratios. Subsequently, a line graph was drawn on the basis of this table (see graph 3), which determined the optimal combination of debt and cost of capital. In the chapter discussing the results, previously asked scientific questions were answered. The relationship between the cost of capital and the debt ratio of the company was found to exist. The question was also answered: What is the optimal capital structure. We answered it using a table (see Table 1). The results were also compared with other results from previous research by others.

We recommend for entrepreneurs of agricultural enterprises in the Czech Republic that their debt ratio be in the range of 20 to 25%. If an entrepreneur had a debt ratio of above 25% or below 20%, it would be inefficient for their business. But this interval may not be optimal for all agricultural companies. Because there are many research limits, which could then change the optimal interval. For example, an unbalanced data file, missing data in a data file, other factors that affect the business, such as the size of the business and so on. Also, this debt interval may not be optimal in a year or in 5 years. Therefore, the entrepreneur should take note of this.

References

1. J. Horák, P. Šuleř, J. Kollmann, J. Mareček, Credit absorption capacity of businesses in the construction sector of the Czech Republic – Analysis based on the difference in values of EVA Entity and EVA Equity. *Sustainability*, **12**(21), 1-16 (2020)
2. J. Horák, V. Machová, Business enterprises and generators of their value. *18th International Scientific Conference on Globalization and Its Socio-Economic Consequences*, 547-554 (2018)
3. J. Steklá, M. Gryčová, J. Homolka, Evaluation of capital structure of agricultural cooperatives. *AGRIS on-line Papers in Economics and Informatics*, 37-48 (2015)
4. L. Frýd, V. Machová, J. Horák, P. Pardal, The wage impact on the net value-added in the agricultural sector. *Littera Scripta*, **13**(1), 186-193 (2020)
5. J. Belás, B. Gavurová, P. Tóth, Impact of selected characteristics of SMEs on the capital structure. *Journal of Business Economics and Management*, **19**(4), 592-608 (2018)
6. J. Kučera, M. Vochozka, Z. Rowland, The ideal debt ratio of an agricultural enterprise. *Sustainability*, **13**(9), (2021)
7. L. Strýčková, Factors determining the corporate capital structure in the Czech Republic from the perspective of business entities. *Economics and Management*, 40-56 (2015)
8. M. J. P. Lampreia Dos Santos, N. Ahmad, Sustainability of European agricultural holdings. *Journal of the Saudi Society of Agricultural Sciences*, **19**(5), 358-364 (2020)
9. G. Chmelíková, Possibilities and limits for capital structure optimalizing model design of Czech dairy industry. *Agriculture Economics – Czech,* **48**, 321-326 (2002)

10. C. Russo, D. Weatherspoon, C. Peterson, M. Sabbatini, Effects of managers' power on capital structure: a study of Italian agricultural cooperatives. *The International Food and Agribusiness Management Review*, **3**(1), 27-39 (2000)
11. A. Wolz, J. Fritzsch, The impact of structural social capital on farm income in the Czech Republic. *Agriculture Economics – Czech,* **52**, 281-288 (2006)
12. H. Hudečková, M. Lošťák, Social capital in the change of the Czech agriculture. *Agriculture Economics – Czech*, **49**, 301-309 (2003)
13. G. F. Ortmann, R. P. King, Agricultural Cooperatives I: History, Theory and Problems. *Agrekon*, **46**(1), 18-46 (2007)
14. J. Stekla, M. Grycova, The relationship between profitability and capital structure of the agricultural holdings in the Czech Republic. *Agriculture Economics – Czech*, **62**, 421-428 (2016)
15. D. Beyer, J. Hinke, European benchmarking of determinants of profitability for companies with accrual accounting in the agricultural sector. *Agriculture Economics – Czech*, **66**, 477-488 (2020)
16. V. Fenyves, K. Pető, J. Szenderák, M. Harangi-Rákos, The capital structure of agricultural enterprises in the Visegrad countries. *Agriculture Economics – Czech,* **66**, 160-16 (2020)
17. R. Kieschnick, R. Moussawi, Firm age, corporate governance, and capital structure choices. *Journal of Corporate Finance*, **48**, 597-614 (2018)
18. M. Yen, M. Miranda, A. Katchova, The effects of capital constraints on the growth of agricultural cooperatives. *Agriculture Economics – Czech*, **66**, 27-33 (2020)
19. M. Wysokiński, J. Domagała, A. Gromada, M. Golonko, P. Trębska, Economic and energy efficiency of agriculture. *Agriculture Economics – Czech*, **66**, 355-364 (2020)
20. M. Dimitrova, Financial liquidity of agricultural holdings. *Trakia Journal of Sciences*, 634-639 (2020)
21. L. Kalusova, P. Badura, Factors determining the financial structure of Czech and Slovak agricultural enterprises. *Agriculture Economics – Czech*, **63**, 24-33 (2017)
22. Y. Wang, The optimal capital structure in agricultural cooperatives under the revolving fund cycles. *Agriculture Economics – Czech*, **62**, 45-50 (2016)
23. J. Špička, P. Dereník, How opportunity costs change the view on the viability of farms? Empirical evidence from the EU. *Agriculture Economics – Czech,* **67**, 41-50 (2021)
24. A. Zeshan, The effect of corporate governance on cost of capital in agriculture sector of Asian countries. *Scientific Papers-Series Management Economic Engineering in Agriculture and Rural*, **1**, 33-40 (2020)
25. Ł. Kryszak, M. Guth, B. Czyżewski, Determinants of farm profitability in the EU regions. Does farm size matter? *Agriculture Economics – Czech*, **67**, 90-100 (2021)

Evaluation of market demand for a specific product

Zuzana Rowland[1], *Jiří* Kučera[2,*], and *Leona* Martínková[1]

[1]Institute of Technology and Business in České Budějovice, School of Expertness and Valuation, Okružní 517/10, 370 01 České Budejovice, Czech Republic
[2]University of Žilina, The Faculty of Operation and Economics of Transport and Communications, Department of Economics, Univerzitná 8215/1, 010 26 Žilina, Slovakia.

Abstract. The aim of this paper was to evaluate the demand for mobile phones in the world at the time of the coronavirus pandemic and to predict future development up until the end of 2022. During the covid-19 pandemic, markets in all sectors were subdued due to limited production and sales. As the production of chips for electronics was one of the deeply affected areas in terms of production slowdown, along with the rapid purchase of electronics at the beginning of the pandemic, there was also a slowdown in sales. The method of describing time series and calculation using linear regression is used. In this work, it is found that the decline in demand for mobile phones was caused by a global pandemic and consumers' uncertainty in maintaining their jobs. Overall, during the covid-19 pandemic, we saw a drop in the number of mobile phone units sold by more than 40,000. Furthermore, it was found that the entire mobile phone market is clearly recovering after the coronavirus crisis.
Keywords: demand, coronavirus, market, smartphones

1 Introduction

The 2019 coronavirus pandemic (COVID-19) according to Majerova and Fernandes [1] disrupted supply chains, exposing gaps and vulnerabilities. Retailers are challenged to address the risks and organize themselves to fit into a new scenario [2]. This global epidemic has created a volatile environment for supply chains, raised the question of market survival and forced companies to rethink resilient strategies to be adopted for the post-pandemic situation in order to mitigate the long-term effects of the virus [3]. These articles point to the issue of market functioning in the current coronavirus era [4-6]. The decisive factor for building interorganizational relations is the interdependence between members. When one of the members does not fully oversee its management, there is a mutual dependence [7]. We need to think about what features the market interpretation allows us to do. Here we also use duality theory, for example for the introduction of demand constraints or the provision of services that result in artificial pricing behavior and economic losses. To overcome these problems, we examine market incentive mechanisms that use supply. We introduce product representation for the supply chain and show that this representation allows us to determine the minimum supply and these subsequently activate the markets. These results then provide

* Corresponding author: kuceraj@mail.vstecb.cz

guidance for the formulation of a supervisory committee that rewards stakeholders who support market transactions. Tominac and Zavala [8] state that customer demand for a product is closely linked to their likelihood of purchase. This probability is determined by their purchase benefit, which depends on the reference price, the selling price and the risk factors. According to their asymmetric perception based on the difference between the reference and the selling price, customers in the market are divided into three categories. Namely, aversion to losses, search for profit and neutrality to losses. Zhao et al. [9] consider the mobile phone market to be very dynamic. It is constantly undergoing innovations and a great competitive environment forces manufacturers to come up with more sophisticated sales methods. In the second quarter of 2019, sales of mobile phones fell by 1.7%. Mobile phone sales grew only in China and Brazil [10]. The aim of this paper is to evaluate the demand for mobile phones in the world during the coronavirus pandemic and predict future developments up until the end of 2022. To meet the aim of the paper, the following research questions were set out:

RQ1: How was the demand for mobile phones developing in the world 5 years before the coronavirus pandemic?

RQ2: How has the demand for mobile phones in the world changed since the beginning of the coronavirus pandemic?

RQ3: What is the estimate of the price development of phones in the period from 6/2021 to 12/2022.

Realizing synergies is a decision-making problem. The aim is to reduce the time and effort required to gather information and assess synergies by industry subjects and intermediary stakeholders. Indicators represent the technical feasibility, economic viability and environmental benefits of a potential resource exchange [11]. However, determining a match between supply and demand and assessing its economic and environmental benefits remains a challenge in view of potentially conflicting objectives. Included are direct waste to resources, indirect waste to resources, by products to resources of higher value and waste to new materials. The cost-benefit analysis, together with the necessary environmental impact assessment, can be used to evaluate and optimize ecological networks so that they are commercially viable [12]. The characteristics of the spot electricity market and a financial model based on the net present value indicator are also discussed. The current document presented a technical and economic assessment of five thermal energy storage technologies in the integration of concentrated solar power system production into the SPOT electricity market [13]. For stated preference methods, products with certain attributes are identified or placed on experimental "markets!" to give respondents preferences for the product considered as a combination of attributes. Revealed preferences can be obtained from actual purchases, such as past market data or experimental procedures. Both methods have their advantages and disadvantages. For mobile phones, market shares were influenced by features related to functionality and appearance, such as battery life and unused appearance. At the same time, full repairability has proved to be less desirable than the ability to replace critical parts [14]. A four-part questionnaire was designed to analyze the potential demand for eggs produced companies that practice social farming. However, according to the respondents, two factors appear to be decisive for ensuring the future development of social farming, namely an adequate certification system and proper communication on the added value of social farming products [15]. The analysis is based on annual data on energy production until 2019 from BP public limited company. Ember's data on the energy sector for the first half of 2020 for a subset of countries, representing about three-quarters of global energy production, are shown in Figure 1b for comparison. The data on CO2 emissions for the period 2011 to 2018 comes from International Energy Agency and data for 2019 is derived from production data contained in the BP public limited company data set assuming constant emission factors by fuel. Uncertainties about the outlook for the near future are considerable, but it is possible that CO2 emissions from the energy sector will not return to the level of 2018. Various policy instruments could effectively support the accelerated reduction of emissions over the next few years [16]. Impact of COVID-19 and national restrictions on vegetable prices with evidence from wholesale markets. Price data and market information were collected from the National Database of Agricultural Product Prices by method of web-browsing [17]. Influence of COVID-19 on real estate prices. Using a revised method of differences with nonparametric smoothing, which more intensively controls the unobservable, we are able to model the effects of the outbreak on real estate prices as a continuous time function. The results show that after the outbreak of the Honolulu epidemic, it is the only place where real estate prices have fallen, with the largest decline approaching 6.69% in April 2020. For the other four areas recording real estate prices, the largest increase was 9.97% in Santa Clara, followed by Irvine with a growth of 5.80%. Houston and Des Moines have also generally seen an increase in property prices [18]. For our paper we will also utilize the description of time series and also calculation using linear regression.

2 Data and methods

2.1 Data

We collected the utilized data from the website [19], where we searched for an article containing the global share of the smartphone market listing quarterly data from 2014 to 2021. We compared the smartphone brands Apple, Huawei, Xiaomi and Samsung. First of all, we will deal with the question of how the demand for mobile phones in the world has been developing 5 years before the coronavirus pandemic outbreak. For this, we will compare the quarterly data of selected smartphone brands from 2015 to March 2021, which we can see in Table 1.

Table 1. Quarterly demand in mil. of pieces 2015-March 2021

Year	Apple	Huawei	Xiaomi	Samsung	Total
31.03.2015	61.2	17.3	15.3	83.2	177
30.06.2015	47.5	3.5	19	74.5	144.5
30.09.2015	48	27.4	18.2	84	177.6
31.12.2015	74.8	32.6	18	81.3	206.7
31.03.2016	51.2	28.4	14.5	78.6	172.7
30.06.2016	40.4	32	14.5	77	163.9
30.09.2016	45.5	33.6	16	76	171.1
31.12.2016	78.3	44.9	15.8	77.5	216.5
31.03.2017	50.8	34.6	13.4	80	178.8
30.06.2017	41	38.5	23.1	80.4	183
30.09.2017	46.7	39.1	28.5	83.3	197.6
31.12.2017	77.3	41	31	74.4	223.7
31.03.2018	52.2	39.3	28.1	78.2	197.8
30.06.2018	41.3	54.2	32	71.5	199
30.09.2018	46.9	52	33.3	72.3	204.5
31.12.2018	65.9	59.7	25.6	69.8	221
31.03.2019	42	59.1	27.8	72	200.9

https://doi.org/10.1051/shsconf/202213201009

30.06.2019	36.5	56.6	32.3	76.3	201.7
30.09.2019	44.8	66.8	31.7	78.2	221.5
31.12.2019	72.3	56.2	32.9	70.4	231.8
31.03.2020	40	49	29.7	58.6	177.3
30.06.2020	37.5	54.8	26.5	54.2	173
30.09.2020	41.7	50.9	46.2	80.4	219.2
31.12.2020	81.9	33	43	62.5	220.4
31.03.2021	59.5	15	48.5	76.6	199.6

Source: [19].

According to Table No. 1, it is clear that the biggest players in the smartphone market are Apple and Samsung. Every year, however, the total number of units sold shows that the largest sales of mobile phones are always in the last quarter of each year. Overall, it is also evident that mobile phone sales are rising worldwide.

2.2 Methods

In the second part of the data comparison, we compare the demand before the coronavirus pandemic, i.e. the quarterly data from March 31, 2018 to March 31, 2019, with the demand during the pandemic from March 31, 2020 to March 31, 2021.

Our next step is to use simple linear regression to calculate the assumed prediction. Here we will use the overall calculation of quarterly data of all selected smartphone brands from 2015 to March 2021 and then calculate the estimate up until December 2022.

Last but not least, we will use the method of exponential alignment of the time series to smooth the short-term prediction of time series. This will utilize all the quarterly series from a given time period plus added prediction.

3 Results

To compare the demand before the coronavirus pandemic, i.e. quarterly data from March 31, 2018 to March 31, 2019, with the demand during the pandemic, i.e. quarterly data from March 31, 2020 to March 31, 2021. For this comparison, we used a line graph that shows the data with a blue line before the pandemic and then an orange line for data for the period during Covid-19; see Figure 1.

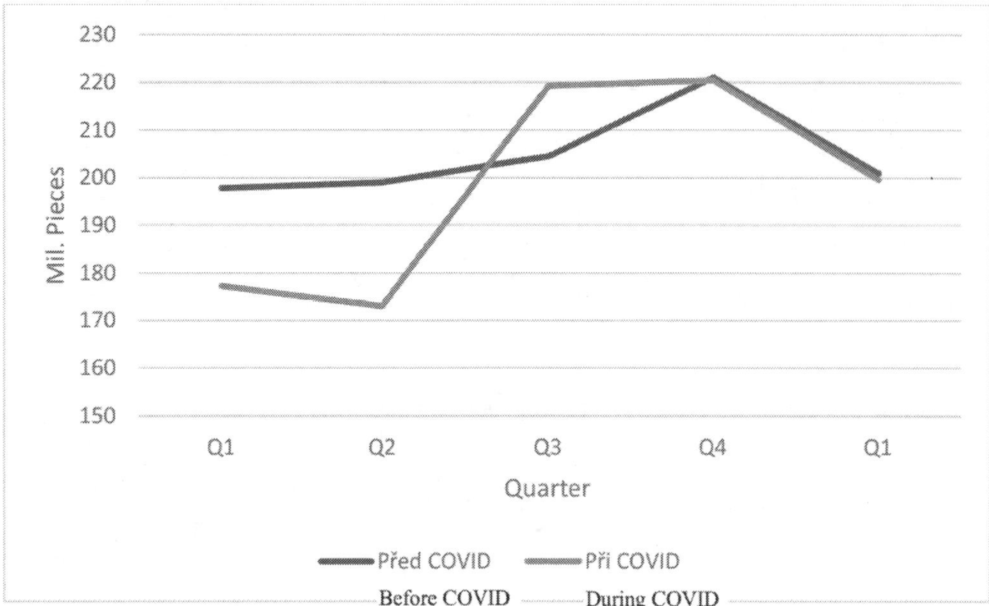

Figure 1. Quarterly data before and during Covid-19
Source: Own processing.

Figure 1 shows that in the period from March 31, 2018 to March 31, 2019, the trend in the number of units of mobile phones sold was usual with a gradual increase in the number of units sold during the year, with highs in the last quarter. During 2020, when the Covid-19 pandemic began, the largest increase in cell phone sales was recorded as anti-epidemiological measures were introduced. With the duration of these measures, the number of units sold remained at the upper end of 2020.

Using a simple linear regression to calculate the assumed prediction, we used the overall calculation of quarterly data of all selected smartphone brands from 2015 to March 2021 and then calculated the estimate up to December 2022. First, we used scatter plot in figure 2 to show the total values from 2015 to March 2021.

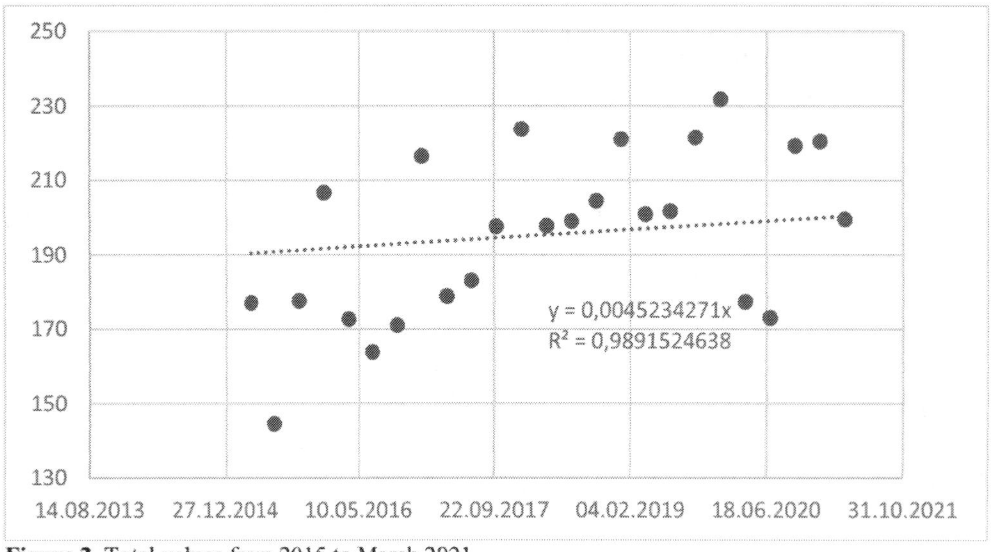

Figure 2. Total values from 2015 to March 2021
Source: Own processing.

Furthermore, when calculating simple linear regression, we used a line graph in figure 3 with a prediction from June 2021 to December 2022.

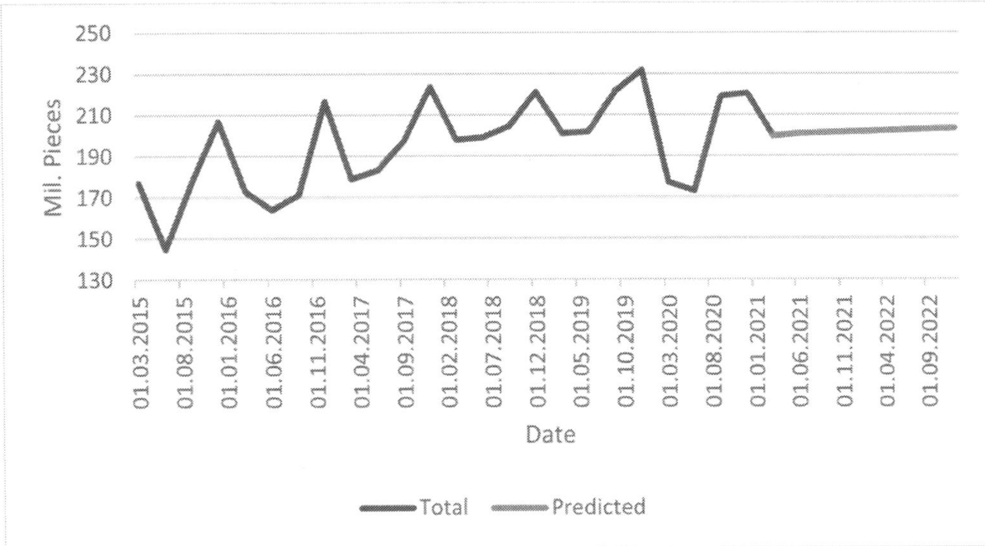

Figure 3. Prediction from June 2021 to December 2022
Source: Own processing.

From Figure 3 it is evident that in the period from March 31, 2021 to Decembed 2022, a gradual increase in the number of ordered pieces can be expected throughout the year 2022.

Finally, we used the method of exponential alignment of time series to smoothen the short-term prediction of time series. We used all quarterly series from the given time period here plus added prediction. To display this data, we again used a simple line graph in figure 4, which shows the total value of the blue line and the estimated prediction of the orange line.

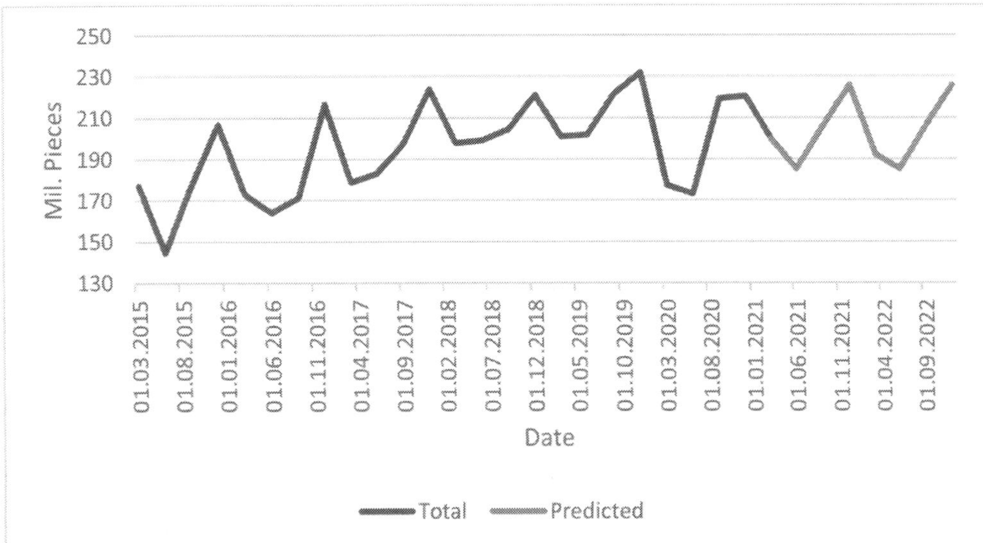

Figure 4. Total quarterly data and smoothened forecast
Source: Own processing.

Finally, the method of exponential alignment of time series was used to refine the prediction using linear regression to smooth the short-term prediction of time series. Based on the achieved results, it is evident that in 2022 the trend in the number of sold pieces of mobile phones from the period before the covid-19 pandemic will be copied again.

4 Discussion

Based on the obtained results, we are able to answer the determined research questions: *How has the demand for mobile phones been developing in the world 5 years before the coronavirus pandemic?* Demand was usually the highest in the 4th quarter, i.e. always at the end of the year. The end of the year means the most profitable time of the year for most retailers, because people buy Christmas presents. This also applies to the smartphone market, as electronics companies present smartphones as an ideal gift for Christmas. For example, Apple introduces new models mostly at the end of the year.

How has the demand for mobile phones in the world changed since the beginning of the coronavirus pandemic? According to the obtained results, it can be seen that at the beginning of the coronavirus pandemic, people saved money. It is possible that the reason for saving was fear and uncertainty in maintaining jobs. However, as we can see in figure 2, the demand for mobile phones increased again at the end of the year. The increase in demand may have been due to the upcoming Christmas holidays or the stabilization of fear from the coronavirus pandemic. About 360 million smartphones were sold in the third quarter, which is only 4 percent less than last year. This growth may be an interesting indicator for the marketing industry in the coming years.

What is the estimate of the price developments of mobile phones in the period from June 2021 to December 2022? Using exponential time series alignment to smoothen the short-term time series prediction, it has been shown that the entire smartphone market is clearly recovering from the coronavirus crisis. According to this data, it can be assumed that the price development of phones in the coming years will be similar to previous years.

5 Conclusion

The aim of the paper was to assess the demand for mobile phones in the world during the coronavirus pandemic and predict future developments up until the end of 2022. Using determined research questions and quarterly data [19] from 2014 to 2021, we compared the demand for Apple, Huawei, Samsung and Xiaomi. Among other things, we found that while sales of Samsung smartphones stagnated, Huawei and Apple phones experienced a year-on-year decline. By comparing the pre-coronavirus pandemic demand using a simple linear regression between March 31, 2018 and March 31, 2019, and then between March 31, 2020 and March 31, 2021, the following was found: before the coronavirus pandemic, almost always in the 4th quarter, demand was highest probably due to the Christmas holidays. During the coronavirus pandemic, the smartphone market declined slightly, but demand increased again in the fourth quarter. Furthermore, using the method of exponential adjustment of the time series for smoothing the short-term prediction of time series, it is assumed that the price development of mobile phones will be similar to previous years. As we have already summarized in the discussion of the results, the decline in demand for mobile phones was caused by a global pandemic and uncertainty of consumers maintaining their jobs. If the situation regarding the coronavirus crisis stabilizes and the smartphone market, and with it the economy as a whole, picks up activity as part of the recovery, mobile phone manufacturers do not need to feel threatened.

References

1. J. Majerova, C. Fernandes, How to make Phoenix to arise from the ashes: Brand loyalty as a prospective pillar of branding in tourism after crisis COVID-19. *Littera Scripta,* **13**(2), 49-58 (2020)
2. P. Pardál, R. Dias, P. Šuleř, N. Teixeira, T. Krulický, Integration in Central European capital markets in the context of the global COVID-19 pandemic. *Equilibrium - Quarterly Journal of Economics and Economic Policy*, **15**(4), 627-650 (2020)
3. J. Horák, K. Mlsová, V. Machová, Impact of the coronavirus pandemic on the tertiary sector. *Littera Scripta*, **14**(1), 28-39 (2021)
4. M. Sharma, S. Luthra, S. Joshi, A. Kumar, Accelerating retail supply chain performance against pandemic disruption: adopting resilient strategies to mitigate the long-term effects. *Journal of Enterprise Information Management* (2021)
5. M. Sharma, S. Luthra, S. Joshi, A. Kumar, Developing a framework for enhancing survivability of sustainable supply chains during and post-COVID-19 pandemic. *International Journal of Logistics Research and Applications* (2020)
6. R. Dias, N. Teixeira, V. Machová, P. Pardál, J. Horák, M. Vochozka, Random walks and market efficiency tests: Evidence on US, Chinese and European capital markets within the context of the global Covid-19 pandemic. *Oeconomia Copernicana,* **11**(4), 585-608 (2020)
7. A. Kamalaldin, L. Linde, D. Sjödin, V. Parida, Transforming provider-customer relationships in digital servitization: A relational view on digitalisation. *Industrial Marketing Management*, **89**, 306-325 (2020)
8. P. A. Tominac, V. M. Zavala, Economic properties of multi-product supply chains. *Computers & Chemical Engineering*, **145** (2021)
9. N. Zhao, Q. Wang, P. Cao, J. Wu, Pricing decisions with reference price effect and risk preference customers. *International Transactions in Operational Research*, **28**(4), 2081-2109 (2019)
10. M. Framingham, Smartphone shipments decline 2.3% in the second quarter on continued challenges across most major regions, According to IDC [online]. Available at: https://www.businesswire.com/news/home/20190731005812/en/Smartphone-Shipments-Decline-2.3-in-the-Second-Quarter-on-Continued-Challenges-Across-Most-Major-Regions-According-to-IDC (2019)
11. L. Kosmol, M. Maiwald, C.Pieper, J. Plötz, T. Schmidt, An indicator-based method supporting assessment and decision-making of potential by-product exchanges in industrial symbiosis. *Journal of Cleaner Production*, **289** (2021)
12. S. Bin, Y. Zhiquan, L.S.C. Jonathan, D.K. Jiewei, D. Kurle, F. Cerdas, C. Herrmann, A big data analytics approach to develop industrial symbioses in large cities Procedia CIRP. *The 22nd CIRP Conference on Life Cycle Engineering*, **29**, 450-455 (2015)
13. I. Khamlich, K. Zeng, G. Flamant, J. Baeyens, C. Z. Zou, J. Li, X. Y. Yang, X. He, Q. C. Liu, H. P. Yang, Q. Yang, H. P. Chen, Technical and economic assessment of thermal energy storage in concentrated solar power plants within a spot electricity market. *Renewable and Sustainable Energy Reviews*, **139** (2021)
14. A. D. Hunka, M. Linder, S. Habibi, Determinants of consumer demand for circular economy products. A case for reuse and remanufacturing for sustainable development. *Business Strategy and the Environment,* **30**, 535-550 (2021)

15. T. Tempesta, D. Vecchiato, F. Nassivera, M. Bugatti, B. Torquati, Consumers demand for social farming products: An analysis with discrete choice experiments. *Sustainability*, **11**(23), (2019)
16. Ch. Bertram, G. Luderer, F. Creutzig, N. Bauer, F. Ueckerdt, A. Malik, O. Edenhofer, COVID-19-induced low power demand and market forces starkly reduce CO_2 emissions. *Nature Climate Change,* **11**(3), (2021)
17. J. Ruan, Q. Cai, S. Jin, Impact of COVID"19 and nationwide lockdowns on vegetable prices: Evidence from wholesale markets in China. *American Journal of Agricultural Economics,* **103**(5), 1574-1594 (2021)
18. B. Wang, How does COVID-19 affect house prices? A cross-city analysis. *Journal of Risk and Financial Management*, **14**(2), (2021)
19. Counterpoint, Global Smartphone Market Share: By Quarter. Counterpoint USA: Counterpoint team [online]. Available at: https://www.counterpointresearch.com/global-smartphone-share/ (2021)

Czech companies will have to deal with import of labour

Eva Kalinová[1,*], *Yaroslava* Kostiuk[1], and *Denisa* Michutová[2]

[1]University of Žilina, The Faculty of Operation and Economics of Transport and Communications, Department of Economics, Univerzitná 8215/1, 010 26 Žilina, Slovakia
[2]Institute of Technology and Business in České Budějovice, School of Expertness and Valuation, Okružní 517/10, 370 01 České Budějovice, Czech Republic

Abstract. Labour market or supply and demand for labour is determined by how individuals demand work (supply) and how the offer of jobs is from the side of companies (demand). This is a very important issue, as it is a part of the main factor markets. The data on labour market are used for analysing the movement of labour market. Thanks to this, it is possible to forecast its approximate development. The basic source are data from the database of the Ministry of Labour and Social Affairs of the Czech Republic. The objective of this paper is to analyse the development of supply and demand for labour in the years 2010-2020 and forecast its development until 2025. The analysis of time series is performed using the method of artificial neural networks, which enables the analysis of the development between 2010 and 2020 and forecasting the further development of supply and demand for labour until the year 2025. The research shows that the development until the year 2025 will not be very favourable. The demand will be much greater than the supply, which means there will be more vacancies than workers. To fill the vacancies and be able to further operate, companies will try to solve this situation by hiring workers from other countries. The results of the paper being submitted may serve for other labour market research.
Keywords: labour market, supply and demand for labour, labour market forecasting

1 Introduction

The labor market is, like many other markets, made up from supply and demand. Firms through demand for their products and services "derive" demand for work. Households, on the other hand, create labor supply as one of the factors of production. The flexibility of the job offer necessarily depends on the types of job positions. In the short term, it is generally possible to state a lower flexibility of the labor supply, especially for highly qualified job positions due to higher transfers and annuities. It is the transitions between job positions that create frictional unemployment in its natural expression. On the other hand, the highest problem is involuntary unemployment, which occurs most often during cyclical and structural fluctuations.

* Corresponding author: kalinova@mail.vstecb.cz

High unemployment rate in many European countries raises concerns of both individuals directly affected by these figures and government and politicians [1], [2].

Labour markets are a good tool to evaluate the performance of the economy, since their observation enables assessing the source of the existing problem, such as unemployment, as well as the success rate of the policies implemented [3].

Svoboda and Applova [4] say, that if jobs offered by various employers are not perfect substitutes, employers have the power to set wages. Its extent can be described by the elasticity of the job supply for the companies [5]. A significant discrepancy between the estimates of the direct elasticity of job offer to changes in wages (lower) and estimates converted from inverted elasticities (higher) indicates that the institutions of labour market can reduce significantly the amount of fixed force determining wages [6]. Tightening of financial conditions leads companies to reduce wages and the number of hours worked per employee [7]. Companies publish fewer vacancies and offer lower wages [8].

A common hypothesis of labour market polarization is that technological progress has reduced the price of machines which are replaceable by medium-skilled jobs and complementary to higher and lower-skilled jobs [9]. Competition between groups of workers and within those groups takes place in the labour markets, which are segmented according to various, often unobservable dimensions [10]. Wage and unemployment rate largely depends on the distribution of unemployment expectations frequency among workers, which is a key feature of the labour market [11].

In macroeconomic literature, most empirical studies that analyse the effects of financial shocks does not address the issues related to labour market [12]. On the one hand, there is an increasing number of theoretical publications analysing the effects of financial shocks on several aspects of the labour market; on the other hand, at the macro level, empirical works include only one variable on the labour market [8].

The objective of the paper is to assess the development of supply and demand for labour in the Czech Republic in the years 2010-2020 and forecast their further development until 2025. In order to achieve the objective, two research questions are formulated:

RQ1: How did supply and demand for labour develop in the years 2010-2020?
RQ2: What will be the development of supply and demand for labour from 2021 to 2025?

Companies decide on how many and which workers to hire. Nevertheless, in most models of the labour market, all workers have the same probability to be hired. Merkl and Rens [13] argue that selective hiring significantly affects the analysis of welfare. They created a model which is isomorphic to a search model in random hiring but enables selective hiring. In selective hiring, the positive predictions of the model change very little but the consequences for welfare are different for two reasons: first, the hiring externality occurs in random hiring but not in selective hiring. Second, the costs of social security in unemployment are much higher in the case of selective hiring, since the risk of unemployment is unevenly distributed among employees [13].

Fachelli and López-Roldán [14] aimed to confirm the general hypothesis claiming that there is no single labour market that would regulate supply and demand for labour but it can be assumed that several segments will structure hierarchical positions on the labour market in accordance with the specific profiles of job offer. The authors had expected that despite contextual and socio-historical differences they would achieve high correspondence between the individual employment segments.

Rkman [3] specifies a simple search and matching model of the labour market and how well it is able to describe the dynamics of the labour market. The model searches vacancies using two types of shocks: productivity shocks and separation shock. The model estimates the data on unemployment and vacancies using Bayesian methods. The model fits well in the data and the estimates show that productivity shocks are the main drivers of fluctuations of the labour market, while separation shocks represent high fluctuating unemployment.

Corbellini, Magnani and Morelli [15] used the method of discriminant analysis and linear or generalized linear models. These models do not consider the presence of possible outliers. At the microeconomic level, theoretical analysis of the functional distribution of income implicitly expressed this assumption by accepting Cobb-Douglas (CD) function for describing production and distribution of income between labour and capital in a sample company. In this context, it holds true that if factor markets are acceptably competitive and the distribution of income reflects the limits of these two factors, the shares of factors are fixed [15].

Nackerdien and Yu [16] examined the relationships on the labour market between the formal and informal sectors, focusing on three main groups of employees: those who work in the formal sector, those who work in the informal sector, those who worked in both sectors. The authors used econometric analysis, based on which older people and people with higher education are more likely to work in the formal sector.

Benda, Koster and Van Der Veen [17] point to the fact that the dependent variable is measured on a continuous scale; therefore, a suitable analytic method appears to be linear regression. Nevertheless, due to the clustered data structure, where people are clustered into countries, the basic assumption of independent observation is violated, which would result in distortion of the estimates. The common method for analysing clustered data is multilevel modelling (MLM), which considers clustering by means of including random variables.

Molnarova and Rostekova [18] used the methodological model of discourse analysis (DIMEAN). For better understanding of the issue, it was necessary to analyse the socio-political development in the last decade with a focus on legislative measures taken in the area of education and to determine the basic theoretical foundations of the research. The research results clearly show that professionalism of secondary education and the need for labour market forecasting are addressed sufficiently [18]. Speckesser, Gonzalez Carreras and Kirchner Sala [19] used aggregate impact analysis to examine the impact of politics on macroeconomic variables, such as employability and youth unemployment. This is based on the assumption that programmes which are efficient in terms of improving individual job opportunities, may affect the balance of youth unemployment. The results show that both wage subsidies and creation of jobs reduce aggregated youth unemployment, which is in the contradiction with the research in several microeconomic studies, which indicate that programmes for creating jobs are not efficient. This points to the importance of helping young people to acquire valuable work experience, which is beneficial for creating jobs, although the work experience is acquired outside regular employment or in the commercial sector [19].

Flek, Hala and Mysikova [20] dealt with the description of the data obtained from the longitudinal database EU-SILC, selected specific countries and a period of the analysis. They explained the concept of the so-called gross flows in the labour market, which can be used to derive the probabilities of the individual transition between employment, unemployment, and economic inactivity. The results focused on the average probabilities of the transition between the situations in the labour markets in individual age groups generally qualitatively coincide with the reference economy of the United Kingdom.

In accordance with the formulated research questions, Flek, Hala and Mysikova [20] document the marginalized status of young people below 24 years in terms of the risk of losing job, which is relatively the highest one of all age groups, followed by unemployment. The so-called "IMF-OECD consensus" indicates that the deregulation of the labour market increases employment and reduces unemployment. Brancaccio, De Cristofaro and Giammetti [21] present a meta-analysis of research on this issue based on the MAER-NET guidelines. The authors analysed the relationship between the indexes of legislation for protecting employment on the one hand and employment and unemployment on the other hand. The decrease of support for the consensus view is notable especially in the last decade [21].

2 Data and methods

2.1 Data

The data on supply and demand on the labour market were obtained from the official website of the Ministry of Labour and Social Affairs [22]. The data are available for the period from 31 January 2010 to 31 December 2020. On their basis, a prediction was made for the period between 31 December 2020 and 31 December 2025.

2.2 Methods

Statistical software TIBCO Statistica was used to evaluate the data obtained. The analysis of the time series (regression) is performed using the method of neural networks, where supply and demand for work is a continuous target variable, and date is a continuous predictor. Predictions will be made using three datasets – training, testing, and validation. The training dataset will contain 70% of the input data, while both testing and validation datasets will contain 15% of the input data each. Each value will be calculated on the basis of the previous one. There will be used MLP (multilayer perceptron networks). The LP activation functions, which are divided into the neurons in the hidden and output layer, the following functions are used: identity, logistics, atanh (hyperbolic tangent), exponential, and sine.

For the selected type of network (MLP), 2-8 neurons in the hidden layer will be used. The second selected type of network is RBF, for which 17-24 hidden neurons will be used. For the training, 1,000 networks will be generated, out of which 5 will be retained.

Based on the date, the development of the variable will be predicted. The network will be the neural network of the radial basis function. Its performance will be marked by a correlation coefficient, with the ideal value being as close to 1 as possible.

In accordance with the selected methodology, the results will be presented in the following form:

- Analysis of the input data,
- Overview of retained neural networks,
- Graph of supply and demand for labour in the years 2010-2020,
- Graph of predictions.

3 Results

Table 1 presents the overall data showing the development of the supply and demand on the labour market in the years 2010-2020. The data were obtained from the website of the Ministry of Labour and Social Affairs [22]. It is a sum of the supply and demand on the last day in December of a given year. This table is compiled only for the comparison of supply and demand in individual years.

https://doi.org/10.1051/shsconf/202213201010

Table 1. Data on supply and demand on the last day of the year

Date	Supply	Demand
31 December 2010	561,551	30,803
31 December 2011	508,451	35,784
31 December 2012	545,311	34,893
31 December 2013	596,833	35,178
31 December 2014	541,914	58,739
31 December 2015	453,118	102,545
31 December 2016	381,373	132,496
31 December 2017	280,620	216,629
31 December 2018	231,534	324,410
31 December 2019	215,532	340,950
31 December 2020	291,977	318,581

Source: Authors.

In accordance with the methodology, 1,000 neural networks (MLP and RBF) were created, out of which 5 with the best values were retained. The results obtained from the three data samples and the aforementioned networks are presented in Table 2. This table shows only the demand for labour. Four out of five retained networks are the MLP networks, one is RBF network. The table also presents correlation coefficient, which is the performance of the network in individual datasets (training, testing, validation), where the values should be as close to 1 as possible. For further analysis, at least one suitable network should be selected. As already mentioned, the closer its value is to 1, the better; therefore, the model of the network with the highest value is selected most often.

Table 2. Retained neural networks - Demand

Index	Network	Training performance	Testing performance	Validation performance	Training error	Testing error	Validation error	Training algorithm
1	MLP 1-7-1	0.999110	0.997401	0.998857	12530932	11992015	20494544	BFGS 206
2	RBF 1-22-1	0.997065	0.995037	0.998912	40797090	83687524	53756542	RBFT
3	MLP 1-4-1	0.999044	0.997411	0.998797	13427109	11368737	19782813	BFGS 309
4	MLP 1-7-1	0.999176	0.997343	0.998786	11468719	13328887	21053133	BFGS 137
5	MLP 1-6-1	0.999339	0.997198	0.998969	9216361	16678819	18668172	BFGS 180

Source: Authors.

All networks show very good performance in all datasets, which is close to the ideal value. However, when summing up the performance in all datasets, it is still possible to determine the most suitable neural structure, 5. MLP 1-6-1. Table 3 below shows 5 retained neural networks from the 1,000 networks generated by the MLP and RBF models. In this case, it is about supply, where the best performance was achieved by the RBF networks. Even in this case, we used training, testing, and validation datasets.

Table 3. Retained networks - Supply

Index	Network	Training performance	Testing performance	Validation performance	Training error	Testing error	Validation error	Training algorithm
1	RBF 1-23-1	0.990547	0.984797	0.985460	165727676	202189905	234466494	RBFT
2	RBF 1-24-1	0.995834	0.991442	0.990744	75121230	93059739	156086224	RBFT
3	RBF 1-24-1	0.989719	0.988636	0.987465	180117147	139846032	202774460	RBFT
4	RBF 1-21-1	0.990859	0.990833	0.987876	160238521	112071333	200791417	RBFT
5	RBF 1-22-1	0.992161	0.982959	0.985830	137499454	258499494	229484576	RBFT

Source: Authors.

The values of correlation coefficients are excellent, achieving almost maximum performance. Out of the five retained networks, the best values are achieved by 2.RBF 1-24-1. Graph 1 shows supply and demand in the years 2010-2020. Each year is represented by the first day in January and July. Supply of labour means that there is an offer from individuals and the number of workers available in the economy. Demand means companies are in demand for labour and want to fill vacancies.

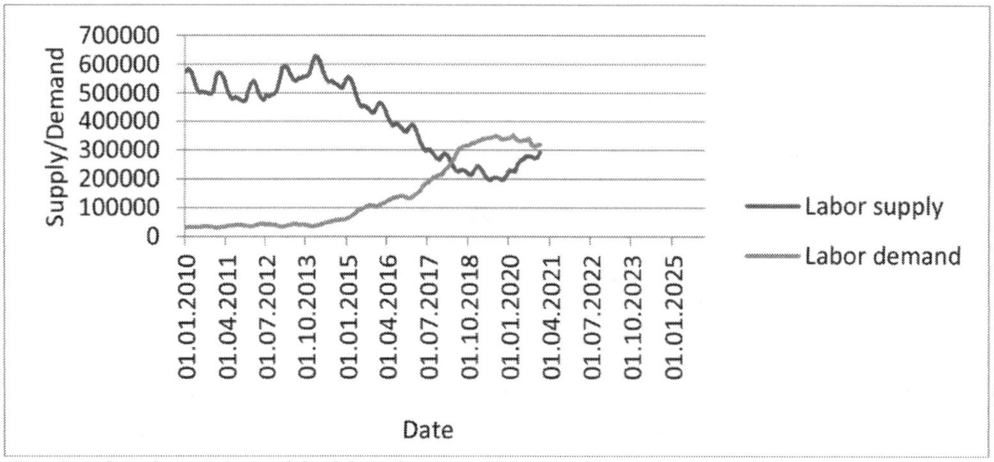

Graph 1. Supply and demand for labour between 2010 and 2020
Source: Authors.

As seen from the graph, supply was very high in the year 2010, achieving almost the value of 580,000; in contrast, at the beginning of 2010, demand was relatively low, not exceeding 32,000. During the decade, approximately at the end of the second half, demand for labour started to grow, exceeding supply in 2018. The peak was achieved in 2014, achieving the value of 630,000. The biggest growth of demand was recorded in 2020, supply and demand started to balance, with the difference between them being nearly 30,000.

Graph 2 shows supply and demand for labour in the years 2010-2020, which is described in more detail in Graph 1. Graph 2 focuses rather on the prediction made until the year 2025 inclusive.

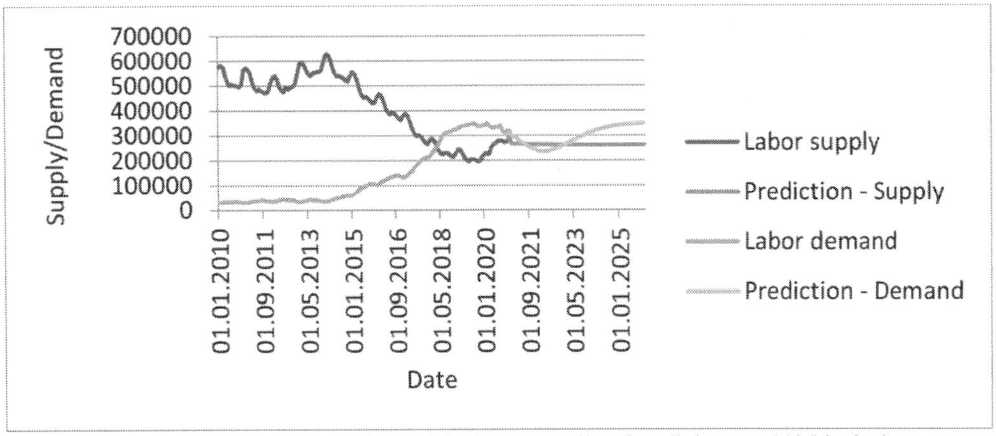

Graph 2. Time series of supply and demand for labour predicted until the year 2025 inclusive
Source: Authors.

In 2020, demand starts to balance supply; however, in 2021, the demand begins to decrease from nearly 300,000 almost to 237,000 in 2022; this is also the lowest value in the period between 2020 and 2025. In 2023, demand starts to increase. At the beginning of 2025, the curve stabilizes remains approximately the same for the whole year, with the values fluctuating around 345,000. Supply is at the same level from the beginning of 2021 to 2025, fluctuating around 260,000.

Graph 3 shows only the prediction for the period 2021-2025, which was described in more detail in Graph 2 above. It is just a more detailed graph of predicted supply and demand for labour in the labour market. The graph presents the data as of the first day in January, April, and July, with the exception of the year 2025, where also October is included.

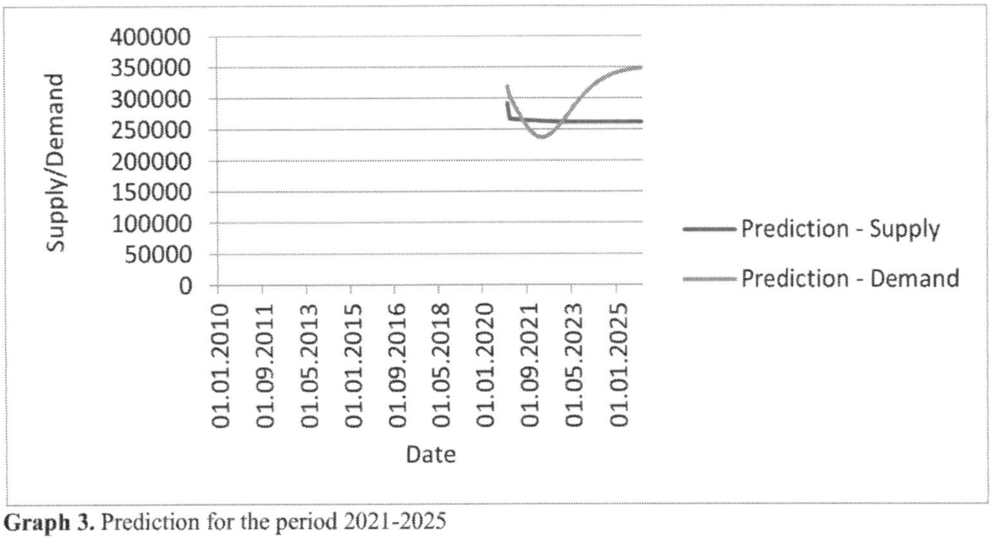

Graph 3. Prediction for the period 2021-2025
Source: Authors.

The predicted supply thus remains approximately at the same level, achieving the values of about 265,000, while demand is predicted to both decrease and increase. At the beginning of 2025, it stabilizes and remains at the same level for the whole year.

4 Discussion

Based on the results obtained, it is possible to answer the research questions formulated in order to achieve the main objective of the paper being submitted:

How did supply and demand for labour develop in the years 2010-2020?

In the labour market, the decline of supply and the increase in demand for labour slowed down. The effects of the global financial and economic crisis, which started to appear in the labour market at the end of 2008, still continued in 2009. The impacts of crisis were visible even in the first months of the year 2010 [22]. The decline in economic performance from the year 2012 (decrease by 0.9%) slowed down slightly in 2013 (decrease by 0.5%). In 2014, there was a positive turnaround in the economic development. According to Labour Force Survey (LFS), there was a year-on-year growth (by 1.4%) in employment in the year 2015 [22]. Along with the growth of the economy, the growth of employment was recorded as well, achieving its peak from the beginning of the economic and financial crisis in 2016. The increase was 1.9%. Despite the positive development of the economy, the lack of suitable labour force became a barrier to more significant growth of employment. The imbalance of employer demand for labour and supply caused a slowdown in employment growth, which fell to 1.6% in 2017, 1.4 % in 2018, and 0.2% in 2019. Currently, the number of registered job seekers is still lower than the number of vacancies announced. In the first quarter of the year 2020, there were 0.7 job seekers per vacancy [22].

The labour market fluctuated between the years 2010 and 2020. In 2010, there was a big difference between supply and demand for labour. At the beginning of 2010, supply fluctuated around the value of 580,000, while demand for labour achieved almost the value of 32,000. At the end of the decade, the difference between supply and demand was smaller, achieving the value of nearly 30,000, with demand exceeding supply.

What will be the development of supply and demand for labour from 2021 to 2025?

Until the year 2025, a slight increase in the number of vacancies is expected in the CR. Compared to the year 2014, the increase accounts for ca. 1.5%, i.e. nearly 80,000 vacancies. However, this does not mean that in the following years, only these new vacancies need to be filled, since this is just the resulting balance between the number of new vacancies on the one side and cancelled jobs on the other side [23].

According to NUV [23], in addition to the independent movement of workers between individual jobs, it will also be necessary to fill the jobs vacated by people leaving the labour market e.g. due to the retirement etc. Moreover, it states that for these reasons, between 2014 and 2025, approx. 1 million of jobs can be vacated in the Czech Republic, which will need to be filled. It is further stated that a lot of jobs will disappear due to the technological progress and ever-increasing efficiency.

By 2025, the fastest decline in the number of employed people is expected in the primary sector. Compared to 2014, in 2025, the number of jobs in this sector will decrease by 6%, i.e. by nearly 17,000. In the secondary sector, the decrease will be about 2.3%, i.e. 41,000 jobs. Other sectors of the Czech economy are expecting growth in the number of jobs. In the tertiary sector, in 2025, it shall be an increase by 14,000 (about 1%) compared to the year 2014. In public quaternary sector, the increase is nearly 36,000 (3.5%), while in quaternary business sector, it will be more than 83,000, which accounts for 10% [23]. At the beginning of the year 2021, supply fluctuates at the same level around the value of 266,000. Demand will fluctuate more and starts to decrease at the beginning of the year 2021. The year 2022 will show a continuous decrease in demand, but at the turn of 2022/2023, demand will increase, exceeding supply. At the end of the year 2025, its value will achieve approx. 348,000.

5 Conclusion

The objective of the paper was to assess the development of supply and demand for labour in the Czech Republic in the years 2010-2020 and to predict their further development until the year 2025. Using time series, the method of artificial neural networks, it was possible to identify the development of supply and demand for labour in specific years and to predict its further development. The objective of the paper was thus achieved.

In the years 2010-2020, supply and demand fluctuated. Due to the economic crisis in 2008, demand for labour was lower than supply at the beginning of the year 2010, with demand achieving the value of approx. 33,000 and supply approx. 580,000. Demand and supply were thus unbalanced until 2018, when demand for labour balanced and even exceeded supply until the end of the year 2020.

As for the predicted supply and demand for labour, supply is expected to remain at the same level until the end of the year, while demand will fluctuate slightly. A decrease is expected at the beginning of the year 2021, which will be followed by an increase with expected stabilization until the year 2025.

This means that the number of vacancies will exceed the number of workers; any individual interested in finding a job will be able to find it without any problems. In contrast, some companies will lack employees. If they are not able to find enough employees soon, it will result in the reduction of production and this would mean bankruptcy. For the state, the fact that companies will not make profit and will not be able to pay taxes, it would mean loss of financial resources used to finance social care, pensions, and other social benefits.

To improve the situation, I would recommend companies to hire labour from other states so that they are able to fill the vacancies and are able to operate. This would also enable people from economically less developed countries to get a job and higher salary. Companies would fill the vacancies, which would enable them to continue operating and the state would thus have sufficient financial resources for the state budget.

On the other hand, it is appropriate to look at the results from a qualitative view. Although the epidemic increase in jobs is evident, it seems that the national economy is generating one that, in essence, is of little interest. They can be defined as supporting activities within production processes and especially also craft activities. The element of manualness is therefore highly undesirable on the labor market in the Czech Republic.

The ongoing migration in the labor market, which is currently somewhat limited by the situation around COVID-19, is a completely natural economic phenomenon. However, there is still a large social group of the economically productive population that perceives migration as a threat to their job certainty. After all, many political actors are aware of this fact. The result is the determination of appropriate party and election programs aimed at obtaining political rents.

Based on the above results and given the current situation concerning COVID-19, the following questions may arise: How will the labour market, or more specifically supply and demand for labour develop after this turnover? Will the Czech Republic be able to recover as soon as possible? Will it be visible in the following years that the world was affected by this disease, due to which many people have lost or will lose their jobs? To answer these questions, further research shall be conducted.

References

1. D. Stroukal, The short-term relationship between unemployment and home ownership. *Littera Scripta*, **9**(1), 140-153 (2016)
2. S. Haskova, P. Volf, V. Machova, Economic convergence of Czech regions in terms of GDP and unemployment rate in response to FDI flows: Do businesses and regions flourish? *Ad Alta: Journal of Interdisciplinary Research,* **9**(1), 326-329 (2019)
3. J. Rkman, An estimated search and matching model of the Croatian labor market: Post-crisis analysis. *Croatian Economic Survey*, **22**(2), 35-72 (2020)
4. O. Svoboda, P. Applova, Determinants of employment and GDP resilience in the context of an economic crisis: Evidence from EU countries and regions. *Littera Scripta*, **9**(2), 140-155 (2016)
5. V. Novak, M. Vokoun, F. Stellner, M. Vochozka, Institutional analysis of the contemporary regional labour market in the Czech Republic. *E+M Ekonomie a Management,* **19**(3), 4-15 (2016)
6. A. Sokolova, T. Sorensen, Monopsony in labor markets: A meta-analysis. *ILR Review*, **74**(1), 27-55 (2021)
7. J. Deggans, T. Krulicky, M. Kovacova, K. Valaskova, M. Poliak, Cognitively enhanced products, output growth, and labor market changes: Will artificial intelligence replace workers by automating their jobs? *Economics, Management, and Financial Markets*, **14**(1), 38-43 (2019)
8. F. Corsello, E. Valerio Nispi Landi, Labor market and financial shocks: A time-varying analysis. *Journal of Money, Credit and Banking*, **52**(4), 777-801 (2020)
9. Ch. Vom Lehn, Labor market polarization, the decline of routine work, and technological change: A quantitative analysis. *Journal of Monetary Economics*, **110**, 62-80 (2020)
10. N. M. Wiener, Labor market segmentation and immigrant competition: A quantal response statistical equilibrium analysis. *Entropy,* **22**(7), (2020)
11. J. J. Silveira, G. T. Lima, Can workers' increased pessimism about the labor market conditions raise unemployment? *International Review of Economics & Finance*, **72**, 125-134 (2021)
12. I. Kmecova, J. Stuchly, L. Polanecky, M. Suta, Analysing structure of employed and unemployed population of Czech Republic as part of human capital on labour market. *Littera Scripta,* **12**(1), 121-141 (2019)
13. Ch. Merkl, T. Van Rens, Selective hiring and welfare analysis in labor market models. *Labour Economics*, **57**, 117-130 (2019)
14. S. Fachelli, P. López-Roldán, Comparative Perspectives on Social Indicators. *Social Indicators Research: An International and Interdisciplinary Journal for Quality-of-Life Measurement*, **154**(3), 755-765 (2021)
15. A. Corbellini, M. Magnani, G. Morelli, Labor market analysis through transformations and robust multivariate models. *Socio-Economic Planning Sciences*, **73** (2021)
16. F. Nackerdien, D. Yu, A panel data analysis of the formal-informal sector labour market linkages in South Africa. *Development Southern Africa*, **36**(3), 329-350 (2018)
17. L. Benda, F. Koster, and R. Van Der Veen, Active labour market policy as a socialising agent: a cross-national analysis of learning attitudes. *Studies in Continuing Education*, **42**(1), 75-101 (2020)

18. E. Molnarova, M. Rostekova, Educational policy of the Slovak Republic in the context of changes in the labour market: Media discourse analysis. *Politické Vědy*, **23**(4), 47-64 (2020)
19. S. S. Speckesser, F. J. Gonzalez Carreras, L. Kirchner Sala, Active labour market policies for young people and youth unemployment. *International Journal of Manpower*, **40**(8), 1510-1534 (2019)
20. V. Flek, M. Hála, M. Mysíková, Unemployment and age-based labor market segmentation. *Politická Ekonomie*, **66**(6), 709-731 (2018)
21. E. Brancaccio, F. De Cristofaro, R. Giammetti, A meta-analysis on labour market deregulation and employment performance: No consensus around the IMF-OECD consensus. *Review of Political Economy*, **32**(1), 1-21 (2020)
22. MLSA, Ministerstvo práce a sociálních věcí [Ministry of labour and social affairs] [online], Available at: https://www.MLSA.cz (2021)
23. NUV, Národní ústav pro vzdělání [National institute of education] [online], Available at: http://www.nuv.cz (2021)

Underestimating risk leading to the collapse of a market leader in tourism

Jana Janíková[1*], *Marek* Vochozka[2], and *Martin* Votava[2]

[1]University of Žilina, Faculty of Operation and Economics of Transport and Communications, Department of Economics, Univerzitná 8215/1, 010 26 Žilina, Slovakia

[2]Institute of Technology and Business, School of Expertness and Valuation, Okružní 517/10, 37001 České Budějovice, Czech Republic

Abstract. The topic of this paper, underestimating risk leading to the collapse of the market leader in tourism, is demonstrated on the example of the British travel agency Thomas Cook, which at one time was one of the oldest and largest travel agencies in the world. The aim of this paper is to analyze the development of the stock prices of Thomas Cook from May 13, 2018 to May 19, 2019 and the factors that had an impact on the share price of this company in the monitored period. The base source of data are the share prices of the travel agency Thomas Cook in the specified period from May 13, 2018 to May 19, 2019 published by MarketWatch. A statistical description of time series is used, a moving average trend line is displayed, and a cause-and-effect analysis evaluating the impact of the published information on the value of Thomas Cook's stocks is carried out. The general lesson for companies resulting from this contribution is that every negative event, announcement or piece of information has a negative impact on the value of a company's shares and a collapse could happen even to the leader of a given industry. The collapse of Thomas Cook provides lessons for companies doing business in tourism, so that in the event of a planned merger, a suitable company is selected, the company's funds are under control and development trends in the field are monitored.

Keywords: Thomas Cook, stock price, factors affecting stock price, bankruptcy

1 Introduction

Low-cost travel agencies and their unbeatable prices were certainly a blessing for customers, but older travel companies, which were forced to reduce their prices, did not manage to adapt, especially if they diversified their activities too much and lost their advantageous positions [1].

In the field of today's tourism, it is not possible to rely solely on case studies and tried and true practices. It is necessary to come up with new approaches that can deal with changes in conditions, such as when confronting the diversification of tourism risks and economies of scale [2].

* Corresponding author: janikova@mail.vstecb.cz

In order to gain a competitive advantage and maintain high market share, companies must constantly adapt, not only by introducing the latest available technologies, but also by creating their own innovations to increase business profitability and sustainability [3].

Changing the business model by introducing technological and sustainable components exposes companies to high risk. On the one hand, medium-sized companies are more flexible, the decision-making process is faster, and they can more easily adopt changes in management strategies, but on the other hand they are more vulnerable to market turbulence and dependent on constant cash flows [4].

There are many companies on the market that are growing much faster than the economy as a whole. Thus, there must be a number of companies whose growth rate is slower, or even shrinking and some are collapsing [5]. A typical example is Amazon's corporate victims, whose list expands almost every day. In the worst case, similar companies lie to themselves and those around them (including shareholders) that there's nothing wrong until the last moment. In the best case scenario, on the other hand, they admit their situation as soon as possible and try to save what they can.

Survival bias in credit risk modeling is a bias that leads to parameter estimates when a company's survival is ignored [6]. The results presented [7] suggest that systemic risk models should take into account the interconnectedness between economic entities, such as the banking and real estate and financial sectors, in a multi-layered approach. Stock prices have predictive power for future long-term economic activity [8].

I demonstrate on the example of the tourism industry, namely the company Thomas Cook. The world's oldest travel company Thomas Cook, a British travel agency with 19 million annual clients, which after 178 years of service has officially announced that it will liquidate its assets and file for bankruptcy, despite attempts to save the brand [1].

The story is well summed up by, for example, British newspaper The Guardian, stating that the company is collapsing after 178 years, caused by erroneous mergers, high debts and the internet revolution, along with uncertainty about Brexit. As other media point out, the final turning point was when the company failed to obtain a loan of £200 million. In May, the company reported a loss of £1.5 billion, mainly due to the write-off of an investment in MyTravel made in 2007. The merger was set to create a European giant and bring in cost savings of £75 million a year. However, the result was mainly higher debts [9].

The aim of the paper is to find out the development of the stock prices in the length of one year from May 16, 2018 to May 16, 2019 and to find out what factors influenced the share price in the examined period, where the statistical unit will be the travel agency Thomas Cook.

In order to meet the aim of the paper, the following research questions are set:

RQ1: What were the stock prices from May 13, 2018 to May 19, 2019?
RQ2: What factors influenced the stock prices during the monitored period?

The relationship between the price and economic activity of stocks is dealt with in a study [10]. This study show that stock prices have predictive power in future panel long-term economic activity. Empirical research by [11] presents a methodological approach to the transmission of impulses in capital markets and deals with the process of transmission of negative impulses in capital markets during the subprime crisis (contagion, dominance, crisis transmission and shocks). A significant and positive benefit of the research is the demonstration of how wavelet analysis can be used to examine the different reactions of financial markets. Investors consider volatility before investing, address capital market risks and measure market performance. Examining the capital market of the joint movement in Southeast Asia and analyzing the correlation of the conventional and Islamic indices in the regional and global equity sector were focused by [12]. The result shows that the conventional or compound index in Malaysia is less volatile than the Islamic one, but on the other hand both control the correlation movement.

Stock price prediction is becoming an important practical aspect of the economic field of study. A framework for addressing these challenges and effectively predicting stock prices using learning models were proposed by [13]. The results show that the applied models outperformed other models in predicting stock prices under different circumstances based on several valuation metrics. Forecasting stock prices using different models partly suffers from problems such as low accuracy, slow convergence and complex network structures. A study [14] developed an echo state network (ESN) model to alleviate these problems. In terms of accuracy, ESN had a unique advantage in acquiring nonlinear data. The proposed ESN has been found to be an effective model that converges faster than others, predicts more accurately and makes time series analyzes easier. Another study, which deals with the large decline in stock prices in 22 emerging markets, was conducted by [15]. In the case of using the analytical report as a proxy for incoming information, it was found that most of the declines in emerging markets are not accompanied by information events and these declines are followed by price reversals.

A model economy consisting of interdependent real, money and stock markets was proposed [16]. The money market is influenced by the real money market through the standard equation LM. Information about specific companies has a dampened effect on the share prices of companies associated with business groups. The exceptional stock returns of these firms respond less sensitively to eccentric commodity price shocks than the concise profits of otherwise similar unrelated firms in the same country and commodity sensitive industry. The use of global commodity shocks means that we evaluate responses to common idiosyncratic shocks of equal magnitude, frequency, and observability. Further identification results from difference-by-difference tests using successful and closed exogenously failed control block transactions. Experts have concluded that the share prices of companies in business groups provide providers and managers of capital with less specific information about the companies [17]. The success of portfolio construction depends primarily on the future performance of stock markets. Recent developments in machine learning have provided significant opportunities to incorporate prediction theory into portfolio selection. However, many studies show that a single prediction model is not sufficient to achieve very accurate predictions and high returns. A new approach to portfolio construction using a hybrid model based on machine learning for stock prediction and a mean variance (MV) model for portfolio selection was published by [18]. The results obtained show that the proposed method is better than traditional methods (without stock prediction) and benchmarks in terms of returns and risks. Finding turning points in the stock market is a task of great importance and challenge. To achieve this goal, [19] decomposed the trend and cyclical components of stock prices with an autoregressive fractionally integrated moving average model that can simulate stationary fractional difference processes. Most importantly, compared to [15] testing the same data is that the Wavelet method more accurately detects turning points. Disclosure of non-financial information has been a trend in listed companies in recent decades. This implies the need to provide credibility of the information by verifying an independent expert who provides a guarantee for the published information. Ibex-35 companies maintain a growing trend in their social commitment and greater recognition of the transfer of verified information to various stakeholders of companies [20]. In addition, companies' efforts are appreciated positively by investors, especially in response to the level of assurance required of the guarantor in the sustainability report. A thorough study of the relevance of value-at-risk (LVaR) forecasts adjusted according to liquidity and expected shortfall (LES) was performed by [21]. The results provide evidence that liquidity risk is proving crucial for the quality of regulatory risk assessments in times of market turmoil. The Cornish-Fisher approximation was found to describe a reasonable choice for LVaR predictions, while the extreme value approach results in corresponding LES predictions. A study [22] assessed the impact of drug development failures (DDS) on the share price of pharmaceutical companies, taking into account the company's financial situation, size and price trend before DDS. In conclusion, DDSs have a negative impact on a company's share price, but this risk can be minimized by investors choosing companies that meet certain criteria. The impact of tour operators' marketing strategies on the cost of sun and beach stays was quantified by [23]. The results confirmed the impact of travel agency variables on price through their brands, accommodation ownership, specific segments they target, awards and incentives, promotional space and pictorial elements in their brochures. Finally, significant differences were found between the analyzed objectives. The total output of the economy usually follows cyclical movements, which are accompanied by similar movements in stock prices. The common explanation depends on the demand side. It points out that the wealth of stock markets controls consumption, which in turn triggers production. A study [24] focused on the effects through the supply side of the economy. The results show that when a bank assesses creditworthiness by stock price, eccentric fluctuations in stock prices have little effect because they distort selection and hinder growth. If bankruptcy is dependent on the optimism on the stock market, the level of influence of stock owners does not matter. If optimism is widespread among stock investors, investment behavior also correlates through stock prices, leading to huge cycles of the real economy without long-term growth. When volatility is taken into account in managers' decisions, more prudent behavior takes place, which promotes growth.

In order to answer research question no. 1, a statistical description of the time series and causal analysis will be used. Research question 2 will be answered through cause-and-effect analysis (causal analysis).

2 Data and Methods

2.1 Data

The base source of data will be the share price of the Thomas Cook travel agency in the period from May 13, 2018 to May 19, 2019 published by [25]. Thomas Cook Group PLC stock price data presented on the MarketWatch website will be further processed in Microsoft's Excel program. The file will contain columns with the date, the price of the value of the stock stated in USD on the given date. In addition, reports published in the media for the monitored period relating to Thomas Cook will be used, and based on these reports, it will be monitored whether there has been a positive or negative development in the share price. These sources will include [9], [26], [27], [28].

2.2 Methods

We will use a statistical description of time series, calculate mode, median, mean average, variance, standard deviation, minimum and maximum share price of Thomas Cook travel agency in the period from May 13, 2018 to May 19, 2019. This data will be presented in a table. Next, a moving average trend line will be displayed to make it easier to detect the time series trend. It will be calculated as the average of a constant number of consecutive periods. It will be displayed using a chart in MS Excel.

The arithmetic mean of the Thomas Cook stock price is calculated using the arithmetic mean formula, I have a set of n values, I denote them x1, x2,..., xn, so we get the average value by adding all xi and dividing the result by n. I denote the average value by $\bar{x}$. Formula:

$$\bar{x} = \frac{x1 + x2 + \cdots + xn}{n} \quad (1)$$

Mode is the value most frequently represented in the Thomas Cook stock price set in the period from 13 May 2018 to 19 May 2019.

If the data set contains an odd number of items, to calculate the median (Me) of a given data set X of size N, which I denote by xi and at the same time the condition of ascending data sorting must be met to use the general formula:

$$Me(X) = x(N + 1)/2 \quad (2)$$

If the data set contains an even number of items to obtain the median, it is necessary to calculate the arithmetic mean of these values and the resulting value is the median of the data set. In this case, we would use the general formula:

$$Me(X) = \left\{\frac{xN}{2} + x\left(\frac{N}{2}\right) + 1\right\}/2 \quad (3)$$

Variance is defined as the average square deviation of the quantity from the mean. In this case, we will not use the variance of the basic set, but we will focus on the sample variance s2. We have only a sample of data from the base set of the variable. Therefore, we use a modified variance formula for the number of observations n < 30:

$$s^2 = \frac{1}{n-1}\sum_{i=1}^{n}(xi - \bar{x})^2 \quad (4)$$

n ... number of observations, xi ... specific realization of the variable X, x ... simple arithmetic mean of the variable X.

The standard deviation is equal to the square root of the variance, we calculate it as follows:

$$\sigma = \sqrt{Var(X)} \quad (5)$$

We can substitute the value of variance with the formula for calculating the variance and thus obtain the formula:

$$\sigma = \sqrt{\frac{1}{N}\sum_{i=1}^{N}(xi - \bar{x})^2} \quad (6)$$

The minimum will represent the lowest value of Thomas Cook's share price in the monitored period. The maximum will represent the highest value of Thomas Cook's share price in the monitored period.

Calculations will be performed using functions in Microsoft Excel. Research question number 1 will be answered by the methods above.

By analyzing the cause and effect, we evaluate the impact of published information [27], from May 16, 2019, where Thomas Cook CEO Peter Fankhauser warned that the world's oldest travel company is struggling in an uncertain environment across all regions. In the previous year, the European heat wave weakened the demand for winter holidays, while British customers continued to delay travel and reservations from Germany weakened. Another report released by the BBC on September 24, 2018, the Thomas Cook travel agency blames the heat wave from the decline in its annual profit forecast. The result will be a table, where the column will contain the date of the published information mentioned above, the price of the Thomas Cook share as of the given day, and a quantification of the% decrease in the price of the Thomas Cook share in connection with the said report. This procedure will answer research question number 2.

3 Results

Graph 1 presents an overview of the development of Thomas Cook's share prices from May 13, 2018 to May 15, 2019. The horizontal axis shows the date information and the vertical axis shows the value of the Thomas Cook Group PLC ADR stock price denominated in USD as of that date.

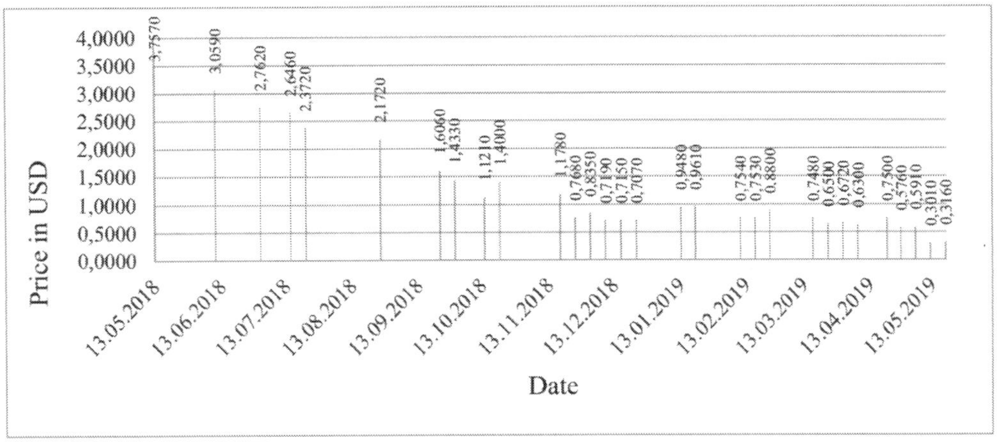

Graph 1. Overview of the development of Thomas Cook's share prices from May 13, 2018 to May 15, 2019
Source: Own processing.

Graph 2 shows the line of the trend of the moving average of the development of Thomas Cook Group PLC ADR share prices in relation to the given date in the period from May 13, 2018 to May 19, 2019. From this chart it can be seen that the link between the trend of the moving average of Thomas Cook stock prices had a predominantly declining trend.

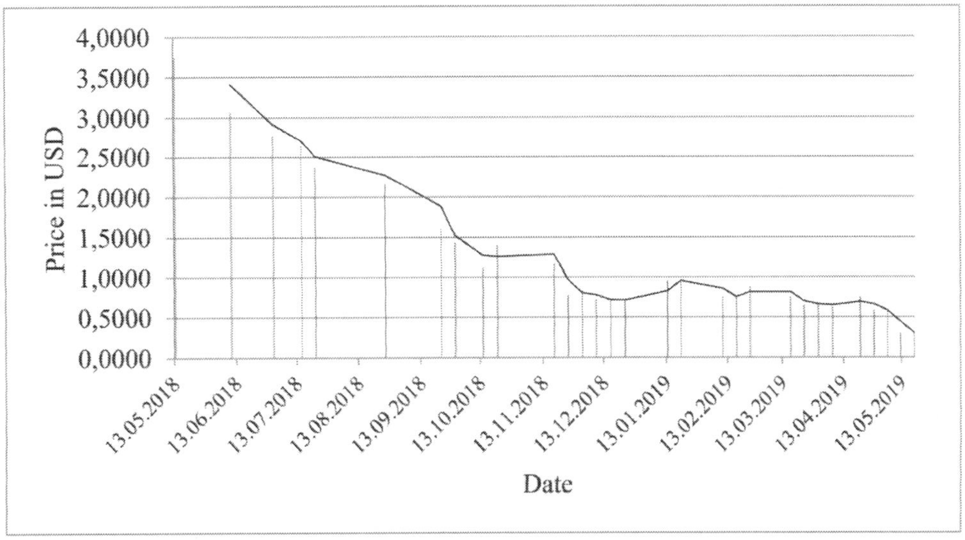

Graph 2. Moving average trend line
Source: Own processing.

Table 1 presents information regarding the price of Thomas Cook shares in USD as of that date. The price of Thomas Cook shares in USD is monitored here in 30 cases in the period from May 13, 2018 to May 19, 2019. Based on the data in Table 1, the mode, median, arithmetic mean, variance, standard deviation, minimum and maximum are calculated.

Table 1. Price of Thomas Cook shares stated in US dollars

Date	Stock price in USD	Date	Stock price in USD
13.05.2018	3.7570	23.12.2018	0.7070
10.06.2018	3.0590	13.01.2019	0.9480
01.07.2018	2.7620	20.01.2019	0.9610
15.07.2018	2,6460	10.02.2019	0.7540
22.07.2018	2.3720	17.02.2019	0.7530
26.08.2018	2.1720	24.02.2019	0.8800
23.09.2018	1.6060	17.03.2019	0.7480
30.09.2018	1.4330	24.03.2019	0.6500
14.10.2018	1.1210	31.03.2019	0.6720
21.10.2018	1.4000	07.04.2019	0.6300
18.11.2018	1.1780	21.04.2019	0.7500
25.11.2018	0.7680	28.04.2019	0.5760
02.12.2018	0.8350	05.05.2019	0.5910
09.12.2018	0.7190	12.05.2019	0.3010
16.12.2018	0.7150	19.05.2019	0.3160

Source: Own processing.

In Table 2, values for mode, median, arithmetic mean, variance, standard deviation, minimum and maximum are calculated using MS Excel. Mode, or in this case the value of the Thomas Cook share, which is most often represented in the data listed in Table 1, is not listed here, as no value of the Thomas Cook share occurs more than once. The median is 0.8015, which means that half of Thomas Cook's shares are less than 0.8015 USD and the other half of Thomas Cook's shares are more than 0.8015 USD. The arithmetic average shows the average value of the price of Thomas Cook stocks, which is 1.2260 USD. The variance is defined as the average square deviation of the Thomas Cook stock price expressed in USD from the mean value, representing the value of 0.772516 USD. The standard deviation represents the square root of the variance, i.e. 0.878928894. The minimum price of Thomas Cook shares in the monitored period is 0.3010 USD and the maximum value of Thomas Cook shares in the monitored period is 3.7570 USD.

Table 2. Formatting sections, subsections and subsubsections

Mode	**Unavailable**
Median	0.8015
Arithmetic mean	1.2260
Variance	0.772516
Standard deviation	0.878928894
Minimum	0.3010
Maximum	3.7570

Source: Own processing.

The cause and effect analysis evaluates the impact of the published information on the price of Thomas Cook stocks. On May 16, 2019 the CEO of Thomas Cook Peter Fankhauser warned that the world's oldest travel company is struggling in a precarious environment across all regions [27]. In the previous year, a heat wave across Europe weakened the demand for winter holidays, while British customers continued to delay travel and reservations from Germany weakened. As a result of the published story, from May 16, 2019, the value of Thomas Cook shares decreased by 40.04% by May 17, 2019. The value of the Thomas Cook shares as of May 16, 2019 was 0.5020 USD, as of May 17, 2019, the value of the Thomas Cook share was 0.3010 USD.

Another report that negatively affected the price of Thomas Cook's shares was published by the BBC on September 24, 2018, when the travel agency Thomas Cook blamed a heat wave for the decline in the annual profit forecast. As a result of the information published above on September 24, 2018, the value of Thomas Cook shares decreased by 10.71% by September 25, 2018. The value of the Thomas Cook share was worth 1.6057 USD as of September 24, 2018, but by the following day, the stock's value was already only 1.4338 USD.

Among the events that affected Thomas Cook's stock price was the fact that Thomas Cook reported a loss of £1.5 billion in May 2019, mainly due to the write-off of an investment in MyTravel made in 2007. The purpose of the merger was to create a European tourism giant and bring cost savings of up to 75 million GBP a year. This merger resulted in no cost savings, only higher debts. This event also had a great impact on the value of Thomas Cook shares, where for demonstration I will show the value from the beginning of the period, on May 13, 2018, when the value of Thomas Cook share was 3.7570 USD and almost a year later, on May 12, 2019 the value of a share of the same company was 0.3010 USD, thus decreasing in value by almost 92%.

The decline in Thomas Cook's stock price was further compounded by the fact that Thomas Cook did not catch up on travel trends in time, when travelers began to make greater use of the Internet and began to organize tours themselves using low-cost airlines and web-based accommodation services around the world.

During the last days of its operation, Thomas Cook had feverish negotiations with creditors and asked the British government for financial assistance. This financial assistance from the government was not approved. The company's management announced at the end of September 2019 that it had entered into forced liquidation.

4 Discussion

Based on the obtained results, we are able to answer research questions that were set out:

RQ1: What were the stock prices from May 13, 2018 to May 19, 2019?

Thomas Cook's share price from May 13, 2018 to May 19, 2019 ranged from 3.7550 USD to 0.3160 USD. The median was 0.8015 USD, meaning that half of Thomas Cook's share price was less than 0.8015 USD and the other half of Thomas Cook's share price was greater than 0.8015 USD. The arithmetic mean shows the average value of the price of Thomas Cook, which was 1.2260 USD. The variance is defined as the average square deviation of the Thomas Cook stock price expressed in USD from the mean value, which was 0.772516 USD. The standard deviation represents the square root of the variance, i.e. 0.878928894 USD. The minimum price of Thomas Cook shares in the monitored period was 0.3010 USD and the maximum value of Thomas Cook shares in the monitored period was 3.7570 USD. The declining trend in the price of Thomas Cook shares in the monitored period is also described by [29] and [30].

In the monitored period from May 13, 2018 to May 19, 2019, the value of Thomas Cook shares had a declining trend, as share price ranged from 3.7550 USD to 0.3160 USD.

RQ2: What factors influenced Thomas Cook's stock prices during the monitored period?

Factors that influenced Thomas Cook's stock prices during the period from May 13, 2018 to May 19, 2019 include the information released by Thomas Cook CEO Peter Fankhauser, who warned that the world's oldest travel company is struggling in an uncertain environment across all regions. The result of the published information was a decrease in the value of stocks. Another report that negatively affected the price of Thomas Cook's shares was the press statement where the travel agency Thomas Cook blames a heat wave for the decline in the annual forecast of its profit. Other events that affected the price of Thomas Cook shares included the fact that Thomas Cook reported a loss of £1.5 billion in May 2019, mainly due to the write-off of the investment in MyTravel, which was made in 2007. Thomas Cook stock value was also damaged by the fact that Thomas Cook did not catch up to current trends in travel. Another factor that affected Thomas Cook's stock prices was the government's rejection of its request for financial assistance. About the aforementioned factors that affected the stock prices of Thomas Cook was published by [30].

During the monitored period, factors affecting Thomas Cook's stock prices included the release of information on Thomas Cook's problems, heat waves affecting the company's annual earnings forecast, loss reporting, failure to catch up to travel trends and the government's refusal of financial assistance. These factors had a negative impact on Thomas Cook's stock prices.

5 Conclusion

The aim of the paper was to analyze the stock prices of Thomas Cook from May 13, 2018 to May 19, 2019 and the factors that had an impact on the stock prices of this company in the monitored period. The goal of the paper was achieved. In the Results section, the share price of Thomas Cook was presented from May 13, 2018 to May 19, 2019, the time series of the share price in the monitored period was described, and the line showing the trend of the moving average of Thomas Cook share prices was shown. Furthermore, a cause and effect analysis was performed, which assessed the impact of the heat wave on the company's annual profit forecast, loss reporting, failure to capture travel trends and rejection of government financial assistance on Thomas Cook's share price.

The British company Thomas Cook was one of the oldest and largest travel agencies in the world and its services were used by millions of clients every year. In 2018, the company's revenues were reduced by a heat wave, which in May 2018 engulfed Europe and significantly reduced the demand for stays abroad. The following year, the British had little interest in holidays, mainly because of Brexit, because they feared a weakening of the pound sterling, which would make trips abroad more expensive. In 2018 and 2019, Thomas Cook faced other external problems; a change in the way people spent their holidays. Instead of organized beach vacations by travel agencies, they preferred to explore cities, destinations on their own, and used the services of low-cost airlines and private accommodation companies. The collapse of Thomas Cook at the end of September 2019 was mainly due to financial difficulties, erroneous mergers and the development of the Internet, as the company no longer had the funds to pay receivables from airlines and hotels.

Thomas Cook is a case in point that a collapse could happen to an industry leader. The collapse of Thomas Cook provides lessons for tourism businesses to select a suitable company in the event of a planned merger, to have the company's finances under control and to monitor developments in the industry. Another lesson is that any negative event, notice, piece of information has a negative impact on the value of a company's stock.

These conclusions apply generally to all companies from different fields and at different times.

The Results section listed the factors that influenced Thomas Cook's share price during the monitored period. In addition to these factors, there are a number of other factors that have influenced Thomas Cook's share price. Further research should focus on the cash flow in the company, as this indicator shows the viability of a company and further focus on the movement of working capital, because this item is able to shock otherwise profitable companies.

References

1. T. Macák, J. Hron, M. Jadrná, Diverzifikace nabídek ubytování v závislosti na kritických úsporách z rozsahu pro cestovní ruch [Diversification of accommodation offers depending on critical economies of scale for tourism]. *Kogentní obchod a řízení [Mandatory business and management]* [online], Available at: doi:10.1080/23311975.2020.1746170 (2020)
2. J. Majerová, C. Fernandes, How to make phoenix to arise from the ashes: Brand loyalty as a prospective pillar of branding in tourism after crisis COVID-19. *Littera Scripta*, **13**(2), 49-58 (2020)
3. M. Vochozka, V. Machová, Determination of value drivers for transport companies in the Czech Republic. *Nase More,* **65**(4), 197-201 (2018)
4. M. Vochozka, J. Vrbka, P. Šuler, Bankruptcy or success? The effective prediction of a company's financial development using LSTM. *Sustainability*, **12**(18), 2020
5. S. Wang, X. Wei, Relationships between exchange rates, economic growth and FDI in China: An empirical study based on the TVP-VAR model. *Littera Scripta,* **10**(1), 166-179 (2017)
6. D. Amaya, M. Boudreault, D. L. Mcleish, Maximum likelihood estimation of first-passage structural credit risk models correcting for the survivorship bias. *Journal of Economic Dynamics & Control* [online], Available at: doi:10.1016/j.jedc.2018.11.005 (2021)
7. T. Ch. Silva, M. S. Alexandre, B. M. Tabak, Bank lending and systemic risk: A financial-real sector network approach with feedback. *Journal of Financial Stability* [online], Available at: doi:10.1016/j.jfs.2017.08.006 (2018)
8. M. Vochozka, J. Horák, T. Krulický, Innovations in management forecast: Time development of stock prices with neural networks. *Marketing and Management of Innovations*, **2020**(2), 324-339 (2020)
9. J. Soustružník, *Thomas Cook – jedno prosté poučení pro investory [Thomas Cook - One simple lesson for investors]* [online], Available at: https://www.patria.cz/zpravodajstvi/4229342/thomas-cook--jedno-proste-pouceni-pro-investory.html (2021)
10. V. Yilanci, O. Ozgur, M. S. Gorus, Stock prices and economic activity nexus in OECD countries: New evidence from an asymmetric panel Granger causality test in the frequency domain. *Financial Innovation*, [online], Available at: doi:10.1186/s40854-020-00221-1 (2021)
11. M. M. Burzala, The process of transferring negative impulses in capital markets - a wavelet analysis. *Journal of Applied Statistics* [online], Available at: doi:10.1080 / 02664763.2020.1864811 (2021)

12. S. A. Rusmita, L. N. Rani, P. Swastika, S. Zulaikha, Capital market volatility MGARCH analysis: Evidence from Southeast Asia. *Journal of Asian Finance Economics And Business* [online], Available at: doi:10.13106/jafeb.2020.vol7.no11.11 7 (2020)
13. M. Sharaf, E. E. Hemdan, A. El-Sayed, N. Al-Bahnasawy, StockPred: a framework for stock Price prediction. *Multimedia Tools and Applications* [online], Available at: doi:10.1007 / s11042-021-10579-8 (2021)
14. G. Sun, J. Lin, Ch. Yang, Stock price forecasting: An echo state network approach. *Computer Systems Science and Engineering* [online], Available at: doi:10.32604/csse.2021.014189 (2021)
15. M. Bai, Y. Qin, H. Zhang, Stock price crashes in emerging markets. *International Review of Economics & Finance* [online], Available at: doi:10.1016/j.iref.2020.12.007 (2021)
16. F. Cavalli, A. Naimzada, N. Pecora, A stylized macro-model with interacting real, monetary and stock markets. *Journal of Economic Interaction and Coordination* [online], Available at: doi:10.1007/s11403-021-00320-x (2021)
17. M. Faccio, R Morck, M. D. Yavuz, Business groups and the incorporation of firm-specific shocks into stock prices. *Journal of Financial Economics* [online], Available at: doi:10.1016/j.jfineco.2020.09.005 (2021)
18. W. Chen, H. Zhang, M. K. Mehlawat, L. Jia, Mean-variance portfolio optimization using machine learning-based stock price prediction. *Applied Soft Computing* [online], Available at: doi:10.1016/j.asoc.2020.106943 (2021)
19. Z. Tan, J. Liu, J. Chen, Detecting stock market turning points using wavelet leaders method. *PHYSICA a-Statistical Mechanics and Its Applications* [online], Available at: 10.1016/j.physa.2020.125560 (2021)
20. Q. M. Miralles, L. J. Miralles-Quiros, I. J. Daza, *The assurance of sustainability reports and their impact on stock market prices* [online], Available at: doi:10.5295/cdg.201262mm (2021)
21. T. Berger, Ch. Uffmann, Assessing liquidity-adjusted risk forecasts. *Journal of Forecasting* [online], Available at: doi:10.1002/for.2758 (2021)
22. A. Venslauskas, E. Stankevicius, Stock price reaction to the drug development setbacks in the pharmaceutical industry. *Daru-Journal of Pharmaceutical Sciences* [online], Available at: doi:10.1007/s40199-020-00349-6 (2021)
23. P. Picazo, S. Moreno-Gil, Tour operators' marketing strategies and their impact on prices of sun and beach package holidays. *Journal of Hospitality and Tourism Management* [online], Available at: doi:10.1016/j.jhtm.2018.02.004 (2018)
24. P. Assmuth, Stock price related financial fragility and growth patterns. *Economics-The Open Access Open-Assessment E-Journal* [online], Available at: doi:10.5018/economics-ejournal.ja.2020-10 (2020)
25. MarketWatch: Thomas Cook Group PLC ADR [online], Available at: https://www.marketwatch.com/investing/stock/tckgy/charts?mod=mw_qu ote_tab (2019)
26. ČTK, *Proč padl Thomas Cook? Firmu zlomily dluhy a strach lidí z brexitu [Why did Thomas Cook fall? The company was broken by debts and people's fear of Brexit]* [online], Available at: https://byznys.ihned.cz/c1-66646080-proc-padl-thomas-cook-firma-neumela-reagovat-na-moderni-trendy-jako-je-dovolena-pres-airbnb-navic-nadelala-velke-dluhy (2019)

27. T. Novák, *Akcie britské společnosti Thomas Cook silně oslabují [Shares of the British company Thomas Cook are weakening sharply].* Fio banka, a.s. [online], Available at: https://www.fio.cz/zpravodajstvi/zpravy-z-burzy/223200-akcie-britske-spolecnosti-thomas-cook-silne-oslabuji (2019)
28. BBC News Business, Thomas Cook blames heatwave for profit warning [online], Available at: https://www.bbc.com/news/business-45624215 (2018)
29. Patria Finance, *Akcie v ČR: Thomas Cook (-38 %) je tu od 19. století. Teď jsou prý akcie bezcenné [Shares in the Czech Republic: Thomas Cook (-38%) has been here since the 19th century. Now stocks are said to be worthless]* [online], Available at: https://www.kurzy.cz/zpravy/494482-thomas-cook--38--je-tu-od-19-stoleti-ted-jsou-pry-akcie-bezcenne/ (2019)
30. P. Michl, *Pád Thomas Cook: proč k němu došlo a kdo na něm vydělal? [The fall of Thomas Cook: why did it happen and who made money on it?]* [online], Available at: https://finex.cz/pad-thomas-cook-proc-k-nemu-doslo-a-kdo-na-nem-vydelal/ (2019)

Evaluation of development of Apple Inc. stock price time series

Jakub Horák[1*], and *Dominik Kaisler*[1]

[1]Institute of Technology and Business in České Budějovice, School of Expertness and Valuation, Okružní 517/10, 370 01 České Budějovice, Czech Republic

Abstract. The paper deals with the the development of a specific company´s stock price time series. The aim of the paper is to use the time series method for a detailed analysis and evaluation of the development of Apple Inc. stock prices. Daily data from 2000 to 2020, daily data from the period of the economic crisis between 2007 and 2009 and daily data from the Covid-19 pandemic period from March 2020 to the end of the same year are used. The data, from the period of 2000 - 2020 show a gradual increase in Apple's stock prices. The most common factor leading to the increase in stock prices is the launch of a new product or service on the global market. On the contrary, the reason for the decline in stock prices is customer dissatisfaction, the excess of demand over supply, or the political situation. The analysis of time series for the period of the economic crisis points to the fact that thanks to the development, innovation and constant introduction of new products into the market, the company was not significantly affected by the crisis and neither were stock prices. Naturally, there were some fluctuations in prices, but at the end of 2009, the company even reached the highest stock prices in its history to date. . The analysis of time series during the global pandemic of Covid-19 shows a steady rise in stock prices. Currently, the company sells more and more products and introduces new services that help us work, study or entertain ourselves in these difficult times, in the safety of our homes.
Keywords: Apple Inc., stock price, stock price development, time series, investing
JEL codes: C22, B26

1 Introduction

Investing in traditional investment institutions, such as banks, is no longer attractive to investors due to the low return on investment [1, 2]. The stock market is now one of the most important components of the financial system [3]. Investing in stocks is becoming more and more popular for the society today. The desire to multiply one's funds in a relatively short period of time and achieve often above-average returns attracts almost all of us. In turn, companies that offer stocks receive funding for their own development. However, it is often a long haul, and it can take several decades for the money invested to recover. In some cases, the capital invested does not

*Author's contact information: horak@mail.vstecb.cz

result in a profit at all but in a loss. Therefore, each step needs to be carefully considered and the purchase of stocks must be carried out with due consideration.

Any investor entering the financial market wishes to be able to predict how many stocks to buy and when. However, the number of stocks depends also on the value of the investor's assets. The next important step for the subsequent valuation of the funds invested is a proper analysis and evaluation of the historical development of stock prices and the company whose stocks we want to buy. This includes its turnover, bringing new products to market or its overall economic situation. There are many factors that affect the subsequent increase or decrease in value. If we successfully predict the development of a company's stock price, the future price could yield a significant profit [4]. This feeling and the ability to predict correctly distinguishes successful investors from the unsuccessful ones.

Interest in stocks actually determines their value. Stock price develops on the basis of supply and demand. As with any commodity on the market, the higher the demand for stock and the lower the supply, the higher its price is. Supply and demand change depending on various factors; therefore, stock price also changes over time. These factors can be predicted and evaluated using several methods and data approaches. One of the most conclusive statistical methods to examine the dynamics and development over time are the so-called time series. Time series of stock price developments are crucial for assessing the factors that influenced stock price developments and determined their performance in the past, as well as for predicting their future development.

The objective of the article is to analyse and evaluate the historical development of Apple Inc.'s stock prices on the basis of secondary data in the years 2000 - 2020, and to determine in what time horizon, on what occasion, or as a result of what circumstances on the financial market, the company's stocks reach the highest and the lowest prices.

In order to achieve the set objective, the following research questions are formulated:

1. What factors influenced Apple Inc.'s share prices in the years 2000 - 2020?
2. How did the COVID-19 pandemic affect Apple Inc. stock prices?
3. What was the development of Apple Inc. stock prices during the 2007-2009 economic crisis?

2 Literature research

Apple Inc. is one of the best-known and richest companies in the world. As a result of its phenomenal success, the value of Apple´s liquid assets at the end of fiscal year 2012 was $121 billion. In April 2013, the company committed to distribute up to $100 billion to shareholders through stock buybacks and cash dividends until the end of calendar year 2015 [5].

Precisely because Apple Inc. is active in the financial market, the company must have well-established financial management and decision support system. Khan et al. [2] argue that knowledge-based decision support systems for financial management are an important part of investment plans. They concluded that investors avoid investing in traditional investment institutions, such as banks, due to the low return on investment.

Currently, the stock market is one of the main areas of investment. Predicting the price of stocks is becoming a difficult task for day traders, investors and data scientists. These are complex functions of a wide range of interracting factors that influence the dynamics of price movements [6]. Politics, the social prospects, the company's sales and socio-economic factors affect global perception of stocks, which is influenced by the balance of supply and demand [7]. This is confirmed also by Gupta and Chen [8], who add that stock prices and financial markets are often based on sentiment, which leads to research efforts to predict stock market trends using public sentiment expressed on social networks, such as Facebook and Twitter. Predictions can be made more effectively by considering price fluctuations and understanding people's mood [9]. However, Chaudhari and Ghorpade [10] believe that predicting stock return is a significant problem in the financial sector, due to the non-linear fluctuations in the company's return on investment in stocks. Inaba [11] analysed global market commodities for 37 developed and developing countries, stating that international stock price returns in developed countries were higher than in developing countries. On the contrary, return on investment of stocks grew faster in developing countries, due to time-varying factors. These factors include not only the openness of international trade and finance, but also institutional factors representing the development of information and communication technologies, the protection of property rights and the transparency of information disclosure.

Time series of stock prices are crucial for the prediction of stock prices. Time series analysis is a fundamental method for predicting stock prices and remains the most widely used research method [12]. However, this method is not entirely accurate, as it does not include external factors, such as news, events, etc., that may affect stock price developments [13]. Rajesh and Gandy [14] focused on time series prediction, introducing the CashTagNN system, which uses sentiment and subjectivity scores of tweet, including cashtags, to model stock market movements, especially to predict opening stock prices. In this system, they used two machine learning methods, namely feed-forward neural network and the deep convolutional neural network. Ebadati and Mortazavi [15] also use the method of neural networks, applying the hybrid method of genetic algorithm (GA) and artificial neural network (ANN) to develop a method for predicting stock prices and time series. In the GA method, the output values are further converted into the developed ANN algorithm to correct errors at the exact point. This analysis suggests that GA and ANN can improve the accuracy with a fewer number of iterations. Gandhmal and Kumar [16] introduce several techniques for predicting stock market trends. The authors perform an analysis that describes the software tools used in documents that also deal with price prediction. According to their research, the leading software tools include JavaScript, Python and Matlab. Vochozka, Horak and Krulicky [17] also agree that the Matlab system is the most frequently used software for making stock price predictions, and use the neural network method for research. Horak and Krulicky [18] aim to compare the method of exponential smoothing of time series and time series smoothing using neural networks as a tool for predicting the future development of a company´s value. First, they analyse the dataset, which is followed by generating neural networks, retaining five with the best characteristics. Exponential time series smoothing is the next step, followed by obtaining predictions for 62 trading days, which is subseuently compared with the results of two retained neural networks.

3 Data and methods

3.1 Data

For time series research, individual data are obtained from the finance.yahoo.com website. This page presents the historical development of Apple Inc.'s stock prices, specifically daily data for the period of 2000 - 2020. The obtained data are processed in Microsoft´s Excel. The dataset contains 5,283 rows. The next step is to edit the dataset and exclude irrelevant data, such as opening prices or the lowest and highest stock prices. Only the columns containing the date and the closing prices of stocks are kept in the dataset. The next step is the evaluation and comparison of the obtained data.

3.2 Methods

The data are evaluated using the time series analysis. Time series analysis compares comparable data, which are ordered by time. Using statistical methods, values such as median, mode, arithmetic mean, variance, variance range and maximum and minimum value are calculated in Excel for each segment. The obtained data are evaluated both qualitatively and quantitatively. The resulting values are also graphically illustrated. The information obtained provides an analysis of the pace and fluctuations in stock prices and their causes. Excessive fluctuations are analysed in more detail by looking for possible events that have affected them. Finally, the development of time series during the COVID-19 pandemic are examined and evaluated. This pandemic has been sweeping the world since March 2020. For this period, separate charts and tables are created.

4 Results

The time series of Apple Inc.'s stock prices is introduced in the reference period (i.e. 2000-2020), separately for the COVID-19 pandemic period (i.e. from March 2020 to the end of the same year), and for the period of economic crisis (i.e. 2007-2009).

The curve of the stock prices development in 2000 - 2020 (see Figure 1) reached its maximum value in December 2020 ($ 136.69). By contrast, the lowest value was recorded in April 2003 ($ 0.23). At the beginning of 2000, stock prices stagnated, but then began to rise gradually from 2004, and their rise did not stop until early the year 2008, when there was a slight decline. In 2009–2012, the value rose slightly. The decline appeared again in 2013, but the following two years brought an increase in stock prices. At the beginning of 2016, prices fell again; however, in the same year, they also started to rise at a dizzying pace. This increase did not stop until 2018, when there was a sharp decline until 2019. After that, however, stock prices were growing again at a rapid pace, and a sharp decline occurred in 2020. Around the middle of the same year, there was the largest and most notable increase in the period under review, reaching the maximum value at the end of the year.

In the monitored periods, the basic statistical characteristics of the time series were calculated – median, mode, arithmetic mean, variance and variance range. The average share price is $ 9.19. The value that occurs most frequently between 2000 and 2020 is $ 0.35. The calculated average stock price is $ 18.70. The variance is $ 565.48 and the difference between the maximum and minimum value is $ 136.45.

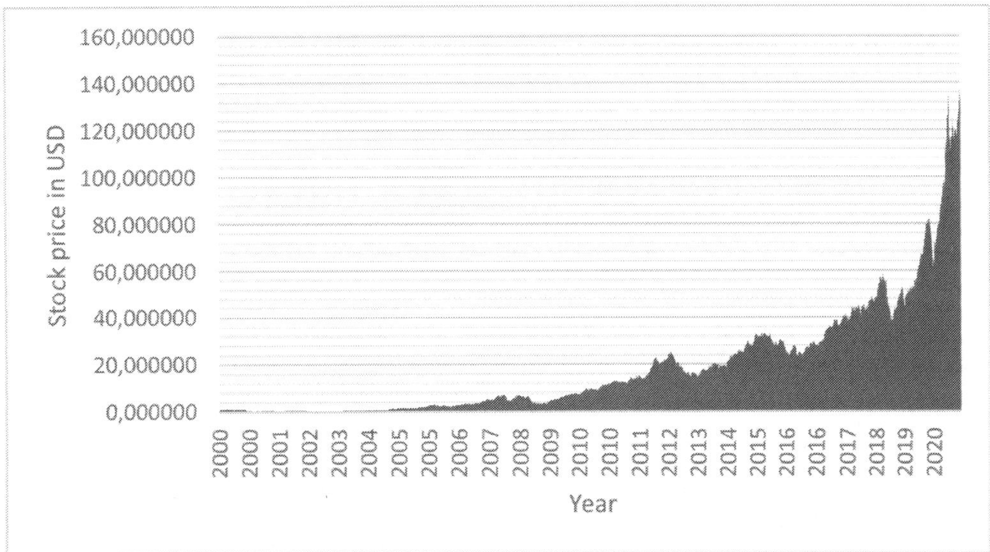

Fig. 1. Historical development of Apple Inc. stock price time series from 2000 to 2020 incl.
Source: Compiled by the authors on the basis of data from finance.yahoo.com

The curve of the development of stock prices during the COVID-19 pandemic (see Figure 2) reached its peak in December 2020. The price was 136.69 USD. It reached its minimum in March 2020 ($ 55.09). With the onset of the COVID-19 pandemic, stock prices fell first, reaching the aforementioned minimum in March. At the beginning of April, the prices grew gradually, without major and demonstrable declines, and the rise did not stop until the end of July. Prices stagnated for a short period, but then there was a sharp rise, which lasted until the early September. After that, prices fell for several days. Between mid-September and early December, stock prices fluctuated rapidly and frequently, falling and rising again.. However, the fluctuations range was not large and always maintained the range of up to ten dollars per stock. At the end of the year, stock prices rose slightly, achieving the maximum before the New Year.

The median value for this period is $ 102.30. The most frequently recorded price is $ 91.02. The average share price in this period is $ 98.46. The variance takes the value of $ 492.03, and the variance range is $ 80.59.

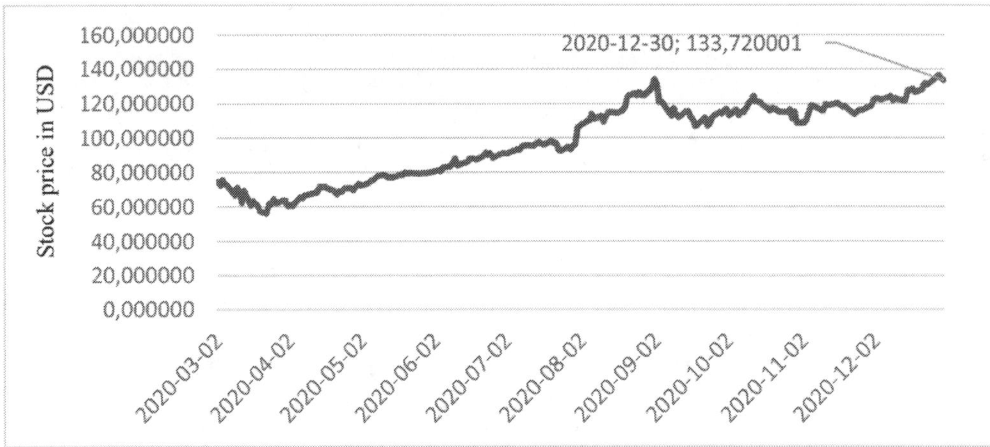

Fig. 2. Development of stock prices during the COVID-19 pandemic
Source: Compiled by the authors on the basis of data from finance.yahoo.com

The curve of the development of stock prices during the economic crisis from 2007 to 2009 (see Figure 3) peaked in December 2009 ($ 7.55). The minimum was recorded in January of the same year, with the price being $ 2.79 per share. At the beginning of the crisis (i.e. from 2007), stock prices rose gradually until the beginning of August, when achieved their maximum. In August, prices decreased slightly, but then rose again at a relatively high rate. Their rise did not stop until November 2007, when there was a short but rather a sharp fall in stock prices. Until the end of 2007, however, prices grew only. At the beginning of 2008, the first major and longer fall in prices came, which lasted until March. However, in early March, stock prices grew at a relatively rapid pace and did not stop until May. Then, in the period May-June, prices stagnated and did not change significantly. During July, there was a slight decline followed by a slight increase; in August, there was the largest and longest fall in stock prices in the period under review; however, the prices did not yet reach their minimum. This fall did not stop until October. In the period from November 2008 to March 2009, prices did not change significantly. However, a significant event that occurred during this period was that in January 2009, share prices reached their minimum for the period under review. After that, however, from April to the end of 2009, prices grew only, reaching their maximum in December.

The median value for this period is $ 4.86. The most frequently recorded value is $ 4.42. The average stock price for the period under review is $ 4.96. The variance is $ 1.75 and the difference between the maximum and minimum is $ 4.76.

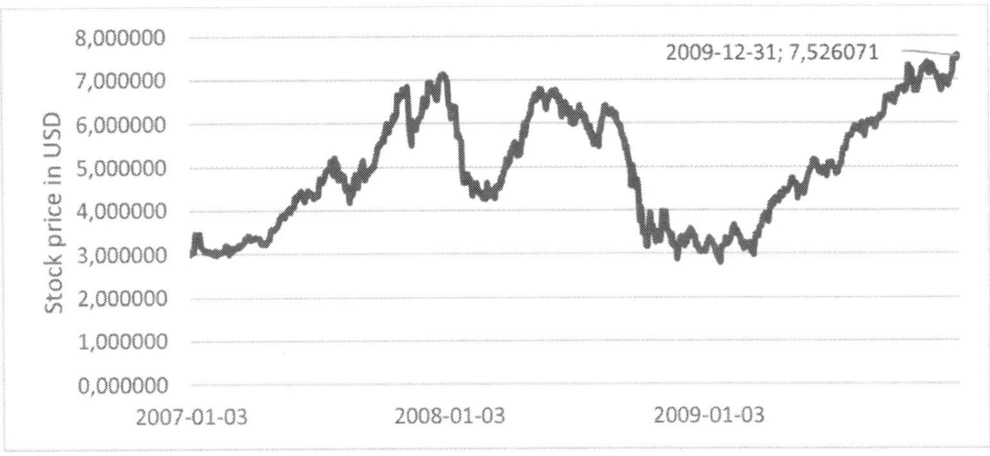

Fig. 3. Development of stock prices during the economic crisis
Source: Compiled by the authors on the basis of data from finance.yahoo.com

5 Result discussion

Based on the obtained results, it is possible to answer the formulated research questions as follows:

1. What factors influenced Apple Inc.'s share prices in the years 2000 - 2020?

In early 2000, Apple Inc. was on the verge of bankruptcy and the price of their stocks fell to less than a dollar per stock. For this reason, the company reached back to its co-founder Steve Jobs and brought him back to the position of CEO. In a few years, Jobs managed to bring the company back to limelight and make it profitable again. Jobs had to turn to the company´s biggest competitor, Microsoft, to borrow $ 150 million to pay off debts and bills. The company introduced iPod in 2001, but this small player did not mean a miraculous breakthrough. In 2004, Apple Inc. introduced iTunes, putting much faith in it. iTunes slowly gained momentum, and thus the rise in stock prices. It soon recorded 200 million downloaded songs and became very popular and well-liked. In 2007, with the advent of the global economic crisis, Apple Inc. released the historically first iPhone. For comparison, the first smart Android phones appeared on the market three years later. The number of stocks purchased climbed to almost $ 24,000 in 2008. In 2009, the financial crisis hit stock markets and Apple Inc. stocks lost up to 70% of their value. However, in 2010, the company introduced new products (iPhone 4 and iPad) and the price of stocks grew to $ 40,000. Even the death of Steve Jobs in 2011 did not caused a fall in stock prices; in contrast, the prices grew slowly but steadily. In 2014, Apple Inc. introduced the first wearable electronics, Apple iWatch. This watch became very popular and used all over the world. This resulted in another growth in the company's stocks prices. In 2016, the turning point came again. The company faced severe criticism for the removal of the 3.5 mm jack connector of the iPhone 7. Despite the alternative offer of the historically first Airpods wireless headphones, stock prices started to fall. In 2018, Apple sold 217 million iPhones, 43 million iPods, 8.2 million Macs and introduced the new state-of-the-art iOS 12 operating system. Therefore, stock prices rose again. In 2019, stock prices fell to the bottom. The main reason are the concerns about the trade war between China and the USA. No one knew what the war would mean for the sale of Apple products in China, Apple Inc.'s third largest market. In 2020, the global pandemic of COVID-19 started. In that year, the latest and best-rated iPhone of all generations was released. At the same time, Apple Inc., for the first time in three years, replaced the South Korean brand Samsung in the position of the world leader in the sale of smartphones, thus beating its biggest rival. Also, for the first time in its history, Apple Inc. reached a record value of $ 136.69 per stock. All key milestones are shown in Table 4.

What was the development of stock prices during the economic crisis of 2007-2009 at Apple Inc.? Between 2007 and 2009, there was a global economic crisis. While most companies faced heavy losses, Apple Inc. showed an opposite trend. The main reason was the release of its flagship in 2007, historically the first iPhone OS mobile phone. The company's profits grew at an incredible pace and phone sales rose to 7.4 million sold devices. However, in 2008, after the release of the second iPhone OS 2, the company faced manufacturing problems due to high demand. The demand exceeded the supply and sales and stock prices began to fall. Because of this, the company was forced to discount the previous generation of the iPhone to $ 99. In addition, in the same year, the company warned against a possible hacker attack. It pointed to a critical error in its cell phone software. The company saw its stock prices rise again in 2009 with the launch of the Macbook, which contributed significantly to the company's high sales, particularly $ 3.05 million. While worldwide laptop sales stagnated during the crisis, Apple Inc. sales jumped by 35%.

How did the COVID-19 pandemic affect Apple Inc. stock prices? Although the world economy and financial markets were rather stagnant during the COVID-19 pandemic, it is surprising that Apple Inc.'s stock prices still grew. This was mainly due to the global lockdown and a change in people's behaviour. Most of them had to start spending much more time at home due to closed workplaces, schools and closure of other services. People began to use the devices for news, consumption, entertainment, but above all for work and study. Due to these factors, the demand for products increased, especially in the case of iPad tablets and MacBook computers. This is evidenced by the fact that worldwide sales of these devices increased by 18.6% to 38.6 million units sold.

Al Aradi and Hewahi [7] argue that stock prices are influenced by supply and demand. I completely agree with this statement. Based on the results achieved, I can confirm that this balance has a significant impact on stock prices. I further agree with the statement of Kaushali and Chaudhary [12], who argue that the method of predicting stock prices using time series analysis is one of the fundamental methods but not entirely accurate one. In my opinion, this method is very effective, which can be evidenced by my research.

This research can serve as support for beginning investors, who can predict future developments in Apple Inc.'s stock prices based on the results. In addition, it can serve as an inspiring article.

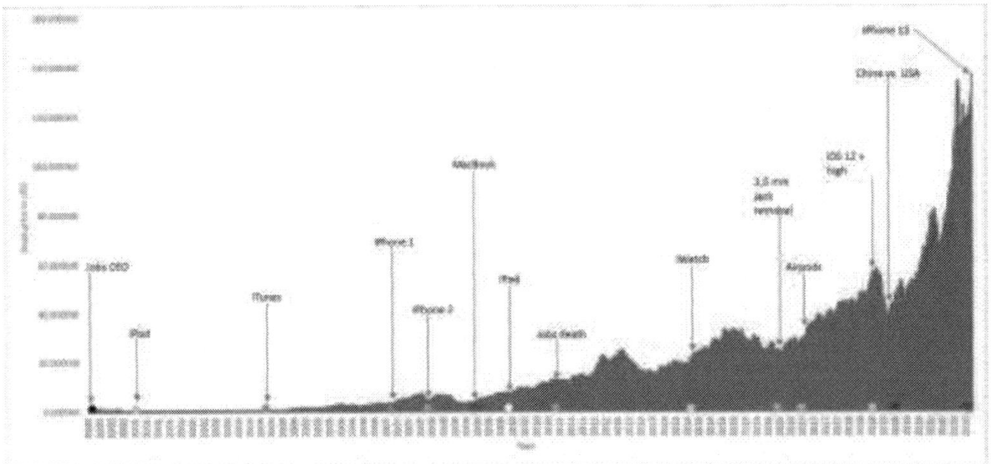

Fig. 4. Historical development of Apple Inc. stock price time series from 2000 to 2020 incl., with the most important milestones
Source: Authors.

Fig. 4 shows the milestones in the history of Apple Inc., (from left): Steve Jobs became CEO, first iPod, origin iTunes, iPhone 1, iPhone 2, MacBook, iPad, Steve Job's death, iWatch, 3,5 mm jack removal, Airpods, iOS 12+ high, trade war between USA and China, and iPhone 12.

6 Conclusion

The objective of the article was to use secondary data to analyze and evaluate the historical development of stock prices of Apple Inc. from 2000 to 2020. Furthermore, it aimed to specify in what time horizon, on what occasion, or as a result of what circumstances on the financial market, the company's stocks reached the highest and the lowest prices. For the period under review, the development of Apple Inc.'s stock prices was analysed and quantitatively and qualitatively evaluated using time series analysis. The objective of the article was thus achieved.

In the results chapter, the maximum and minimum stock prices, mode, median and variance, the average price and the difference between the maximum and minimum value of stock prices were calculated using statistical methods. Based on the data obtained, graphs were created and presented especially for three specific periods: during the COVID-19 pandemic, for the economic crisis in 2007–2009, and the overall development of stock prices from 2000 to 2020. The discussion of the results then described the factors that caused the stock price fluctuations. I can conclude that stock prices are mainly affected by the following factors: the launch of new products, product prices, the introduction of support programs and accessories, customer reactions, external events and the overall operation of the company.

However, the above results raise the question of how long Apple Inc. will remain in the forefront of the stock market? Will Apple announce a ground-breaking innovation in the future that will raise stock prices to an astronomical amount? To what extent can the political situation affect Apple Inc.'s stock prices? In order to answer these questions, further research is needed. Further research could focus on the development of stocks from the establishment of the company or its comparison with the competitors.

Acknowledgement

The article was prepared within the project SVV202109 "Scientific school" – Integration of students and experts from practice in creative and scientific research activities of the School of Expertness and Valuation.

References

1. R. Gottwald, Optimal portfolio of chosen stocks of the Prague stock exchange. *Littera Scripta*, **7**(1), 12-24 (2014)
2. U. Khan, F. Aadil, M. A. Ghazanfar, S. Khan, N. Metawa, K. Muhammad, I. Mehmood, Y. Nam, A robust regression-based stock exchange forecasting and determination of correlation between stock markets. *Sustainability*, **10**(10) (2018)
3. Z. Yeze, W. Yiying, Stock price prediction based on information entropy and artificial neural network. *2019 5th International Conference on Information Management (ICIM)*, pp. 248-251 (2019)
4. S. Arivarasan, A. Kumaravel, Stock market price prediction by 6 datamining techniques anad final decision by comparison. *International Journal of Applies Engineering* Research, **9**(22), 7173-7178 (2014)

5. W. Lazonick, M. Mazzucato, Ö. Tulum, Apple's changing business model: What should the world's richest company do with all those profits? *Accounting Forum*, 249-267 (2013)
6. P. Domanižová, F. Milichovský, K. Kuba, Business models, strategy and innovation in the new world of digization. *Littera Scripta*, **13**(1), 17-31 (2020)
7. M. Al Aradi, N. Hewahi, Prediction of stock price and direction using neural networks: Datasets hybrid modeling approach. *2020 International Conference on Data Analytics for Business and Industry: Way Towards a Sustainable Economy (ICDABI)*, pp. 1-6 (2021)
8. R. Gupta, M. Chen, Sentiment analysis for stock price prediction. *2020 IEEE Conference on Multimedia Information Processing and Retrieval (MIPR)*, pp. 213-218 (2020)
9. M. Jaggi, P. Mandal, S. Narang, U. Naseem, M. Khushi, Text mining of stocktwits data for predicting stock prices. *Applied System Innovation*, **4**(1) (2021)
10. A. Chaudhari, P. Ghorpade, Forecasting a firm's position based on Pitroski's F-score using ARIMA. *2020 International Conference on Data Analytics for Business and Industry: Way Towards a Sustainable Economy (ICDABI)*, pp. 1-5 (2020)
11. K. I. Inaba, Information-driven stock return comovements across countries. *Research in International Business and Finance*, **51** (2020)
12. R. Dias, P. Alexandre, P. Heliodoro, Contagion in the LAC financial markets: The impact of stock crises of 2008 and 2010. *Littera Scripta*, 13(1), 32-45 (2020)
13. A. Kaushal, P. Chaudhary, News and events aware stock price forecasting technique. In: *2017 International Conference on Big Data, IoT and Data Science (BID)*, pp. 8-13 (2017)
14. N. Rajesh, L. Gandy, CashTagNN: Exploiting the use of cashtags to predict stock market prices using convolutional networks. *Proceedings of the 2020 4th International Conference on Algorithms, Computing and Systems*, pp. 1-5 (2020)
15. O. M. E. Ebadati, M. T. Mortazavi, An efficient hybrid machine learning method for time series stock market forecasting. *Neural Network World*, 41-55 (2018)
16. D. P. Gandhmal, K. Kumar, Systematic analysis and review of stock market prediction techniques. *Computer Science Review*, **34** (2019)
17. M. Vochozka, J. Horak, T. Krulicky, Innovations in management forecast: Time development of stock prices with neural networks. *Marketing and Management of Innovations*, **2020**(2), 324-339 (2020)
18. J. Horák, T. Krulický, Comparison of exponential time series alignment and time series alignment using artificial neural networks by example of prediction of future development of stock prices of a specific company. *SHS Web of Conferences: Innovative Economic Symposium 2018 – Milestones and Trends of World Economy, EDP Sciences* (2019)

Is it recession-proof? Masstige purchase intention: the moderating effect of perceived economic crisis

Belay Addisu Kassie[1], and *Jounghae* Bang[2,*]

[1] Graduate School of Business Administration, Kookmin University, Seoul, South Korea
[2] College of Business Administration, Kookmin University, Seoul, South Korea

Abstract. The aim of this study is to investigate the effect of economic crisis on consumers' masstige behavioural intention. The study examines the effects of uniqueness, hedonic value, conspicuous consumption and perceived quality on masstige purchase intention, and as well the moderating effect of perceived economic crisis. The study conducts the hierarchical multiple regression analysis. The results show that uniqueness and hedonic values have significant influences on masstige purchase intention, while conspicuous consumption and perceived quality have no influence on masstige purchase intention. In addition, perceived economic crisis has no moderating effect on the relationship between uniqueness, hedonic value, conspicuous consumption, perceived quality and masstige purchase intention.
Keywords: Masstige, Economic crisis, Uniqueness, Conspicuous consumption, Hedonic value, Perceived quality

1 Introduction

The recent economic crisis, caused by COVID-19, impacted the global market, even well-established brands and businesses experienced tough times, but how did it affect the masstige sector? The luxury industry has been viewed as recession immune and crisis and uncertainty invulnerable. Intriguingly, past luxury studies witnessed a hybrid trajectory of various luxury product categories. As reported by Vogue Business, the luxury sector has sustained the impact of the 2008 recession hit, although the global personal luxury market shrank by 9%. After more than a decade now, according to [1], the personal luxury goods market dropped by 23%, which is the largest contraction the industry has ever experienced. Despite the turbulence, the luxury market is expected to return to the growth trajectory as evidenced by a rise of 2-3% in the first quarter of 2021. What about the masstige market? Will the economic downturn tone down the booming masstige segment? Our fear is that the recent recession could lead the middle-class, the growing purchasing group of the new luxury, to abandon masstige goods and switch to subdued product lines. We undertook this study to examine the influence of the recent economic crisis on the purchase intention of masstige goods.

For years, luxury was the privilege and the signal of some people who have superfluous wealth and it was reserved for selected few. However, in the last decades, the luxury market has experienced a new trend [2]. Now, it is no longer remains accessible only to a niche of richest people. An increasing slice of the world's middle-class population began to experience a glimpse of luxurious life [3]. This change is partly attributed to the growing of richness of the mass-class consumers and the democratization of luxury. The rapid economic progress of the

* Corresponding author: bangjh@kookmin.ac.kr

middle-class gives the consumer the financial freedom to enjoy a limited array of luxury. By consequence, many luxury brands, Louis Vuitton, for example, designed reasonably priced prestige products to allure the middle class segment [2]. These paradigm shift in brand positioning from the conventional view of luxury *'luxury only for the rich'* to *'luxury to the mass'* gives birth to masstige. Masstige is a portmanteau term used to describe Mass and Prestige which holds the charming spot between the mass and the rich [4]. The democratization of luxury or as some researchers prefers to call it, the populence, mass affluence, or accessible luxury, is a recent departure from luxury, which is a downward extension of the traditional luxury. According to the masstige theory, as explained by [2], [4], masstige goods are the finest in terms of quality but they are priced far lower than the super-premium brands. The luxury brand, Giorgio Armani, for instance, has both downward and upward extensions; the brand sells a luxury fashion jeans with Giorgio Armani label to affluent entrepreneurs and professionals and targets the mass youthful audience with Armani Exchange, at a lower price points [5]. Over the last decades, the world has witnessed an upsurge in the demand of both luxury extensions, mainly for less-priced downward luxury extensions. However, in this paper we contend that the Covid-19 pandemic could trim down the private spending of the middle-class population, and may pull back the years of growth of the masstige sector.

When threatened by crisis consumers evaluate the perceived severity of the risk and develop coping mechanism. The Protection Motivation Theory (PMT) may explain individual's protective behavior during crisis situation [6]. According to PMT, individual's decision to engage in threat preventative behavior depends on their evaluation of the two cognitive processes; the threat and coping appraisal. In a threat appraisal individuals evaluate the perceived severity (the degree of seriousness of the potential threat) and the perceived vulnerability (the perceived likelihood of being exposed to the harm). The coping appraisal reflects one's ability to respond to the threat either by adopting or avoiding the threat. The coping appraisal also includes two essential elements; self-efficacy (the capability to perform the coping behavior) and response efficacy (the perceived effectiveness of the coping response in preventing the threat). The model predicts that the motivation to perform protective behavior intensifies with high degree of threat appraisal and risk susceptibility coupled with the likelihood and effectiveness of coping response, which results in a change in behavior.

During periods of economic turndown, people trade-off their choices to what they believe is critically important and look for a new landmarks to challenge the treat [7]. Consequently, businesses change or tone down their target market and tailor their brand positioning to reflect values sought by consumers experiencing recession. Empirical and anecdotal evidences suggest that consumer behave differently during blossoming and economic downturn circumstances. For example, [8] indicated that, over half of the U.S. consumers, either decrease their loyalty or shifted to competitors alternative brands during the 2009 global financial crisis. We undertook this study to examine the influence of the pandemic on the masstige consumer's behavioral intention.

The goal of this study was to examine the impact of uniqueness, hedonic value, conspicuous consumption and perceived quality on masstige purchase intention and further to investigate whether perceived economic crisis moderates the interplay. As such, it is expected that the middle-income consumers, who are targets of the masstige goods, are less likely to desire uniqueness, hedonic value, conspicuous consumption and quality, which may leads to lower purchase intention for masstige brands.

2 Literature Review and Hypothesis development

In their Harvard Business Review article, [7] elaborated the concept of "masstige" by focusing on the U.S. middle consumer behavior. They indicated that with the increasing number of middle-class population and the desire of many to possess high-quality goods, luxury products became reachable and attainable by the large segment of the market. The notion of luxury and its dimensions are well articulated in past researches, whereas the concept of masstige is still at the premature stage. In his study, Paul [2] developed the masstige pyramid by targeting the Louis Vuitton based in Japan which focuses on the middle and upper-lower class consumers. His work identified luxury brands such as Gucci, Apple, Victoria, Versace, Starbucks, and Louis Vuitton as the best examples that applied the masstige model.

In this study we asserted that companies' departure from the traditional luxury to allure the mass into the world of luxury doesn't make the luxury brand conventional. Luxury was unattainable for the mass fashion oriented young consumers and now a new breed of luxury which is well-designed, well-crafted and even with same label became accessible to the less wealthy segment. Luxury superpowers, Luis Vuitton and Gucci, for example, introduced a line of handbags, which still has the company label, although it is a downward extension [5]. Therefore, we argue that although luxury brands extended downward aiming to maximize their profit by catering the middle class consumer, the approach does not dilute the core brand essence. That is, the new luxury offspring maintains a certain level of uniqueness, emotional value, conspicuousness and quality from the luxury brand. Besides, intriguingly, luxury is very subjective, and it is a matter of perception. What is lavish to one might not necessarily be to another [3]. Hence, more precisely we contend that for the middle-class consumer, owning accessible premium goods could provide a sense of uniqueness, emotional feelings, social status and reliability. Figure 1. below is the research model that we propose.

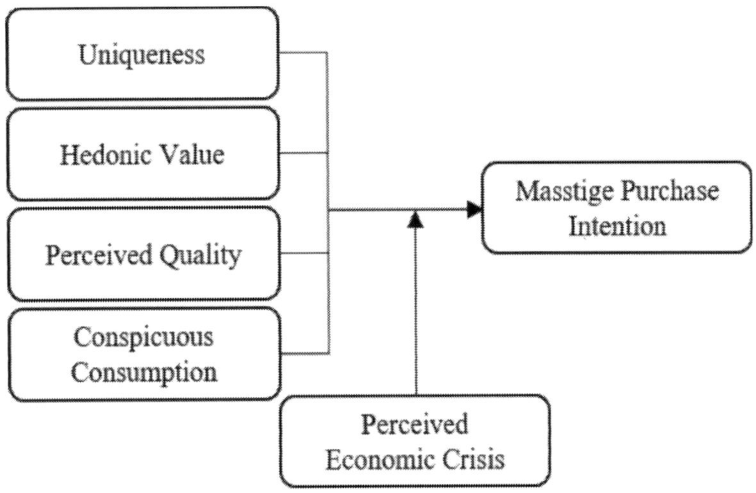

Figure 1. Research Model
Source: Own processing.

2.1 Uniqueness and Masstige Purchase Intention

Uniqueness is the behavior of buying distinctive products with the intention of impressing other and creating a difference among others. According to the uniqueness theory individuals tend to change their attitude when they feel higher similarity with others and engage in behaviors that exhibit their difference [9]. Consumers seek to regenerate a unique use for a product when they experience high similarity (vs. low similarity). Consistent with the theory of uniqueness, the conformity behavior study postulates that when the similarity magnitude among others heightens, the desire for conformity decreases proportionally [10]. For instance, preference for exclusiveness and unavailable products was rated high when the degree of similarity was experienced as high than moderate. When consumers feel similarity with others, they will reach a verdict to recoup their individuality and uniqueness. Individuals with a higher need for uniqueness tend to adopt new brands more quickly. The theory of social comparison [11] argues that an individual is motivated to compare him/herself with significant others and this comparison triggers desire for uniqueness which may influence consumption of consequent brands. Thus, in this study we conjecture that consumer desire for uniqueness may increase the likelihood of purchasing masstige products. That is, people who desire to feel differentiated from their social group may have strong predisposed to buy masstige brands.

H1: Uniqueness will have a positive effect on masstige purchase intention.

2.2 Hedonic value and Masstige Purchase Intention

The psychological value of the brand dominates hedonistic consumer who responds and reacts emotionally to the brand. The emotional attachment, relationship, and feeling such as excitement, warmth, and fun are more important attributes in consumer consumption behavior. Hedonistic consumption behavior is triggered by the affection and emotional values consumers get by using the product [12]. Possession of masstige is a means of obtaining some aesthetic value such as joy and happiness. Previous studies identified hedonistic need as important luxury purchase parameter. Masstige consumers express themselves by owning emotionally rich products and possession enhances their self-esteem and provides a sense of pleasure. However, hedonic is subjective; what is hedonic to one group may not be perceived the same in another society. For example, British consumers are more emotional oriented than Indian consumes. In contrast, concerning the rating of hedonic and utilitarian value in the U.S and Chinese consumers, a very insignificant difference was evidenced [13]. Hence, in this study we expect that hedonic value may influence the purchase intention of masstige products.

H2: Hedonic value will have a positive effect on masstige purchase intention.

2.3 Conspicuous consumption and Masstige Purchase Intention

Veblen [14] in his book 'The Theory of Leisure Class', defined conspicuous consumption as consumers' willingness to pay more for goods that are rich in displaying wealth and signifies status. Conspicuous consumption is the inspirational process by which consumers motivated to enhance their status in the society by purchasing expensive products, displaying and showing off their wealth. As the concept of conspicuous consumption entails, consumers purchase expensive and distinctive products to display their income level and identify their social status in the reference group [5]. From this perspective, the masstige products serve as a way of expressing symbol. In this study we posit that consumers conspicuous consumption behavior influence the intention to purchase masstige products.

H3: Conspicuous consumption will have a positive effect on masstige purchase intention.

2.4 Perceived quality and Masstige Purchase Intention

Perceived quality is the ability of the product to exhibit superior and excellent performance. Quality is a consequence of individuals' judgment and evaluation of a product [15]. Product quality evaluation consists of two spectrums; consumers' subjective and objective recognition of a product. Objective product quality is the use of prior product experience and knowledge to appraise the overall product benefit, performance and trustworthiness, while perceived quality linked with users' feeling, subjective judgment and evaluation of the product intrinsic and extrinsic attributes [16]. In this study we used perceived quality to indicate consumers' subjective evaluation and predisposition to purchase of masstige products.

Masstige brands have better quality and enhance the owner's level of satisfaction, compared to ordinary brands. Even before a direct personal experience of brands, consumers can tell about the superior performance of masstige products. Masstige brands are expected to display the finest brand quality and premium price. Masstige brands such as Starbucks, Coach, and Bath have one thing in common that they emphasis on premium product quality to the mass market and make prestigious brands while maintaining their product affordability [4]. Continuous product quality improvement contributes significantly to masstige purchase intention. The notion is supported by the Louis Vuitton masstige strategy of incessant quality enhancement and modification and a positive result was observed with consumers purchase intention and retention as the company widens the product range and improve its quality [2]. Thus, in this study we posit that consumers' perceived quality may influence the purchase intention of masstige products.

H4: Quality will have a positive effect on masstige purchase intention.

2.5 The moderating effect of Perceived economic crisis

In this study, we argue that perceived economic crisis could play a moderating role in the interplay between the four independent variables and masstige purchase intention. During the great recession period of 2008-2011, many middle-income consumers were extremely hit by the crisis and consumers resist purchasing expensive products, tend to buy low-quality and counterfeit luxury items. Even affluent and high-end consumers avoid luxury during the economic crisis with the view of protecting the psychology of those who used to but no longer affords to buy luxury. Because luxury products are perceived as rich in conspicuous and hedonic value, wealthy luxury consumers considered the consumption of status enhancing products as 'luxury shame' and condemn lavish spending and avoid show off when unemployment intensifies and the social suffering enlarges [8].

The luxury sector was projected to grow in 2020, however, the economic downturn sluggish a decades of progress of various product lines. The pre-pandemic outbreak estimate for both the personal and experiential global luxury was expected to be on a positive trajectory. However, even financially strong and true luxury consumers reserved from making huge investments. Consumers are more frugal and conscious in their spending when experiencing crisis [5]. On the contrary, the financial crisis leads consumers to be selective and purchase single but super quality product than purchasing multiple items [5]. Reaction to economic crisis largely depends on the ratio of the perception of the price of the product and perception of once earnings. Although the world has witnessed an upsurge in the income of the middle-class consumers, the global market and economic volatility caused by COVID-19 may have a significant impact on consumers' willingness to buy masstige products. Therefore, in this study we predicted that recession threat could condition the middle-class consumers' need for uniqueness, hedonic value, conspicuousness, and perceived quality resulting in contraction in the masstige shopping experience.

H5: Perceived economic crisis will negatively moderate the relationship between uniqueness (H1), hedonic value (H2), conspicuous consumption (H3), perceived quality (H4) and masstige purchase intention.

3 Methodology

The data was collected by online survey in South Korea. We assumed that college students who are familiar with the subject under investigation could provide useful information to test our predictions, thus, we systematically targeted business major students. Our survey was conducted from May 28 to June 15, 2021. We prepared a survey questionnaire and collected the response using 'Google Form,' and we shared the form to the respondents through 'Kakao Talk,' which is the most popular SNS in South Korea.

The measurement items were adopted from existing literatures and previously validated scales. The constructs were measured using a five-point Likert scale, ranging from 1= strongly disagree to 5= strongly agree. The uniqueness construct in this study was measured from [17] and [18] five-item scale measurement items. The items for hedonic value were adopted from [19] and [12], which measures respondent's feeling of emotion in owning masstige goods. As for conspicuous consumption, the constructs were adopted from [2] and [20]. Perceived quality scale was measured using [2] and [21] scale, which indicates individuals' perceived consistency, durability, value for money, and confidence in the quality of masstige brands. The five-item scale developed by [22] and [2] was adapted to measure purchase intention, which measure respondents' likelihood to purchase masstige goods.

For those measurement items, reliability and validity are tested. As shown in the Table 1, KMO = 0.822, Bartlett's Test of Sphericity was significant at 0.000 for the factor analysis. The "Total Variance Explained" with the Eigen Value of greater than 1 represents four factors explaining 71.569% of the model. Cronbach's Alpha (α) checks internal consistency of measurements and assures whether items measure the same scale. According to Saunders et al. (2012), if items and constructs Cronbach's Alpha (α) value shows 0.7 or above, items said to have internal consistency. Table 1 explains factor loadings and Cronbach's Alpha (α) score of the constructs and all variables were higher than 0.7, indicating measurements strong internal consistency.

To test the hypothesized relationships, we used regression analysis. Also, to examine the weakening effect of COVID-19 on consumer masstige consumption, we tested the moderating role of perceived economic crisis using hierarchical multiple regression analysis. More detailed results of the hierarchical multiple regression analysis are shown in Table 4. We analyzed the data using SPSS version 27.

Table 1. Reliability and validity

		Factor Analysis			Reliability
Factor	Survey information	Factor Loadings	Eigen value	Variance explained	Cronbach's (α)
Uniqueness	UNI1	0.841	6.944	28.933	0.826
	UNI2	0.689			
	UNI3	0.678			
Hedonic Value	HED1	0.880	3.969	16.539	0.902
	HED2	0.798			
	HED3	0.717			
Conspicuous Consumption	CC1	0.758	2.287	9.529	0.815
	CC2	0.655			
	CC3	0.649			
	CC4	0.543			
	CC5	0.541			
Perceived Quality	PQ1	0.887	1.544	6.431	0.802
	PQ2	0.675			
	PQ3	0.613			

Perceived Economic Crisis	PEC1	0.901	1.343	5.595	0.907
	PEC2	0.852			
	PEC3	0.844			
	PEC4	0.840			
	PEC5	0.721			
	PEC6	0.585			
Purchase Intention	PI1	0.693	1.191	4.963	0.848
	PI2	0.784			
	PI3	0.887			
	PI4	0.799			
	PI5	0.451			
KMO= .822 X^2= 2413.409 df= 276 p=.000					

Source: Own processing.

4 Result and Discussion

4.1 Characteristics of the sample

Out of 184 questionnaires distributed, 177 valid responses were utilized for further analysis. As shown in table 2, 48% of participants were male and 52% were female. The study participants' age ranged from under 20 to 50 years and above. More than half of the respondents were between the age range of 20-30 (66.7%) and about one third of the respondents earn a monthly income of less than KRW 1.5 million. Of the sample population, 137 (77.4%) were single households, while the remaining 39 were married.

Table 2. Demographic profile

Variable	Classification	Frequency	Percentage %
Gender	Male	85	48.0
	Female	92	52.0
Age	< 20	2	1.1
	20-30	118	66.7
	31-40	20	11.3
	41-50	18	10.2
	>50	18	10.2
Income	Less than KRW 1,500,000	104	58.8
	KRW 1,500,000 – KRW 2,500,000	31	17.5
	KRW 2,500,001 – KRW 3,500,000	10	5.6
	KRW 3,500,001 – KRW 4,500,000	7	4.0
	KWN 4,500,001 – KRW 5,500,000	8	4.5
	More than 5,500,000	16	9.0

Source: Own processing.

4.2 Hypothesis testing

We performed hypothesis testing at two stages. First, we performed a regression analysis to examine the effect of independent constructs (H1, H2, H3 and H4) on masstige purchase intention. Next, we performed a hierarchical multiple regression analysis to assess the effect of the moderating construct (PEC) (H5) using SPSS.

As shown in Table 3, the regression analysis shows that the values of Variance Inflation Factor (VIF) for all constructs were less than 5.0 and the Tolerance values ranged between 0.633 and 0.762, indicating the absence of Multicollinearity. The results shown in the table 3 indicates that UNI, HED, CC, and PQ account for 37% (Adjusted R^2= 0.370) of the variance in the MPI. Regarding the goodness of fit, the analysis of variance (ANOVA) presents strong evidence that factors have a non-zero coefficient and considered fit (F-value = 26.868, P=0.000), and R^2=0.385 indicating 38.5% of the observation were fit for the sample regression line.

Table 3. Regression result

Model	Unstandardized Coefficients		Standardized Coefficients			Collinearity Statistics	
	B	S.E.	Beta	t	Sig.	Tolerance	VIF
(Constant)	.683	.280		2.439	.016		
Uniqueness	.249	.067	.267	3.717	.000	.694	1.441
Hedonic Value	.379	.072	.398	5.291	.000	.633	1.581
Conspicuous consumption	.043	.081	.039	.529	.598	.666	1.502
Perceived Quality	.047	.075	.043	.623	.534	.762	1.312
Goodness of fit	F=26.868		P=0.000	R^2=0.385		Adjusted R^2= 0.370	
a. Dependent Variable: Masstige Purchase Intention (MPI)							

Source: Own processing.

The regression table shows uniqueness (b = .267, t = 3.717, P < 0.01) to have a significant effect on MPI thus, the hypothesis 1 is supported. The finding also shows that hedonic value is a more significant contributor to MPI (b = .398, t = 5.291, P < 0.01), which provides strong statistical evidence that there is a significant impact of hedonic value on masstige purchase intention. Thus, H2 can be supported. In contrast, conspicuous consumption and perceived quality were not significant predictors of masstige purchase intention, indicating both constructs to have insignificant impact on MPI, therefore, H3 and H4 are not supported.

Secondly, we examined the effect of perceived economic crisis as a moderator on the relationship of the four constructs and MPI (the dependent variable) (Table 4). In order to measure the role of PEC and to better understand the moderating effect, we introduced a dummy variable (DPEC; high vs. low perceived economic crisis). In order to examine the moderating effect, we performed a hierarchical multiple regression analysis. The result, as shown in table 4, indicates that PEC has no moderating effect. R^2 changes were not significant when PEC was entered in the regression while UNI and HED remained significant to MPI.

Table 4. Moderating effect of PEC

Model Summary									
					Change Statistics				
Model	R	R Square	Adjusted R Square	Std. Error of the Estimate	R Square Change	F Change	df1	df2	Sig. F Change
1	.620[a]	.385	.370	.75721	.385	26.868	4	172	.000
2	.620[b]	.385	.367	.75942	.000	.001	1	171	.980
3	.636[c]	.405	.373	.75553	.021	1.441	4	167	.223

a. Predictors: (Constant), Quality, Uniqueness, Conspicuous, Hedonic
b. Predictors: (Constant), Quality, Uniqueness, Conspicuous, Hedonic, DPEC
c. Predictors: (Constant), Quality, Uniqueness, Conspicuous, Hedonic, DPEC, InteractionQuali_DPEC, InteractionUniq_DPEC, InteractionHedon_DPEC, InteractionConsp_DPEC

Source: Own processing.

5 Discussion

The finding showed that uniqueness and hedonic value are significant predictors and have positive effect on masstige purchase intention. However, conspicuous consumption and perceived quality do not influence purchase intention. It indicates that consumer's emotion or need to possess aesthetically rich materials and feeling of uniqueness triggers the predisposition to buy masstige products. The finding is interesting in that consumers who need to feel different from their social group tend to purchase masstige brands. When a product becomes popular, it ceases to be exclusive and thus no longer treated as a status symbol [14]. But, luxury companies and scholars realized the fact that luxury is not about sophistication alone. Louis Vuitton, for example, has managed to deploy the masstige strategy by combining the cachet attribute of the core brand with attainable price to make it available to the mass. They designed the new masstige to appeal the middle-class while maintaining the prestige image and exclusive sentiment of the brand [23, 2].

Notwithstanding the change in the income of the middle-class consumers caused by COVID-19 pandemic does not affect the masstige purchase intention. COVID-19 has given stressful and traumatic experience which leads to low self-esteem. According to [24] when people perceive low self-concept they encouraged to acquire self-enhancing and unique products. Thus, individuals' low self-esteem caused by the COVID-19 pandemic could have a spillover effect on the need for uniqueness which may lead to masstige behavioral intention. Similarly, there is no moderating effect of perceived economic crisis on the relationship between CC and MIP. The finding of the study shared some similarities with [5] study such that conspicuousness endures regardless of economic turndown and status seekers tend not to change their consumption trend during recession. However, about three decades ago [25] found that Asian consumers cut back their spending and look for less priced products when times are tough. Consumers reduce their spending in the period of crisis and this cut in consumption is mostly proportional to their income. In contrast, this study supports the work of [5] such that consumers need for conspicuousness remains intact during economic downturn. We also predicted that during the pandemic consumers may prioritize health-related issues and become more concerned about safety than aesthetic value resulting in the deferment of masstige spending. However, the result showed the non-significant effect of perceived recession on the link between hedonic value and masstige purchase intention.

6 Conclusion

By employing the Protection Motivation Theory, we posit that the recent economic crisis may weaken consumers' need for uniqueness, hedonic value, conspicuous consumption and desire for quality, resulting consumer to turn away from purchasing masstige goods. That is, when individuals perceive high threat and vulnerability they perform preventive behavior which leads to a change in the consumption behavior. However, intriguingly, quite the reverse to the marketing and psychology literature, our study shows that during the pandemic masstige consumers seems not to turn away from masstige. This implies that consumer's predisposition to purchase masstige is not likely to be mitigated by the economic crisis. Admittedly, COVID-19 has disrupted consumers' consumption pattern.

As with all researches, the findings of this study are not without limitations. We considered masstige handbags, wrest watches and perfume products. Thus, further research is needed to examine the effect of the studied constructs on masstige purchase intention with a large set of product categories. Finally, as discussed in the methodology, because of the pandemic we used students as a sample, and the use of student samples may question the generalizability of the finding to the larger population. Future research with a shopping mall intercept approach may also be needed to address the actual masstige consumer group of the population.

References

1. Bain & Company, Global personal luxury goods market on track for recovery [online], Available at https://www.bain.com/about/media-center/press-releases/2021/Global-personal-luxury-goods-market-on-track-for-recovery/ (2021)
2. J. Paul, Masstige model and measure for brand management. *European Management Journal*, **37**(3), 299-312 (2019)
3. M. Kastanakis, G. Balabanis, Between the mass and the class: antecedents of the "bandwagon" luxury consumption behavior. *Journal of Business Research*, **65**(10), 1399-1407 (2012)
4. A. Kumar., J. Paul., A. Unnithan, 'Masstige marketing': a review, synthesis and research agenda. *Journal of Business Research*, **113**, 384-398 (2020)
5. J. Nunes, X., Drèze, Y. Han, Conspicuous consumption in a recession: toning it down or turning it up? *Journal of Consumer Psychology*, **21**(2), 199-205 (2011)
6. S. Youn, J. Lee, J. Ha-Brookshire, Fashion consumers' channel switching behavior during the covid-19: protection motivation theory in the extended planned behavior framework. *Clothing and Textiles Research Journal*, **39**(2), 139-156 (2021)
7. M. Silverstein, N. Fiske, Luxury for the masses. *Harvard business review*, **81**(4), 48-59 (2003)
8. N. Piercy, D. Cravens, N. Lane, Marketing out of the recession: recovery is coming, but things will never be the same again. *The Marketing Review*, **10**(1), 3-23 (2010)
9. H. Fromkin, C. Snyder, The search for uniqueness and valuation of scarcity. *Social Exchange,* Springer, Boston, MA, 57–75 (1980)
10. I. Abosag, Z. Ramadan, T. Baker, J. Zhongqi, Customers' need for uniqueness theory versus brand congruence theory, the impact on satisfaction with social network sites. *Journal of business research*, **117**, 862-872 (2020)
11. L. Festinger, A theory of social comparison process. *Human Relations,* **7**(1), 117–140 (1954)
12. K. Voss, E. Spangenberg, B. Grohmann, Measuring the hedonic and utilitarian dimensions of consumer attitude. *Journal of marketing research*, **40**(3), 310-320 (2003)

13. P. Shukla, K. Purani, Comparing the importance of luxury value perceptions in cross-national contexts". *Journal of Business Research*, **65**(10), 1417-1424 (2011)
14. T. Veblen, *The theory of the Leisure Class*. Boston, MA: Houghton Mifflin, (1973)
15. V. Zeithaml, Consumer perceptions of price, quality, and value: a means-end model and synthesis of evidence. *Journal of marketing*, **52**(3), 2-22 (1988)
16. D. Aaker, Measuring brand equity across products and markets. *California management review*, **38**(3) (1996)
17. K. Tian, W. Bearden, G. Hunter, Consumers' need for uniqueness: scale development and validation. *Journal of consumer research*, **28**(1), 50-66 (2001)
18. Q. Bian, S. Forsythe, Purchase intention for luxury brands: a cross cultural comparison. *Journal of Business Research*, **65**(10), 1443-1451 (2012)
19. J. Sweeney, G. Soutar, Consumer perceived value: the development of a multiple item scale. *Journal of retailing*, **77**(2), 203-220 (2001)
20. Y. Liang, Exploring chinese consumers' luxury value perceptions: development and assessment of a conceptual model. *Bournemouth University*, (2018)
21. S. Agarwal, R. K Teas, Cross-national applicability of a perceived quality model. *Journal of Product & Brand Management*, **11**(4), 213-236 (2002)
22. A. Arora, N. Kishor, Factors determining purchase intention and behavior of consumers towards luxury fashion brands in India: an empirical evidence. *British Journal of Marketing Studies (BJMS)*, **7**(4), 34-58 (2019)
23. J. Paul, Masstige marketing redefined and mapped: introducing a pyramid model and mms measure. *Marketing Intelligence & Planning*, **33**, 691-706 (2015)
24. M. Lynn, C. Snyder, Uniqueness seeking. *Handbook of positive psychology,* Oxford, Oxford University Press, 395–410 (2002)
25. S. Ang, S. Cheng, A. Lim, K. Tambyah, Spot the difference: consumer responses towards counterfeits. *Journal of consumer Marketing*, **18**(3) 219-235 (2001)

Exploratory Study on Digital transformation: Capabilities and Expected Performances

Byungchul Choi[1,*] and *Seunghyun Kim*[2]

[1]College of Business, Hankuk University of Foreign Studies, Seoul, South Korea
[2]Science and Technology Policy Institute, Sejong, South Korea

Abstract. Nowadays many firms encounter the macro-level changes called digital transformation and research on it has drastically increased since 2014. In response to this emerging phenomenon, this study explores how firms prepare for digital transformation and what do they expect from it. More specially, based on the survey from 439 Korean scaleup firm (high-growth firm), we suggest the concepts to necessary capabilities for and expected performance improvement through digital transformation. Our study illustrates that how a firm's perceptions on those concepts varies upon firm size and industry type. Additionally, our study offers a clue for 'digital divide' that can potentially threaten the survival of numerous firms.
Keywords: digital transformation, capabilities, performance improvement, firm size, industry type

1 Introduction

In recent years, the business world has encountered the macro-level changes sourced from of digital transformation. While digital transformation is not a completely new topic to many entities across the areas, the growing body of research has been shown since 2014 (see Figure 1, Google trend). However, as mentioned above digital transformation comes with unprecedented changes and convergence across various technological domain and industrial sectors, it can be a double-edged sword to both incumbents and new ventures. Although it can potentially provide diverse and constructive collaboration opportunities to many firms (Luo, 2021), it can also threat the business of many firms as existing technological standards and business systems will be drastically either changed or challenged (Oh & Rhee, 2008).

In response to this mixed expectation on digital transformation, this study explore how firms are currently viewing digital transformation in terms of preparation and desirable outcomes. More specifically, by using the Korean high-growth firm data, we first offer the results of survey implying what types of capabilities those high-growth firms pereceive as key capabilities to cope with changes sourced from digital transformation and what types of firm performances they anticipate to achieve. Then, we shift our attention to important firm-specific characteristics, firm size and industry type that can matter when firm adopt marco-level changes (Li et al., 2018; Eller et al., 2020; Siachou et al., 2021; Savastano et al., 2019;

[*]Corresponding author: bchoi@hufs.ac.kr

Vogelsang et al., 2018; von Leipzig et al., 2017; Martín-Peña et al., 2020; Ribeiro-Navarrete et al., 2021), and provides a comparative analysis. The results indicate taht firms' perception of and preparations for digital transformation vary upon firm-specific charateristics. We also find that some firms might suffer from digital divide (bipolarization of digital capability).

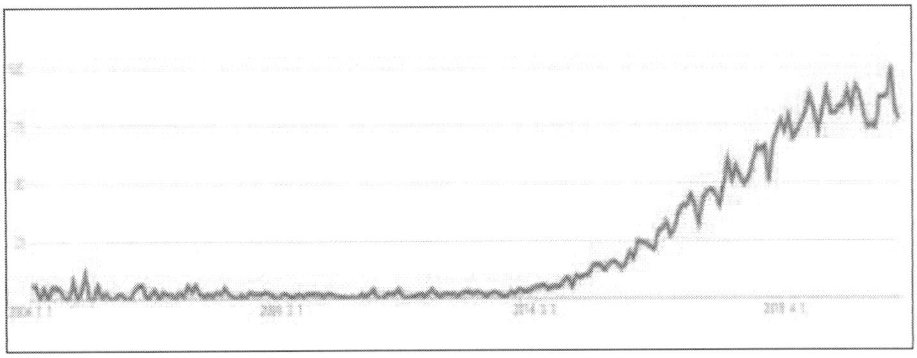

Fig. 1. Interest in Digital Transformation
Source: Google Trend.

2 Research Design

2.1 Selection of capabilities and performance

To design survey, we select the following three capabilities: (1) business capability that refers to a firm's ability for task management, process improvement, and data management (2) digital capability, which is defined as a firm's technological capability for programming, software development, and data analysis (3) soft capability that refers to a firm's capability for problem-solving, collaboration, and creativity management. Those capabilities are necessary to achieve innovation (Teece, 1986, 2018). The reason we focus on these three capabilities is that a firm's ability to cope with macro-level changes ultimately depends on how a firm develop necessary technologies, how a firm apply those technologies to real business activities and how a people in those firms manage these transitional processes. Regarding anticipated performances that firms eventually aim to achieve, we develop three performance dimensions, which are (1) profitability growth (2) new product development and (3) new partnership.

2.2 Sample collection

We implemented a survey based on the list of scaleup firms in Korea. To collect the sample firms, we first draw on the Korea Enterprise Data (KED) database subscribed by Science and Technology Policy Institute (STEPI), and gather the initial sample firms that have shown 20 percent increase in sales with more than 10 employees during the period of 2016-2018. This initial sample include 3,391 scaleup firms. Then, we sent email survey to all these firms and a total of 439 survey was received with response rate of 13%. We classified those 439 firms based on two categories, firm size and industries. Regarding the former category, while 86 firms (20%) belong to a large firm, 353 firms (80%) were Small and Medium Enterprises (SMEs). In relation the latter category, while manufacturing industries comprised 44% of the sample (194 firms), non-manufacturing firms such as service consist 56% of the sample (245 firms). We designed a survey questionnaire by referring to 2019 ScaleUp Survey of Scaleup Institute, the UK private sector, not-for-profit company (s).

2.3 Analysis

Collected data were analyzed with a focus on the following four matters: (1) importance of digital transformation and preparations, (2) expected change digital transfor-mation would bring to the industry and main products, (3) capabilities necessary to prepare for digital transformation, and (4) anticipated performance improvement from digi-tal transformation. Digital transformation is a macro-trend phenomenon that affects all industries and many firms, but individual firms may have different views depending on their characteristics or the industry to which they belong. Therefore, this study applies the classification method according to firm size and industrial characteristics for each of the aforementioned matters.

3 Results

Table 1 shows firms' percption on "capabilities considered important to prepare digital transformation" depending on firm size and industry typel. In all capabilty types, small-medium size firms show more than 50% o "neutral" views and low level of 'Positive' views, indicating that small and medicum size firms have pessimistic perspective on needs of those capabilites. Simiarily, compare to manufaturing firms, non-manufacturing firms have more 'Neutral' and 'Negative' perspective on the value of each capability.

Table 1. Do you think this capability is important to prepare digital transformation?

		Business capability (%)			
		N	Negative	Neutral	Positive
Category 1	Large	86	5.9	43.0	51.2
	Small-Medium	353	18.1	51.6	30.3
Category 2	Manufacturing	194	12.4	51.0	36.6
	Non-Manufacturing	245	23.2	61.9	14.9

		Digital capability (%)			
		N	Negative	Neutral	Positive
Category 1	Large	86	5.9	44.2	50
	Small-Medium	353	22.7	50.1	27.1
Category 2	Manufacturing	194	14.4	52.6	32.9
	Non-Manufacturing f	245	29.4	58.2	12.4

		Soft capability (%)			
		N	Negative	Neutral	Positive
Category 1	Large	86	7	43.0	50
	Small-Medium	353	19.9	51.8	28.3
Category 2	Manufacturing	194	13.4	53.6	33
	Non-Manufacturing	245	25.7	59.8	14.5

Source: Own processing

Table 2 shows firms' expectation on "performance impovement through digital transformation. Compared to small-medium firms, large firms show more optimistic view and less pessemistic view on each performance dimenstion. Similarily, compared to non-manufacuting firms, manufacturing firms show more optimistic view on each performance dimenstion.

Table 2. What types of performance you want to improve through digital transformation?

		Profitability growth (%)			
		N	Pessemistic	Neutral	Optimistic
Category 1	Large	86	24.5	54.7	20.9
	Small-Medium	353	54.1	35.7	10.2
Category 2	Manufacturing	194	45.8	38.7	15.5
	Non-Manufacturing	245	37.1	50.5	12.4

		New product development (%)			
		N	Pessemistic	Neutral	Optimistic
Category 1	Large	86	25.6	59.3	15.1
	Small-Medium	353	55.8	36.5	7.6
Category 2	Manufacturing	194	45.9	41.8	12.4
	Non-Manufacturing	245	40.7	51.0	8.3

		New partnership (%)			
		N	Pessemistic	Neutral	Optimistic
Category 1	Large	86	26.7	55.8	17.4
	Small-Medium	353	56	37.1	6.8
Category 2	Manufacturing	194	47.4	40.2	12.3
	Non-Manufacturing	245	40.1	52.1	7.8

Source: Own processing

After explore the general perspective of firms on each dimenion of capability and perforamnce, we conduct sets of OLS regression to get further insights on the relationship between those capabilites and performances. Table 3 shows summary statisctics and indicate that each dimension of capabilites and performances is highly correlated. Hence, we maily focus on the direct relationship between each capability and performance.

Table 3. Summary Statistics

		Mean	S.D.	1	2	3	4	5
1	Business capability	3.24	0.99					
2	Digital capability	3.15	1.00	0.89				
3	Soft capability	3.21	1.00	0.89	0.91			
4	Profitability growth	2.41	1.05	0.41	0.42	0.40		
5	New product development	2.34	1.01	0.38	0.39	0.37	0.90	
6	New partnership	2.34	1.00	0.37	0.38	0.36	0.89	0.96

N = 439
Source: Own processing

Table 4-7 show the result of OLS regression. As expected, while each capability has significant and positive relationship with each anticipated performance, Table 5 shows interesting results. In case of large firms, no capability has significant impact on new product development. This may indicate that large firms view digital transformation as instrument to renovate business model and network rather than merely developing new product. The findings presented in Table 4-7 does not necessarily mean the impact of those capabilities on three types of performances are confirmed. Rather, it reflects how people in firms perceive or believe the association between capabilities and performances. However, it is still meaningful and valid to understand how various firms interpret encountering situations and changes brought by digital transformation.

Table 4. OLS regression on firm performances (Small & Medium Firms)

	Profitability			
Independent variables				
Business capability	0.417***			0.256*
	0.050			0.119
Digital capability		0.403***		0.173
		0.050		0.126
Soft capability			0.392***	0.001
			0.051	0.127
N	353	353	353	353
	New product development			
Independent variables				
Business capability	0.404***			0.244
	0.049			0.116
Digital capability		0.387***		0.129
		0.049		0.123
Soft capability			0.385***	0.051
			0.050	0.124
N	353	353	353	353
	New partnership			
Independent variables				
Business capability	0.371***			0.211+
	0.048			0.114
Digital capability		0.362***		0.169
		0.048		0.121
Soft capability			0.352***	0.009
			0.049	0.123
N	353	353	353	353

Source: Own processing

Table 5. OLS regression on firm performances (Large Firms)

	Profitability			
Independent variables				
Business capability	0.291**			-0.569
	0.107			0.348
Digital capability		0.386***		0.679+
		0.109		0.366
Soft capability			0.332**	0.258
			0.102	0.371
N	86	86	86	86
	New product development			
Independent variables				
Business capability	0.068			-0.50
	0.101			0.3361
Digital capability		0.134		0.451
		0.105		0.353
Soft capability			0.106	0.178
			0.097	0.358

N	86	86	86	86
	New partnership			
Independent variables				
Business capability	0.122			-0.665+
	0.103			0.338
Digital capability		0.212*		0.611+
		0.106		0.356
Soft capability			0.172+	0.246
			0.099	0.360
N	86	86	86	86

Source: Own processing

Table 6. OLS regression on firm performances (Manufacturing Firms)

	Profitability			
Independent variables				
Business capability	0.388**			0.101
	0.076			0.171
Digital capability		0.393***		0.021
		0.076		0.181
Soft capability			0.377***	0.268
			0.073	0.189
N	194	194	194	194
	New product development			
Independent variables				
Business capability	0.356***			-0.509
	0.074			0.336
Digital capability		0.352		0.451
		0.074		0.353
Soft capability			0.106	0.178
			0.097	0.358
N	194	194	194	194
	New partnership			
Independent variables				
Business capability	0.337			0.036
	0.074			0.170
Digital capability		0.360***		0.182
		0.074		0.181
Soft capability			0.356***	0.161
			0.073	0.188
N	194	194	194	194

Source: Own processing

Table 7. OLS regression on firm performances (Non-Manufacturing Firms)

	Profitability			
Independent variables				
Business capability	0.461**			0.23
	0.058			0.149
Digital capability		0.462***		0.347*
		0.056		0.158
Soft capability			0.427***	-0.104
			0.058	0.155
N	245	245	245	245
	New product development			
Independent variables				
Business capability	0.398***			0.206
	0.056			0.145
Digital capability		0.401***		0.324*
		0.055		0.154
Soft capability			0.366***	-0.117
			0.057	0.151
N	245	245	245	245
	New partnership			
Independent variables				
Business capability	0.385***			0.197
	0.055			0.144
Digital capability		0.386***		0.283+
		0.054		0.153
Soft capability			0.357***	-0.081
			0.056	0.149
N	245	245	245	245

Source: Own processing

We coduct additional analysis to explore more detailed picture of how firm size and industry type conjontly influence the relationship between capabilities and expected performance. While most of main effect of individual capability show significant and positive effect of each expected performance, full model illustrate more interesting findings. (We do not report the main effect of individual capability here due to the space limit). Table 8 shows that large-manufacturing firms and smal-medium non-manufacturing firms demonstrate some statistically signifianct results. Interestingly, in case of large-manufacturing firms, business capabiliy and firm profitability are negatively related. This finding does not mean business capability for digital transformation leads firms experience to lower profitability. Instead it implies that people in large-manufacturing firms does not view business capability as a key determinant of profitability in the era of digital transformation. However, they consider digital capability as a positive and key determinant of it.

Table 8. OLS regression on three dimensions of firm performance

	Profitability			
Independent variables	SMEs MFG	Large MFG	SMEs non-MFG	Large non-MFG
Business capability	0.229	-1.257**	0.260	-0.0365
	(0.182)	(0.446)	(0.160)	(0.497)
Digital capability	-0.023	1.000*	0.303+	0.456
	(0.198)	(0.401)	(0.166)	(0.882)
Soft capability	0.210	0.500	-0.115	0.030
	(0.212)	(0.363)	(0.161)	(0.993)
N	160	34	193	52
	New product development			
Independent variables	SMEs MFG	Large MFG	SMEs non-MFG	Large non-MFG
Business capability	0.178	-0.642	0.270+	-0.495
	(0.178)	(0.437)	(0.153)	(0.488)
Digital capability	-0.114	0.628	0.280+	-0.105
	(0.195)	(0.393)	(0.159)	(0.865)
Soft capability	0.347+	0.093	-0.126	0.737
	(0.208)	(0.356)	(0.154)	(0.975)
N	160	34	193	52
	New partnership			
Independent variables	SMEs MFG	Large MFG	SMEs non-MFG	Large non-MFG
Business capability	0.171	-1.423**	0.230	-0.137
	(0.175)	(0.453)	(0.154)	(0.471)
Digital capability	0.046	1.172**	0.249	-0.0132
	(0.191)	(0.407)	(0.160)	(0.836)
Soft capability	0.161	0.457	-0.0814	0.342
	(0.204)	(0.369)	(0.155)	(0.941)
N	160	34	193	52

Source: Own processing

Discussion

Our study provide unique insights in several ways. First, we offer useful concepts to evaluate capabilities for and expected outcome of digital transformation. In our survey, three important capabilities for digital transformation (business, digital, soft capabilities) and three expeceted performanace improvement (profitability, new product development, new partnership) are addressed. More importantly, our study demonstrate how those concepts are differently perceived to an individual firm. While large size firms and manufacturing firms show more positive perspective on digital transformation in terms of both capabilities and expected outcomes, those firms demonstrate some interesting patterns such as negative or no effect (in fact, perceived expectation) of businss capability on performances.

More importantly and interestingly, by investigating firm size effect and industry effect, our study offer a clue for 'digital divide' that can potentially threaten the survival of numerous firms that fail to adopt relevant skills and capabilities to prepare digital transformation. As digital transformation will come with unprecedented degree of convergence across industrial sectors, the negative effect of digital divide can creat trenmendous ineuqlity among firms (e.g., large firm vs. small firm). This may recall for policy maker considering some policy efforts for balanced development.

References

1. L. Li, F. Su, W. Zhang, JY. Mao, Digital Transformation by SME Entrepreneurs: A Capability Perspective. *Information Systems Journal*, **28**(6), 1129–1157 (2018)
2. R. Eller, P. Alford, A. Kallmünzer, M. Peters, M. Antecedents, Consequences, and Challenges of Small and Medium-Sized Enterprise Digitalization. *Journal of Business Research*, **112**, 119–127 (2020)
3. E. Siachou, D. Vrontis, E. Trichina, Can Traditional Organizations Be Digitally Transformed By Themselves? The Moderating Role of Absorptive Capacity and Strategic Interdependence. *Journal of Business Research*, **124,** 408–421 (2021)
4. M. Savastano, C. Amendola, F. Bellini, F. D'Ascenzo, Contextual Impacts on Industrial Processes Brought by the Digital Transformation of Manufacturing: A Systematic Review. *Sustainability,* **11**(3), 891-929 (2019)
5. K. Vogelsang, K. Liere-Netheler, S. Packmohr, S, U. Hoppe, Success Factors for Fostering a Digital Transformation in Manufacturing Companies. *Journal of Enterprise Transformation*, **8**(1-2), 121–142 (2018)
6. T. von Leipzig, M. Gamp, M, D. Manz, K. Schöttle, P. Ohlhausen, G Oosthuizen, D. Palm, K. von Leipzig, Initialising Customer-Orientated Digital Transformation in Enterprises. *Procedia Manufacturing*. **8**, 517–524 (2017)
7. ML. Martín-Peña, JM. Sánchez-López, E. Díaz-Garrido, Servitization and Digitalization in Manufacturing: The Influence on Firm Performance. *Journal of Business & Industrial Marketing*, **35**(3), 564–574 (2020)
8. S. Ribeiro-Navarrete, D. Botella-Carrubi, D. Palacios-Marqués, M. Orero-Blat, The Effect of Digitalization on Business Performance: An Applied Study of KIBS. *Journal of Business Research*, **126**, 319–326 (2021)
9. Y. Luo. New OLI advantages in digital globalization. *International Business Review*, **30**(2) (2021)
10. J. Oh, S. Rhee, The influence of supplier capabilities and technology uncertainty on manufacturer supplier collaboration: a study of the Korean automotive industry. *International Journal of Operations & Production Management*, **28**(6), 490-517 (2008)
11. DJ. Teece, Profiting from technological innovation: Implications for integration, collaboration, licensing and public policy. *Research Policy*, **15**(6), 285-305 (1986)
12. DJ. Teece, Profiting from innovation in the digital economy: standards, complementary assets, and business models in the wireless world. *Research Policy,* **47**(8), 1367-1387 (2018)

Design Crowdsourcing Supply Chain in Short Life Cycle Products

Dolgion Boldbaatar[1], and *Daeheon* Choi[1*]

[1]College of Business Administration, Kookmin University, 77, Jeongreung-ro, Seongbuk-gu, 02707 Seoul, Korea

Abstract. Design crowdsourcing is the largest open innovation model that can create value with potential consumers. It offers an opportunity to quickly respond to the market by obtaining instant designs from the crowd, freelancing designers with fundamental skills. In addition, it can secure globally innovative competitiveness without financial burden, which is more effective to start-up companies and small and medium enterprises (SMEs) in the field of seasonal product industry. Developing standardized design crowdsourcing supply chain processes and mathematical models is essential to respond to market trends and customer needs in the seasonal product industry. This study has been carried out to determine the best mechanism in the design crowdsourcing supply chain and coordinate each supply chain member whose desires meet each other. Thus, we identify contracts under which conditions can coordinate the crowdsourcing supply chain by a newsvendor model with a manufacturer and a retailer with a crowdsourcing platform. To see that, we study the coordination of the crowdsourcing supply chain through the following contracts: wholesale price, buy-back, and revenue sharing contract. For the forecasting, we present a framework of the design crowdsourcing supply chain and compare supply chain performance under crowdsourcing supply chain coordination. The summarized result shows that the wholesale price contract cannot coordinate the crowdsourcing supply chain efficiently. In contrast, buy-back and revenue sharing contracts can coordinate the crowdsourcing supply chain.
Keywords: crowdsourcing, seasonal products, supply contracts, supply chain coordination, newsvendor model

1 Introduction

Manufacturing, which requires specialized technology and facility investment, has long been recognized as a large corporation and professional's domain. On the other hand, digital manufacturing such as 3D printers and CNC machines are now lowering manufacturing barriers. Now is the time for crowds to design their ideas on computers, share design files, and produce using digital manufacturing [23]. These days with the help of digital manufacturing and production systems, manufacturers can produce products designed by the crowds. Design crowdsourcing is a key action if companies want to continue to be successful.

* Corresponding author: dhchoi@kookmin.ac.kr

There are many types of crowdsourcing platforms, but their goals are the same: they connect people and use collective wisdom to tackle innovation challenges. The 3D printing and digital manufacturing industries have very close neighbors for crowdsourcing platforms. Design crowdsourcing is not just about acquiring new ideas but also improving and fine-tuning ideas. It provides the ability to collaborate with the broader community at a high level [1].

Meanwhile, the life cycle of various goods has been shortening since consumers' lifestyles and trends keep changing at a fast pace. And bringing new products to market with innovation is one of the significant issues in business research today [2]. Recently, in the fashion industry, as customer needs are diversified and supply lead time has been greatly shortened, designs that immediately reflect the latest trends and fast fashion, which compete with rapid product turnover, are in the spotlight. In addition, as competition between companies intensifies, how quickly and efficiently satisfy customer needs is being emphasized as an important success factor in the fashion industry [3]. As the speed of product design development with short life cycle products is getting faster, enterprises need to secure competitiveness through open innovation that can quickly reflect trendy designs [4]. Design crowdsourcing increases the resources available for new product development. In research, managers said that crowdsourcing design as a resource replenishment strategy could simplify and speed up the new product development process [1].

After all, social technologies raise new sensitivities, seeking to breach organizational walls and instill more collaborative mindsets. The multiplication of knowledge channels drove process improvements such as faster time to market and improved product and service quality [5]. Thus, global companies are applying crowdsourcing as a business model of open innovation to product development and production process [6]. Companies in the fast-fashion sector that deal with trendy products with uncertain demand apply a new concept, crowdsourcing supply chain (CSC, Crowdsourcing Supply Chain), to their business models to respond to the market quickly [7]. The crowdsourced design business model refers to how the user becomes the subject of design by opening the product development and design process to the public. It can share and reproduce the created design [6]. In addition, the incremental value of design crowdsourcing is high when the quality of the idea is low. When the initial quality of raw ideas is high, the company can be better with in-house design [1]. Crowdsourcing is the largest open innovation model that can create value by taking a closer look at the opinions and reactions of potential consumers. Also, it is making another form of partnership among companies and consumers.

In the case of SMEs, it is generally challenging to maintain competition without innovation [8]. Crowdsourcing can secure globally innovative competitiveness without financial burden, which is more effective to small and medium enterprises (SMEs). Using the help of Internet users, the company is delegating tasks in which they have no experience. Otherwise, it can cost them a lot of resources [9].

Despite its enormous potential, the crowdsourcing supply chain has yet to be studied theoretically. Although it is being applied to various business fields such as idea-sharing and problem-solving, research on using design crowdsourcing in the supply chain has not been conducted yet. Companies are concerned about the complexity of the crowdsourcing supply chain as it involves the public or consumers to the supply chain online, unlike traditional supply chains. Involving consumers in the production process of new products does not mean that companies lack creativity. What they want is to be closer to consumers, hear their voices, create unique value for them, and replace the traditional way of thinking [10]. In this paper, we intend to suggest a design crowdsourcing supply chain business model that outsources the design of short-term seasonal products whose value decreases after the season passes. In addition, in the process of outsourcing the design from the crowd by the manufacturer, we intend to analyze the impact of design crowdsourcing on the entire supply chain profit. However, when design crowdsourcing is applied to the supply chain, the relational complexity and problems between each entity arise.

Therefore, we will try to find out which supply chain structure is optimal under design crowdsourcing. We examine crowdsourcing supply chain coordination with the following contracts, wholesale price, buyback, and revenue sharing contract. There are three related research questions addressed in this paper. Firstly, how does design crowdsourcing affect the total supply chain profit? Secondly, which contract can coordinate the design crowdsourcing supply chain? Thirdly, which contracts are worth adopting in it? In the next phases, we perform the numerical analysis to conduct the suggested model.

2 Literature review

2.1 Crowdsourcing

The term crowdsourcing is a combination of crowd and outsourcing, and entities can access experts faster and lower than traditional channels [24]. The concept of crowdsourcing was introduced in 2006 by Jeff Howe in his article "The Rise of Crowdsourcing" published in the Wired Magazine journal [25]. With the advent of Web 2.0, the mass of people, whose skills are close to experts, increased on platforms, and an economic principle appeared that applied collective intelligence to business [26]. Anyone who can connect to the Internet, whether an expert or an expert, can provide solutions and answers for open outsourcing [9]. Unemployed, enthusiastic publics with overflowing talent and creative imagination generate professional-level results beyond amateurs [28]. Web 2.0 applications take full advantage of the inherent benefits of the platform: the software is delivered as a continuously updated service, the more people use it, the better it gets, and consumption and remixing come from multiple sources. While providing its data and a form of service, it allows others to mix, create network effects through "participatory architecture." and go beyond the metaphors of Web page 1.0 to provide a rich user experience [11]. Hence, crowdsourcing has the potential to remarkably transform the business processes by incorporating the knowledge and skills that globally dispersed experts, at shorter cycles and low cost [12].

At the heart of every R&D, there is innovation, and the innovation process is conspicuously similar across industries [13]. If the current trend is based on the principle of high prices for a single product, then the crowdsourcing process and consumer personalization provide consumers with an opportunity to obtain a unique product that perfectly meets customers' needs [9]. Several well-known venture companies have used crowdsourcing as their business model, successfully leveraging crowd participation and collaboration, and this business model is rapidly changing the industry [14]. In addition, product ideas with significant improvement needs may benefit from an iterative process, allowing for improvements to ideas driven by crowdsourcing [1]. Innovation strategies moving from closed to open innovation, from R&D to connection and development, and from competition to cooperation are paving the way for establishing new forms of collaborative production, particularly suitable for SMEs [8]. Therefore, SMEs and global companies can achieve great results by allowing the crowd to perform various activities and functions previously only available to large corporations. Using a crowdsourcing strategy provides companies with essential advantages, such as establishing a closer relationship with consumers and facilitating the optimization of their expectations, without forgetting the potential communication between them [9]. It also allows SMEs to engage in new projects within a complementary approach, allowing them to develop new competencies in the long term [8].

2.2 Seasonal product

Recent trends in consumer personalization, diversity, and technological development are accelerating the change from existing products to new products. Due to this change, many products are classified as seasonal products. Seasonal products have a low residual value in a market with uncertain demand and have a long manufacturing and procurement period and a short sales period [15]. Rapidly growing electronic products and products in the fashion field, where trends frequently change, rapidly decline in value after a certain period has passed. Even social phenomena are classified seasonally. According to this social phenomenon, the demand of medical device manufacturers and pharmaceutical companies changes with the season.

The seasonal product in this paper is about products that always have to deal with changing trends with innovation and new designs. Also, seasonal products are sold only in the peak season due to factors such as seasons and trends and then form data in the form of a time series that calms down again. As such, it is most important to maintain an appropriate inventory for seasonal products that take the form of a time series that rises and falls due to trend fluctuations and seasonal factors. When calculating the initial supply quantity by intuition, there is a shortage of inventory or bad stock, which leads to promotional activities such as discounts to exhaust the bad stock, hindering profit activities and creating a vicious cycle in the supply chain [17]. Demand for seasonal products has a short peak season throughout the year and changes in trends every year, so past sales information is not very useful [18]. In this paper, seasonal products is categorized where demand plummets after the peak season.

2.3 Design crowdsourcing

In design crowdsourcing, any customer and user can also become a designer for a company. Crowdsourcing offers many benefits, especially in the graphic design industry, by providing multiple-choice, creativity, cost, and speedy solutions. The tools used for development, design, and collaboration have undergone a fundamental shift in the past decade. It becomes more powerful and easier to use, even if their prices plummet [22].

Startups selling seasonal products that become unprofitable once a season passes or are out of vogue adopt crowdsourcing for 'designing new products' because they are busy preparing various designs for the next season [8]. As product life cycles are getting shorter, companies need to read customer needs through crowdsourcing quickly. Furthermore, companies are under pressure to produce more innovative products while constrained by internal resources. The needs of modern users are increasingly diversified and advanced because enterprises cannot sufficiently satisfy the needs of these advanced users. It is because enterprises have to pursue profits by producing mass products that satisfy the consumers [23]. On the other hand, upon request, enterprises can easily collaborate with artists and designers to create logos, branding, marketing materials, and more for their business or ideas with design crowdsourcing. When companies know what products or solutions, they want but are looking for executable designs, design crowdsourcing helps make development happen.

The diversity of the crowd makes any activity more attractive. Beyond Prosumers and the Cresumers, the general crowds can now participate in corporate manufacturing activities. The crowd broadens access to a wider variety of competition and ideas and resources than a company can find internally. As the crowd participates from the product planning stage, it is possible to create the maximum effect at the minimum cost by developing products that reflect the needs of consumers but also promoting products from the beginning [19]. Although design requires expertise, many companies outsource freelancers because it is often carried out as a project. The open-minded experts and the system that supports it are essential to realize the public's support rather than a few experts and lead to the right product. Moreover, differentiated print patterns always give fashion manufacturers a competitive advantage in the market [20]. For brands, crowdsourcing is the way of proposing new and original products that will help them organize a brand image to offer unique products, especially in fashion or ready-to-wear products [9]. Companies may be concerned that the open design process will lose control over it, the research interviews show that companies can better control the design process while creating slack for their R&D teams, design through participation in the process, and choosing the right design crowdsourcing platform to iterate with users or designers [1].

There are several companies that utilize design crowdsourcing in their business model. As a representative example of design crowdsourcing, Threadless in the United States is a design-based crowdsourcing platform that makes and sells T-shirts with designs created by the crowd. Anyone can create and upload designs by signing up for the Threadless website. Internet-born design shopping, known as screen-printed graphic T-shirts, was once known (at the time it was thought impossible) as the best example of how to make money on the Internet [29]. The public uploads their designs according to the design guide provided by Threadless. On average, more than 1,000 designs are uploaded a week, and an average of about 50,000 designs are uploaded annually. Selected designers receive 20% of sales as royalties. The fashion companies generally produced T-shirts' print designs using expensive print patterns from internal experts or external freelancers.

Lego, a Danish toy company, collects ideas through the 'LEGO Ideas' platform launched in 2007 and proceeds to commercialization after user voting and screening. 1% of sales is provided as a reward to providers whose ideas are adopted on the crowdsourcing platform [28]. This method allows Lego to get unique product designs for free and understand customers' needs. When users try to promote their product ideas and gain likes, they create more buzz around Lego. This, in turn, adds to the mindshare and serves as free promotion for the brand. Lego transforms consumers into active participants to innovate beyond the traditional mass-market model [14]. After collecting creative ideas from the public through their 'LEGO Ideas' platform, they commercialize products with new designs through customer voting and screening processes.

Designhill, a graphic design market based in New Delhi, created infographics as part of its digital media strategy. Rahul Aggarwal founded Designhill with his brothers. As a startup, Aggarwal understands the need for interactive guides, mini-quizzes, and other content. The platform connects clients and designers all over the world. A client who needs a logo, website, and graphic design launches a contest and chooses from designers' work.

Table1. Examples of Design crowdsourcing platforms

	Description	Company (platform)	Category	Designer's Revenue
Type 1	The third-party platform connects crowds and clients for design crowdsourcing	99 designs	Logo, Web page, product package, and other categories	Prize money
		Crowdspring	Custom design, logo	Prize money
		Designcrowd	Logo, app, t-shirts, art, etc.	Prize money (minimum $200)
		Fiverr	Graphic design, Digital marketing, video, and animation, etc.	Price is set by freelancer
		Cadcrowd	3D design and 3D printing ideas	Prize money

Type 2	Companies utilize design crowdsourcing and sell creativity	Away to Mars	Clothing designs	Profit-sharing (20% for the designer)
		Threadless	Screen graphic designed T-shirts by the crowd	Profit-sharing (20% for the designer)
		Istockphoto	High-quality Images and Illustrations	Approximately 25-45 cents per image, per month
		Spoonflower	Design crowdsourced fabric	10% of the retail price
Type 3	Brands using crowdsourcing for open innovation and new design	LEGO	New Lego product sets	1% of sales
		Unilever	New product design ideas	Prize money
		Localmotors	3D printed car design	Prize money

Source: Own Processing.

We classify the chart according to platform types. And Type1 is the crowdsourcing platform in this research work model.

Type 1. The third-party platform connects crowds and clients for design crowdsourcing

In type1, there are third-party platforms that connect freelance designers to manufacturers. In their platforms, they usually classify freelance designers according to their skills and working experience. They offer cost-effective design solutions, and this kind of platform is popular with small businesses. Access dozens of new design ideas from a diverse pool of talented and capable designers, both professional and amateur, from around the world. Moreover, creative design concepts are available at very affordable prices. So, if companies are looking for a creative logo design, they need to start a contest on the site and enter their design requirements. Soon they will start receiving dozens of design concepts according to their design papers.

Type 2. Companies utilize design crowdsourcing and sell creativity

Companies that utilize design crowdsourcing and sells creativity are mentioned in type2. The platform companies collect new design ideas from freelance designers. This approach may be more common in start-ups, using and trying collaboration solutions for the first time. Crowds participate in a new designing process without prepayment, then after selection, the designer, whose idea selected, gets certain revenue from profit. As you can see, the freelancer designer's position in this section is higher than the manufacturer and platform owner. It is because organizations that adopt a passive approach to managing crowdsourced IP do not care about crowdsourced IP and are less active in limiting liability. Moreover, the designer has the copyright of product design rights acquired by non-exclusive license [21].

Type 3. Brands using crowdsourcing for open innovation and new design

Companies using crowdsourcing for open innovation and new design are classified in type 3. They open projects or contests on their own platform to gather new designs and ideas. Large companies usually facing complex challenges turn to crowdsource to find ideas and possible solutions. This is a collaborative model that prioritizes good ideas and new designs and even rewards freelance designers with prize money. They also use crowdsourcing as a marketing tool to reach out to more creative customers. Employees also gain specific experience and knowledge in the narrow areas where the company is concentrated. In contrast, crowds can attract global participants in various companies, territories, and industries that have their interests and motives.

3 Data and methods – basic order model

In this paper, we consider the supply chain structure of design crowdsourcing as a two-level supply chain consisting of a supplier and a retailer. An independent platform aimed at crowdsourcing serves as connecting freelancer designers and providers. Because of its characteristics, the products in this research work have a short product life cycle (PLC) and produce small quantity batch production of various types and are sold only for a single period. We assumed that the manufacturer produces multiple items of product on the same production line. The supplier is at the retailer's request in the original design manufacturing (ODM) method in which the supplier takes charge of the design and production. The operation process of the crowdsourcing supply chain proceeds through the following procedures in Figure1.

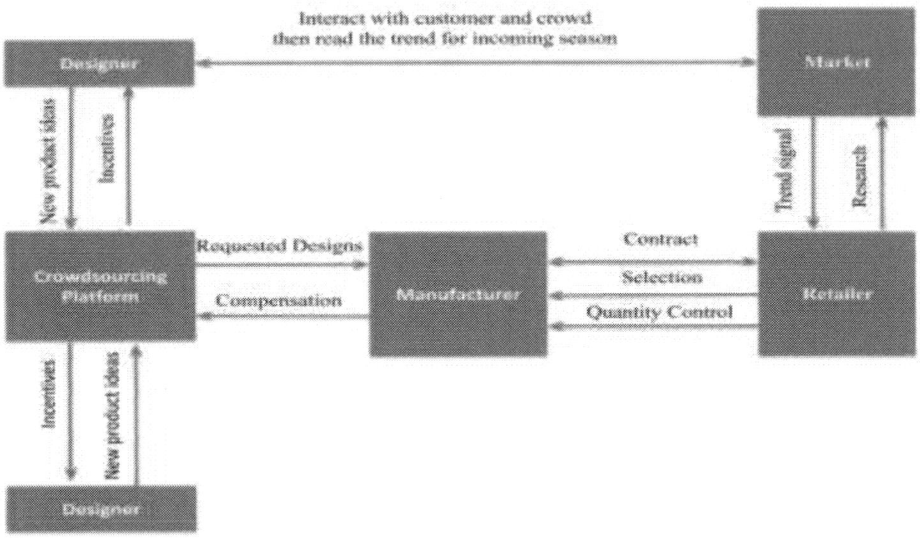

Fig. 1. Design crowdsourcing supply chain process
Source: Own Processing.

4 Results

4.1 Model

There is one manufacturer and a retailer with a crowdsourcing platform for design in the supply chain. The retailer chooses an order quantity before the start of a single season that has a stochastic demand. Let D be in demand during the selling season. ($D > 0$). Let F be the distribution function of demand and f be its density function: F is differentiable, strictly increasing, and $F(0) = 0$. Let $\mu = E[D]$. D follows the independent and identical distribution. The manufacturer rewards crowdsourced designer j with financial incentives, r_j, if a product designed by designer j in the crowdsourcing platform is selected. The unit production cost, wholesale price, and selling price are denoted by c_m, p_w , and p. At the end of the season, any unsold product will be cleared as salvage value v when demand does not exceed sales. Lost sales occur as value ω when demand exceeds sales. $S(q)$ is expected sales, min (q, D), expressed by the following.

$$S(q) = q(1 - F(q)) + \int_0^q xf(x)dx = q - \int_0^q f(x)dx \tag{1}$$

Let $I(q)$ be the expected leftover inventory and $L(q)$ be the lost sales function. At the end of season, each function for two possible cases can be defined by actual demand occurred during the season, $I(q) = (q - D)^+ = q - S(q)$ *and* $L(q) = (D - q)^+ = \mu - S(q)$. Using these terms, the retailer's profits are:

$$\pi_r(q) = pS(q) - p_w q + vI(q) - \omega L(q) \tag{2}$$

In this case, the retailer's optimal order quantity is:

$$q_r^* = F^{-1}\left(\frac{p - p_w + \omega}{p - v + \omega}\right) \tag{3}$$

Meanwhile, crowdsourcing supply chain's profits are:

$$\pi_{sc}(q) = (p - z_j r_j)S(q) - c_m q + vI(q) - \omega L(q) \tag{4}$$

Where $z_j = \alpha\tau_j$, $(\alpha \in (0,1), \tau_j \in [0,1])$. τ_j is a binary variable, taking the value of 1 if a product designed by designer j in the crowdsourcing platform is selected, otherwise, taking the value of 0. Selected designer j is rewarded at rate α based on sales volume.
Total supply chain's optimal order quantity is:

$$q_{sc}^* = F^{-1}\left(\frac{p - c_m + \omega - z_j r_j}{p - v + \omega - z_j r_j}\right) \tag{5}$$

In the next section, we examine the mechanism of crowdsourcing supply chain coordination in the form of several supply contracts with the optimal order quantities of two entities above.

4.2 Crowdsourcing supply chain coordination

The following sequence of events occurs in this coordination. The manufacturer provides the contract to the retailer. The retailer accepts or rejects the contract. Assuming the retailer accepts the contract, the retailer submits the order quantity to the manufacturer. The manufacturer produces after selecting the designer's item on the platform. The manufacturer delivers to the retailer before the selling season. Seasonal demand arises. Finally, payments are made between the companies based on the agreed contract. If the retailer rejects the contract, the game will end.

In design crowdsourcing supply chains, the newsvendor model is not complex, but it is fully to study the coordination mechanism in supply chains. First, which contracts coordinate the design crowdsourcing supply chain? Second, which contracts have sufficient flexibility to allow for any type of the crowdsourcing supply chain's profit? Third, which contracts are worth adopting in it? As a result, contract designers may prefer to propose simple contracts, even if those contracts do not optimize the performance of the supply chain. We examine crowdsourcing supply chain coordination with the following contracts, wholesale price, buyback, and revenue sharing contract. In the next phases, we perform the numerical analysis to conduct the suggested model.

Conclusions

This paper analyzes the design crowdsourcing supply chain in seasonal products. In the basic order model, we established a design crowdsourcing supply chain with retailer, manufacturer, and design crowdsourcing platforms. Supposing the manufacturer produces multiple items of product on the same production line, and the platform has fundamental designs to pick. In order to coordinate the total members of the supply chain, analyzed contracts; wholesale price, buy-back and revenue sharing contracts. The summarized results are; (1) Wholesale price contract cannot coordinate the crowdsourcing supply chain efficiently. (2) Buy-back contract and revenue sharing contract can coordinate crowdsourcing supply chain as reducing the retailer's loss and hedging the risk under volatile price scheme and fluctuating demand. Using the results, we seek the mechanism to fully coordinate all participants of the design crowdsourcing supply chain, designers, platform, manufacturer, and retailer.

For future research, we discuss several types of future research suggestions. First, it is worth investigating the three-level supply chain in design crowdsourcing including manufacturer, retailer, and the crowds with platforms. This could help the optimization more realistic way when crowds get incentives from retailer. Second, it would be interesting to empirically examine how the design crowdsourcing affects seasonal products' supply chain performance.

References

1. B. Allen, D. Chandrasekaran, S. Basuroy, Design Crowdsourcing: The Impact on New Product Performance of Sourcing Design Solutions from the "Crowd". *Journal of Marketing,* **82**, 106-123 (2018)
2. J. Hauser, G. J. Tellis, A. Griffin, Research on innovation: A review and agenda for marketing science. *Marketing Science*, **25**(6), 687–717 (2006)
3. H. Park, K. Park, A Study on the Intelligent Quick Response System for Fast Fashion (IQRS-FF). *Journal of intelligence and information systems*, **16**(3), 163-179 (2010)
4. E. Lee, A Study on the Process Innovation of Fast Fashion -Focusing on Developing Management Algorithm of Zara. *Journal of the Korean society of design culture*, **17**(3), 641-651 (2011)
5. M. Harrysson, D. Schoder, A. Tavakoli, The evolution of social technologies. *McKinsey Quarterly,* 8-12 (2016)
6. H. Park, A Study on Crowdsourcing Design Appearing in Contemporary Fashion Industry. *The Research Journal of the Costume Culture*, **25**(6), 893–912 (2017)
7. J. Li, C. Liu, X. Zeng, N. Zhang, Optimization and Coordination of Crowdsourcing Supply Chain in Fast Fashion Industry. *Mathematical Problems in Engineering* (2018)

8. R. Maiolini, R. Naggi, Crowdsourcing and SMEs: Opportunities and Challenges. In: D'Atri A., Ferrara M., George J., Spagnoletti P. (Eds) *Information Technology and Innovation Trends in Organizations*. Physica-Verlag HD. pp 399-406 (2011)
9. J. Dugout, M. Azouri, J. Decaudin, S. Rochard, Crowdsourcing, Outsourcing to Obtain a Creativity Group. *Arab Economic and Business Journal*, **8**(1/2), 6-15. ISSN 2214-4625 (2013)
10. A. Bujor, S. Avasilcai, L. Alexa, Co-Creation In Fashion Industry: The Case Of Awaytomars. *Annals of the Oradea University: Fascicle Management and Technological Engineering* (2017)
11. T. O'Reilly, What Is Web 2.0: Design Patterns and Business Models for the Next Generation of Software, MPRA. *Communications & Strategies*, 4578 (2007)
12. M. Vukovic, C. Bartolini, Towards a Research Agenda for Enterprise Crowdsourcing. *Leveraging Applications of Formal Methods, Verification, and Validation* 425-434 (2010)
13. C. Terwiesch, W. Xu, Innovation Contests, Open Innovation, and Multiagent Problem Solving. *Management Science,* **54,** 1529-1543 (2008)
14. T. Kohler, Crowdsourcing-Based Business Models: How to Create and Capture Value. *California Management Review*, **57**. 63-84 (2015)
15. J. Li, Y. Zhou, W. Huang, Production and procurement strategies for seasonal product supply chain under yield uncertainty with commitment-option contracts. Intern. *Journal of Production Economics,* **183**, 208-222 (2017)
16. S. Noh, S. Gu, S. Jang, A Study on the Improvement of Inventory Management in the Supply Chain Using Seasonal Demand Forecasting. *Proceedings of the Spring Joint Conference of the Korean Society of Industrial Engineers*, 4179-4205 (2019)
17. H. Park, K. Park, T. Kim, A Computation Model of the Quantity Supplied to Optimize Inventory Costs for the Fast Fashion Industry. *Journal of Society of Korea Industrial and Systems Engineering,* **35**(1), 66-78 (2012)
18. B. Ahn, Sales Forecasting for Inventory Control on Seasonal Fashion Product, *Proceedings of the Korean Operations and Management Science Society Conference*, 953-959 (2002)
19. J. Park, E.Choi, M. Park, Strategy of Crowdsourcing for Open Innovation. Sejong: *Korea Institute for Industrial Economics and Trade* (2016)
20. C. Lee, Roles and Possibilities of Collective Intelligence in Fashion Industry - Centered on Crowd Sourcing. *Fashion Informationa and Techonlogy*, **12**(0), 71-79 (2015)
21. J. de Beer, I.P. McCarthy, A. Soliman, E. Treen, Click here to agree: Managing intellectual property when crowdsourcing solutions. *Business Horizons*, **60**(2), 207-217 (2017)
22. K. J. Boudreau, R. L. Karim, Using the Crowd as an innovation Partner. *Harvard Business Review* [online], Available at: https://hbr.org/2013/04/using-the-crowd-as-an-innovation-partner (2013)
23. N. Kim, J. Yu, Individuals are also emerging as agents of innovation. *LG economic research institute* [online], Available at: http://www.lgeri.com/report/view.do?idx=18381 (2013)
24. M. K. Pratt, C. Gonsalves, *Crowdsourcing* [online], Available at: https://searchcio.techtarget.com/definition/crowdsourcing (2006)
25. https://www.wired.com/2006/06/crowds/
26. M. Fashionn, *Crowdsourcing, gathers the wisdom of the public*

https://doi.org/10.1051/shsconf/202213201015

[online], Available at: https://m.fashionn.com/board/read.php?table=1004&number=5213 (2011)

27. N. Titus, *Threadless: Crowdsourcing Creativity* [online], Available at: https://www.racked.com/2017/11/6/16551468/threadless-t-shirts-ecommerce (2018)
28. L. Hyung-Seok, Lego became a hit by introducing the 'crowd source' model, *Sisa Journal* [online], Available at: https://www.sisajournal.com/news/articleView.html?idxno=178825 (2018)
29. Racket, *What Happened to the Internet's Favorite T-Shirt Company?* [online], Available at: https://www.racked.com/2017/11/6/16551468/threadless-t-shirts-ecommerce (2017)

Semi-supervised generative adversarial networks for anomaly detection

Juan Montenegro[1], and *Yeojin* Chung[1,*]

[1]Department of Data Science, Kookmin University, 77 Jungrungro, Sungbukku, Seoul, 02707 South Korea

Abstract. Advancements in security have provided ways of recording anomalies of daily life through video surveillance. For the present investigation, a semi-supervised generative adversarial network model to detect and classify different types of crimes on videos. Additionally, we intend to tackle one of the most recurring difficulties of anomaly detection: illumination. For this, we propose a light augmentation algorithm based on gamma correction to help the semi-supervised generative adversarial networks on its classification task. The proposed process performs slightly better than other proposed models.
Keywords: Generative adversarial network, Gamma correction, Computer vision, Anomaly detection

1 Introduction

The role of anomaly detection is identifying certain events, observations or data characteristics that diverge from a supposed norm [1]. The video anomalies such as anomalous activities and anomalous entities are defined as the abnormal or irregular patterns present in the video that does not follow the normal trained patterns [2]. An anomaly detection system aims to prompt a signal for a certain activity that diverges from a pattern that is considered normal. On a video, anomaly detection can subtract the anomalies from the video and categorize them according to a certain classification list [3].

Anomalous activities such as fighting, riots, traffic rule violations, and stampede as well as anomalous entities such as weapons at the sensitive place and abandoned luggage should be detected automatically in time. However, video anomalies detection is a complex task due to different reasons such as, the ambiguous nature of searched anomaly, various environmental conditions, the complex nature of human behaviours, and/or the lack of proper datasets.

For the computer vision field, one of the more recurring and greatly challenging problem to address is the one of anomaly detection. Until now, many attempts have been done to tackle the detection of violence on surveillance footage [3]. However, the present investigation focuses on the recent works to detect anomalies on videos based on the biggest crime related video anomaly detection dataset to date: the UCF crime detection dataset from the University of Central Florida of the United States.

* Corresponding author: ychung@kookmin.ac.kr

Numerous methods have been proposed to approach this dataset. Waqas et all.[3] proposed a deep Multiple Instance Learning (MIL) ranking loss. This method of anomaly detection treated the classification of the video as a regression problem. Adding to this, the authors also conducted experiments for the classification of the crimes committed on the videos. For this, they used two models in order to detect actions in videos: 3 Dimensional Convolutional Neural Networks (CNN) and a Tube-Convolutional Neural Network(T-CNN). Other investigation like Dubey et al. [4] proposed a deep-network with Multiple Ranking Measures (DMRMs) framework. Additionally, Doshi et all. [5] suggest the use of a framework called Multi-Objective Neural Anomaly Detector (MONAD). The proposed MONAD consists of a Generative Adversarial Network (GAN) a lightweight object detector (YOLOv3) to extract meaningful features.

The present investigation tries to tackle one of the most recurring obstacles in the UCF dataset: light issues on videos. Sensors on a camera perceive colour and luminance differently than the human eye. For example, as the number of photons captured by a camera's sensor increase, the signal will also increase, which in return makes the image brighter [6]. However, when human eyes perceive double the amount of light, it is processed as only a fraction brighter than the original amount of light met. On the "Light-based data augmentation" section, a method called "gamma correction" is introduced to counter the light issue on the videos of the UCF crime dataset.

To the investigator knowledge, there has not been any efforts to solve this problem, and so, the present study not only focuses on detecting anomalies on videos but also to regulate the low and high luminosity of the videos in question. The model is expected to perform better than previous studies and offer a state-of-the-art detection task. The contributions of the present investigation are the following:

- Use a novel framework composed of an extension of a Generative Adversarial Network for multi-label classification.
- To extensively evaluate the proposed framework on the largest publicly available video anomaly detection data set: The UCF crime detection dataset.
- Use a light-based data augmentation process to give allow the model to perform a better feature extraction for high and low light videos.

2 Related works

Over the recent years, the number of studies focused on anomalous detection algorithms has increased and recent studies have developed techniques, such as image processing [7]. From this point onwards, other studies have directed their attention to action recognition [8, 9], detection of objects [10], tracking objects [11, 12], and other type of visual focused tasks. Consequently, some recent studies [13-16] were proposed mostly using auto-encoder methods and Recurrent Neural Networks (RNN) [17, 18] for more advanced and better performance models on anomaly detection.

Nonetheless, the focus of anomaly detection has been diverted to crime detection. For this area, the quantity of studies is not plenty. The reason for this is the limits of the available datasets, such as, their overwhelming difficulties (obscure, low quality, and short length videos) and small number of samples.

It was until 2018, that the UCF crimes detection dataset was published. This dataset not only presents numerous examples of crimes but also classifies them in 14 crime categories. It, in turn, presents a challenging task of detecting a crime when it occurs and classifying it correctly to the type of crime it is being committed. Due to the recentness of the UCF crime dataset, the number of studies that try to tackle this new repository is limited.

The people responsible of the publication of the UCF crime dataset also carried a study to test the capture of anomalous events. Waqas et al. [3] proposed an anomaly detection algorithm with weakly labeled videos for training. The authors introduced sparsity and smoothness limitations applied in temporal spaces when applying the ranking loss function to improve the detection of anomalies during training. The anomaly detection model was trained using the proposed deep MIL ranking loss. In the proposed approach, the method of anomaly detection was treated as a regression problem. Adding to this, the authors also conducted experiments for the classification of the crimes committed on the videos. For this they used two models: 3-dimensional CNN and T-CNN for action detection in videos.

Dubey et al. [4] proposed a DMRMs framework, which addresses context-dependency using a joint learning technique for motion and appearance features. Afterward, the authors extracted features and fused them for joint learning. However, due to the complexity of these anomalies, using only normal data was not optimal for anomaly detection. First, an extensive and annotated dataset for both normal events and abnormal events is necessary to train the proposed framework since it performs in a semi-supervised manner. A large-scale dataset requirement is an inherent problem in almost all existing visual recognition methods using deep learning. Second, the proposed framework is not robust to certain noises (occlusion, camera jitters, and illumination variations).

Doshi et al. [5] introduces a framework MONAD, which unifies deep learning-based feature extraction and analytical anomaly detection by incorporating two modules. The first module consists of a GAN-based future frame predictor and a lightweight object detector to extract meaningful features. The second module consists of a nonparametric statistical algorithm which uses the extracted features for online anomaly detection. Nevertheless, this proposed framework sacrifices the normalization (preprocessing) of new videos for in time online detection. This makes the framework more prone to mistakes. Additionally, the MONAD framework does not attempt to tackle various noises on videos (blurriness, illumination and sudden movements).

3 Proposed method

3.1 Semi-Supervised GAN

The GAN is a type of generative model that uses deep learning methods to automatically discover and learn recurrent characteristics or patterns on the input data for the GAN model to generate or output artificial examples that could plausibly been on the original dataset [19].

The SGAN model is a branch of the original GAN architecture which is characterized by involving the training of a supervised discriminator and, at the same time, also training an unsupervised discriminator and a generator model [19]. The model serves for both a supervised classification model that performs well to unobserved examples and a generator model that is in charge of outputting realistic examples of images from the main dataset. The discriminator model from the GAN is made to act as a multi-label classification model by classifying the images as real or fake and also, to identify the respective labels of certain images.

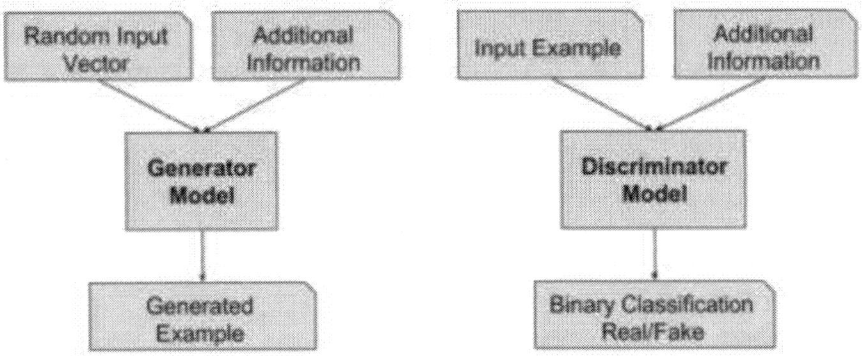

Fig. 1. Example of a Conditional Generative Adversarial Network Model Architecture [18]
Source: J. Brownlee, Generative Adversarial Networks with Python: Deep Learning Generative Models for Image Synthesis and Image Translation. Machine Learning Mastery (2021), pp. 12.

The resulting model is a classifier that can generalize well and offer optimal results on some classic classification tasks (such as MNIST), when trained on a small number of labeled examples, as tens, hundreds, or just one thousand images [19]. As an example, 100 images were taken as input for two models: CNN and the SGAN. The accuracy for the CNN after training was 0.895 but for the SGAN was 0.928. Additionally, the SGAN is capable of generating plausible artificial images of numbers from the MNIST dataset.

The SGAN training algorithm [20] works as follow: For each input of I (number of total iterations), the network will:

- Draw m noise samples $\{z^{(1)}, \ldots, z^{(m)}\}$ from the noise prior $P_g(z)$.
- Draw m examples $\{(x^{(1)}, y^{(1)}), \ldots, (x^{(m)}, y^{(m)})\}$from data generating distribution $P_d(x)$.
- Perform gradient descent on the parameters of the Discriminator (D) and image classifier (C) and then, pass those values to the outputs on the combined minibatch of size $2m$.
- Draw m noise samples $\{Z^{(1)}, \ldots, Z^{(m)}\}$ as for the noise prior $P_g(z)$.
- Perform gradient descent on the parameters the Discriminator (D) and image classifier (C) and then, pass those values to the outputs on the minibatch of size m.

3.2 Light-based data augmentation

A light-based data augmentation approach was used to help the classification process since numerous videos are presented at nighttime which makes the crime undistinctive and difficult to classify. For this, the gamma correction technique was used. The implementation process was applied as follows: The image's pixels intensity has to be converted from the range [0, 255] to [0, 1.0]. Afterwards, the gamma correction is applied on the image by following the equation:

$$O = I^{\left(\frac{1}{G}\right)}$$

Where ***I*** is the input image and G is the gamma value. The output, ***O***, is scaled again to the range [0, 255]. Gamma values of less than one will shift the image towards the darker end of the spectrum while gamma values of greater than on can change the image to appear brighter. A gamma value of one will have no effect on the input image.

3.3 Anomaly detection process with Gamma correction

Fig. 2 illustrates the proposed anomaly detection process. The input images are preprocessed by the gamma correction algorithm and then fed to both the supervised and unsupervised discriminator model. The output from the supervised discriminator model results on the classification of the images into their predicted classes. The output from the unsupervised discriminator is fed to the generative model and its output is a new plausible artificial image.

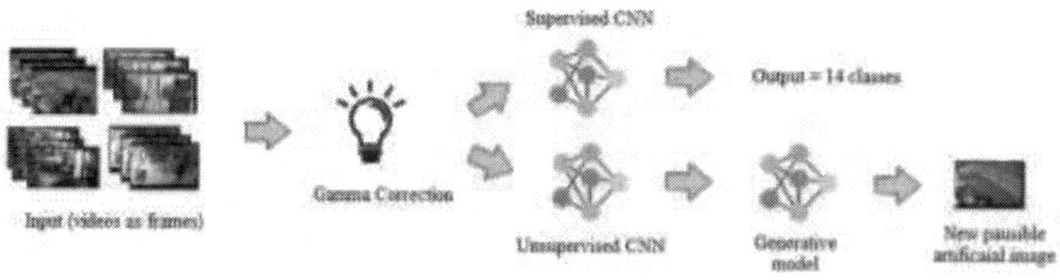

Fig. 2. Flow diagram of the proposed model
Source: Own processing.

4 Dataset

The UCF-Crime dataset was used to evaluate our method. The dataset is composed of a group of lengthy surveillance videos which can be divided in 13 real world anomalies, such as, Abuse, Arson, Road Accident, Burglary, Fighting, Robbery, Shooting, Stealing, Shoplifting, and others. It also includes footage of normal activities. The dataset is 128 hours long.

The UCF crime dataset serves two purposes: In one hand, it can be used as a general anomaly detection dataset for capturing all anomalies and differentiate them from normal events. On the other, it proposes the more challenging task of classifying any of the detected anomalies into the 13 anomalous activities categories.

5 Experiment results

In this work, the trained GANs were designed in two types of variations: The first SGAN was introduced as a normal model without any type of data augmentation process. The second SGAN was modified to deal with the light issues with the gamma correction process mentioned earlier. The output layer of both SGANs is the 14 classes from the different crimes that comprise the UCF dataset.

The videos used for training the models were trimmed in order to contain relevant footage and avoid additional visuals that could increase the noise in the model (credits, police logos, introduction sequences for news sites, etc.). A total of 3,976 extracted frames at an interval of one second were used. From this number, 284 random frames for each type of crime were used.

For both the unsupervised and supervised discriminator models a 2-dimensional CNN was used. The CNN consisted of 4 convolutional layers with 16, 32, 64 and 128 filters. Each layer down sampled the images with a 2x2 stride and used a Leaky-RELU activation function with an alpha of 0.2. Batch Normalization was also used within layers. Also, according to GAN literature and best practices, the kernel initializer corresponds to a random normal distribution generator with the standard deviation of 0.02.

For the generator model, due to the output of the unsupervised discriminator model, the structure follows 4 upsampling layers[†] of Transposed 2-dimentional Convolutions. The properties of the generator model are the same as the supervised and unsupervised discriminator model. The output of the generator model is a 2-dimentional convolutional layer with Tanh activation.

The optimizer used for training both models was Adam (with a learning rate of 0.0005) and the loss functions used were the sparse categorical crossentropy for the supervised discriminator model and the mean squared error for the unsupervised one. The results of the model after 50 epochs are in Fig. 3 and it shows that the model trained is overfitted and producing fluctuating output with poor testing and training accuracy.

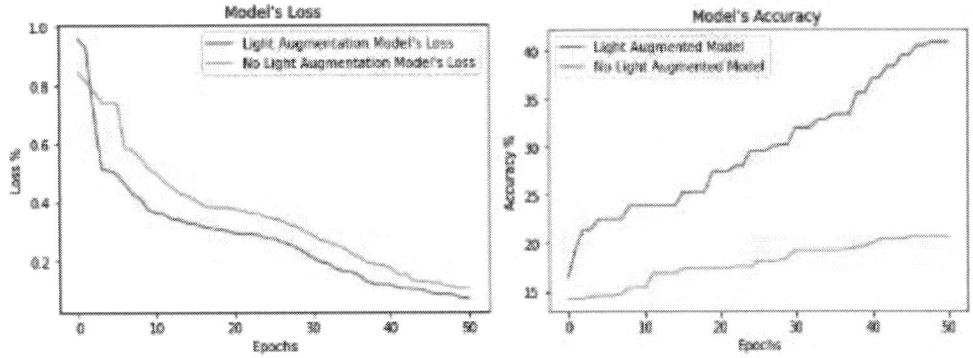

Fig. 3. Different evolutions of the accuracy score and loss of the supervised discriminator as it classifies the training video frames over iterations. The model is tested with and without the light augmentation algorithm
Source: Own processing.

According to the results, a better score on accuracy and lower loss are achieved with the proposed light augmentation algorithm. The gamma correction discussed previously. With this technique, the accuracy score is almost doubled compared to running the model without using it. The final accuracy score with light augmentation is found to be 40.9%. Meanwhile, for a non-light augmented process, the score is only 20.7%.

The light augmented model performs significantly better than the model suggested in [3] and demonstrates the effectiveness of using a gamma correction to change the luminosity inside the video frames for both the anomalous and normal videos. Light adjustment appears to be an indispensable method for a robust anomaly detection system.

Table 1. Accuracy results from [3] with the C3D and T-CNN models compared to the SGANs used

Method	C3D	t-CNN	SGAN (No Light Augmentation)	SGAN (with Light Augmentation)
Accuracy (%)	23.0	28.4	20.7	40.9

Source: Own processing.

Additionally, the use of SGANs also provides the generation of artificial images of crimes. Due to the low number of epochs the model was trained on, the images are not as convincing as they could be. Nonetheless, on the final epochs, the generated images have

[†] An upsampling layer, in this case and UpSampling2D layer, is used to double the dimensions of the input image. [19]

some small characteristics that could resemble those of real crimes images. For example, on epoch 50 there is a resemble to a frame of an arson video.

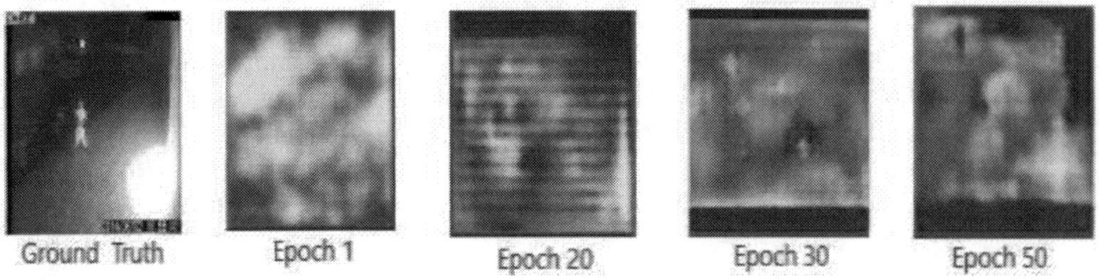

Fig. 4. Pictures generated by the generator model from the SGAN
Source: Own processing.

6 Discussions

A SGAN with a light augmentation method based on gamma correction is proposed to detect and classify real world anomalies in surveillance videos according to the type of crime that is occurring on the video. The results of the light augmented SGAN on the UCF crime detection dataset show that the anomaly detection process performs significantly better than other proposed methods. Moreover, the model demonstrates the usefulness of correcting the light on the videos.

Due to computational limitations, the model was trained only for 50 epochs. Due to this fact and the complexity of the dataset, the accuracy of both models is bound to be low. Therefore, and taking into consideration past investigations, is advised that for future implementations and/or experiments, is recommended that the model should be run for more than the number of epochs used in this investigation.

References

1. I. Cohen, Anodot. [online], Available at: https://www.anodot.com/blog/what-is-anomaly-detection/(2020) (2020)
2. R. Nayak, S. Das, U. A. Pati, Comprehensive review on deep learning-based methods for video anomaly detection. *Image Vision Computer* (2021)
3. S. Waqas, C. Chen, S. Mubarak, Real-world Anomaly Detection in Surveillance Videos. *Computer Vision Foundation* (2019)
4. S. Dubey, A. Boragule, J. Gwak, M. Jeon, Anomalous Event Recognition in Videos Based on Joint Learning of Motion and Appearance with Multiple Ranking Measures. *Appl. Sci.* (2021)
5. K. Doshi, Y. Yilmaz, Online anomaly detection in surveillance videos with asymptotic bound on false alarm rate. *Pattern Recognition* (2021)
6. A. Rosebrock, Pyimagesearch, [online], Available at: https://www.pyimagesearch.com/2015/10/05/opencv-gamma-correction/ (2015)
7. K. Hara, H. Kataoka, Y. Satoh, Can Spatiotemporal 3D CNNs Retrace the History of 2D CNNs and ImageNet? *IEEE Conference on Computer Vision and Pattern Recognition* (CVPR), Salt Lake City, UT, USA, 18–23, pp. 6546–6555 (2018)

8. L. Wang, Y. Xu, J. Cheng, H. Xia, J. Yin, J. Wu, Human Action Recognition by Learning Spatio-Temporal Features With Deep Neural Networks. *IEEE Access, **6***, pp. 17913-17922 (2018)
9. J. Yu, D. Y. Kim, Y. Yoon, M. Jeon, Action Matching Network: Open-set Action Recognition using Spatio-Temporal Representation Matching. *Visual Computer, 36,* pp. 1457-1471 (2020)
10. S. Ren, K. He, R. Girshick, J. Sun, Faster R-CNN: Towards Real-Time Object Detection with Region Proposal Networks. *IEEE Transactions on Pattern Analysis and Machine Intelligence, **39***(6), pp. 1137-1149 (2017)
11. S. Sun, N. Akhtar, H. Song, A. Mian, M. Shah, Deep Affinity Network for Multiple Object Tracking. *Journal of Latex Class Files*, ***13***(9) (2018)
12. G. Ciaparrone, F. L. Sa ́nchez, S. Tabik, L. Troiano, R. Tagliaferri, F. Herrera, Deep learning in video multi-object tracking: A survey. *Neurocomputing*, 381, pp. 61-88 (2020)
13. J. Yu, K. Yow, M. Jeon, Joint Representation Learning of Appearance and Motion for Abnormal Event Detection. *Computer Science, Machine Vision and Applications* (2018)
14. D. Abati, A. Porrello, S. Calderara, R. Cucchiara, Latent Space Autoregression for Novelty Detection. *IEEE Conference on Computer Vision and Pattern Recognition* (2019)
15. L. Wang, F. Zhou, Z. Li, W. Zuo, H. Tan, Abnormal Event Detection in Videos Using Hybrid Spatio-Temporal Autoencoder. *IEEE International Conference on Image Processing* (2018)
16. D. Xu, Y. Yan, E. Ricci, N. Sebe, Detecting anomalous events in videos by learning deep representations of appearance and motion. *Computer Vision Image Understanding, 156,* 117-127 (2017)
17. Y. S. Chong, Y. H. Tay, Abnormal Event Detection in Videos Using Spatiotemporal Autoencoder. *Computer Vision and Pattern Recognition* (2017)
18. J. R. Medel, A. Savakis, Anomaly Detection in Video Using Predictive Convolutional Long Short-Term Memory Networks. *Computer Vision and Pattern Recognition* (2016)
19. J. Brownlee, Generative Adversarial Networks with Python: Deep Learning Generative Models for Image Synthesis and Image Translation. J. Brownlee. *Machine Learning Mastery* (2021)
20. A. Odena, Semi-Supervised Learning with Generative Adversarial Networks. *Machine Learning* (2016)

A Study on the Mechanism of Social Robot Attitude Formation through Consumer Gaze Analysis: Focusing on the Robot Face

Sangjip Ha[1] , *Eun-ju* Yi[1], In-jin *Yoo*[1], and *Do-Hyung* Park[1,*]

[1]Graduate School of Business IT, Kookmin University, Seoul, The Republic of Korea

Abstract. This study intends to utilize eye tracking for the appearance of a robot, which is one of the trends in social robot design research. We suggest a research model with the entire stage from the consumer gaze response to the perceived consumer beliefs and further their attitudes toward social robots. Specifically, the eye tracking indicators used in this study are Fixation, First Visit, Total Viewed Stay Time, and Number of Revisits. Also, Areas of Interest are selected to the face, eyes, lips, and full-body of a social robot. In the first relationship, we check which element of the social robot design the consumer's gaze stays on, and how the gaze on each element affects consumer beliefs. The consumer beliefs are considered as the social robot's emotional expression, humanness, and facial prominence. Second, we explore whether the formation of consumer attitudes is possible through two major channels. One is the path that the consumer beliefs formed through the gaze influence their attitude, and the other is the path that the consumer gaze response directly influences the attitude. This study made a theoretical contribution in that it finally analysed the path of consumer attitude formation from various angles by linking the gaze tracking reaction and consumer perception. In addition, it is expected to make practical contributions in the suggestion of specific design insights that can be used as a reference for designing social robots.
Keywords: Social Robot, Gaze Analysis, Eye Tracking, Robot Design, Attitude Formation

1 Introduction

Robot is coming to the life of human. Robots initially being installed in the factory for industrial use are being transformed for personal use and spread to the daily life of diverse users. Researchers are investigating how the life with robots would influence to human being from various perspective and background. Personal Robot is largely divided into two: One is Service Robot, which is a type of tools for movement, maintenance, cleaning etc. and the other is Social Robot, which is distinguished from that of former in that it is regarded as a companion or family for its main purpose is to communicate at a high level including emotional exchange [1].

* Corresponding author: dohyungpark@kookmin.ac.kr

Research on Social Robots are specifying them as Social Agent or Social Entity able to move autonomously and verbal conversation and focusing on the interaction between the users and robots (Human-Robot Interaction: HRI). More specifically, studies on social robot design are classified as surface and appearance design, functional and role-based design, and interaction design including verbal-nonverbal expression.

Social robots, which are destined to support human's well-being, are equipped with face, body, leg(s), and arm(s) for better physical help by moving like human and easier interaction by showing their emotion and intends like that of human. Certainly, people showed intimacy and stronger anthropomorphism as robot's appearance gets closer to human. However, researchers also found out that this preference stops and decline dramatically at a certain level for people experienced unpleasant, repulsion and abhorrence [2]. This phenomenon called Uncanny Valley represents that the appearance of robots is one of the key elements in designing robots for it has great impact on people's attitude. Therefore, designing the robot's appearance needs new approach which is not merely drawing the external features but regard it as one of the key factors in determining user's attitude.

To elucidate the mechanism of use's attitude on social robot, we designed an experiment of looking at the picture of social robots on screen, collected the user's gaze tracking data with eye-tracker and surveyed to ask about the attitude to each robot. The tracking indicators used for this study is the number (of times) of the fixed view (Fixation), time (in seconds) to the first visit (First Visit), duration (in seconds) of the totally stayed (Total Viewed), and number (of times) of the visits (Revisits). Segmented AOI (Areas of Interests) is face, eyes, lips and full-body. Collected data from survey is design evaluation on each social robot's emotional expression, humanness, and facial prominence and attitude as a dependent variable.

The theoretical contribution of this research is that we have connected 2 different data between gaze tracking reaction and consumer perception and finally analysed the path of consumer attitude formation with diverse perspective. The result of this research provides specific design guideline in designing the social robot at work.

2 Theoretical backgrounds

2.1 Social Robot

The first robot materialized by scientific technology was installed and fixed on the automobile assembly line to replace humans' simple and repeatable labour force. Nowadays we can find robots are at home, in room and on desk. Unlike industrial robots, they were born to serve person [1]. They work either as a type of material equipment in charge of maintenance, cleaning or as a type of companion or as a type of companion or family members in charge if safe, entertainment and emotional well-being. The latter type of robot is called as Social Robot, which is capable in verbal or non-verbal communication and interact as humans do [3, 4]. The qualification of social robot mentioned by prior studies is that it has the proper embodiment for its use, the feeling of living creature, face recognition function, its own reaction to each situation and motion that is designed to be easily understood by users [1]. Also studies on the guideline in designing social robots suggest that it gratifies the user's expectation, communicate verbally or non-verbally and comply with human's social norms [5].

Meanwhile as conceptual social robots become invented and more and more social robots are being commercialized out of laboratory, designing has become one of the most significant tasks. Nowadays designing means not only drawing visual elements but also creating functional factors. Specifically, the former is modelling, formative and artistic area and applied in producing the surface of robot. The latter is related with functional planning to fulfil user's purpose: planning itself, planning process and status that the planned elements are installed [6].

2.2 Social Robot Design

We classified 3 types of research in social robot design. First is the surface design, which is in artistic area and deals with human-likeness and aesthetic balance of the robot's face, body, arm(s), and leg(s). Most representative jargon in this field is the 'Uncanny Valley' phenomenon which means that the preference for a robot increases as robot's human likeness increases until it reaches at a certain likeness level and paradoxically falls significantly down after the certain likeness level [2].

Second is the functional and role-based design to meet user's needs. Social robot may have its unique and special tasks as intended for example elderly care [7], autistic child screening [8] and therapy [9-a], children education [9], and so on. Also, it is designed to function in the emotional areas as a robot counsellor and social areas to be a good member in the Human-robot-team.

Finally, third field is the designing the engagement between robot and users. Social robot has diverse means to induce interaction with users. It fits out with both verbal and non-verbal communication including facial expression, body motion, effect sound. Unlike the chatbot service which allows text or tags only or AI speakers which allows sounds only, robot has its presence, and this embodiment increments user's social benefit a lot [11]. It may dance or hug for user's pleasure and take or lift user' body in accordance with user's wish. Researchers have commented the elements to consider for a social robot to improve user's gratification such as softness, size, temperature, movement speed, pressure, voice-tone (Friend-like vs. Engineer-like), voice pitch and nudging [12, 13].

2.3 The Effect of Social Robot Design on Consumer Attitudes

There is a difference between industrial robots and social robots. While the appearance design of industrial robots emphasizes the performance and functional aspects of robots, social robots emphasize the cognitive and emotional exchanges through interaction with humans [14]. Furthermore, the appearance design of such social robots affects consumers, and consumers may change their evaluations and attitudes toward robots [15]. Therefore, when considering social robot design elements, individual perception or psychological characteristics should not be overlooked.

As representative factors, "Robot face", "anthropomorphism", and "emotional expression" affect the attitude of consumers. First, the face design of social robots is used as a means of basic communication with humans, and it can also affect the attitude of consumers [16]. For example, in previous studies, preference for the appearance of social robots differs depending on the proportion and type of face, and there is also a study that reported that the more a humanlike face is designed, the greater the preference of consumers [17].

Second, it is a design about the anthropomorphism of social robots. anthropomorphism refers to giving human characteristics to non-human objects such as animals and inanimate objects [14]. In particular, the anthropomorphism trend can be mainly confirmed in robots that require human-like shapes and functions, is applied in various ways in relation to physical shape, motion, and interaction when considering robot design [18]. This level of anthropomorphism affects consumer preferences. For example, it has been found that consumers expect robots used in public services to behave similarly to actual service workers, and the more the verbal interaction is similar to humans, the more consumers prefer [19].

Third, it is a design related to the emotional expression of social robots. For the expression of social robots, non-verbal expressions such as gestures, facial expressions, intonations, gaze directions, and postures are used for effective interaction with users [20]. In particular, when expressing emotions through robot movements, consumers may perceive that the robot has emotions. In other words, non-verbal expressions of robots play an important role in forming social relationships with robots [21]. Furthermore, in a study that considered the emotional expression and expression of the robot, the evaluation of consumers was different depending on the shape of the pupil, which is an important medium for conveying emotions [22].

In previous studies, the factors that can influence the consumer's evaluation of social robot design were subdivided and the influence of each factor was explored. However, these studies have limitations in the objectivity of the methodology because they were conducted based on users' subjective questionnaires and interviews. In other words, objective responses of consumers to social robot design were not collected, and precise measurement tools were not used. Therefore, in this study, the factors affecting the attitude of consumers and the mechanism of forming the attitude are to be verified using objective indicators measured by eye tracking.

2.4 Eye Tracking

Eye tracking identifies the point where a person's gaze has stayed, and it refers to the process of measuring the relative movement of the eyes based on the position of the head This movement of the gaze can be explored through the eye tracking device, and it can be observed by visualizing it based on the data derived through the eye tracking. In particular, research has been mainly conducted to designate an Area of Interest (AOI) and to explore where the focus is mainly on among the areas of interest. Eye tracking is being used in various fields, among which there is a study related to social robots. First, a study that confirmed the interaction between a robot character and a person report that the human gaze has a similar pattern when dealing with a robot and a real person [23]. In other words, when a person interacts with a robot, the result is that the gaze stays in the mouth for a long time as when interacting with a person. Meanwhile, to study the design of the robot, an eye tracking device and an EEG device were also used. As a result of the study measured through two devices, it was reported that the features of the head and face were more important than the body organs such as arms and legs of the robot in the appearance of the robot [24].

Meanwhile, eye tracking utilizes various measuring indicators in research. In this study, we tried to use the indicators that are mainly used among them. The first is the 'fixation', which is information about how long the user's gaze stayed on the object. Fixation is usually the most important indicator used because it is possible to confirm which specific area the user is watching, and to objectively explore the user's attention or interest [25]. The fixation collected by the eye tracking device can be subdivided into fixed point coordinates as well as fixed time, changed time, total time the gaze stayed, the distance from the previous fixed point, and movement speed.

The second is the gaze path. It is an indicator of the overall pattern of gaze movement on the object. The gaze path has the advantage of informing a comprehensive gaze movement pattern as the path the gaze has moved is visualized and expressed according to time sequence. In particular, the gaze path is mainly used in design fields such as marketing, product, and outdoor advertising [26].

In summary, in this study, the fixation and the gaze path were subdivided and collected with an eye tracking device. "Fixation" intends to utilize "Number of the fixed view (Fixation)", "Time (in seconds) to the first visit (First Visit)", and "Duration (in second) of the totally stayed (Total Viewed)". In addition, we want to utilize the "Number of the revisits(revisits)", which is data reconstructed to reflect "gaze path" as a quantitative indicator.

2.5 Attitude Formation Mechanism Based on Consumer Gaze Response

This study intends to analyse it by composing it as one research model. This study deals with the relationship between consumer gaze response, perceived beliefs about social robots, and attitudes toward social robots. Specifically, first, it checks which element of the social robot design the consumer's gaze stays on object, and how the gaze on each element affects the consumer's beliefs. Since the face of the social robot has been treated as important in the previous research, we would like to divide the face of the social robot into eyes and mouth in detail and compare it with the whole body. The consumer's beliefs considered the social robot's emotional expression (Expression), humanness (Human-Like), and facial prominence (Facial-Highlighted). Second, we want to check whether it is possible to form a consumer's attitude through two main ways. One is the way in which the beliefs of consumers formed through the gaze influence the attitude, and the other is the direct way in which the gaze of the consumer affects the attitude. Instead of considering all beliefs about social robots, this study tried to confirm the influence of beliefs not covered in this study through the direct influence path of the gaze. The research model proposed in this study is summarized and expressed in <Figure 1>.

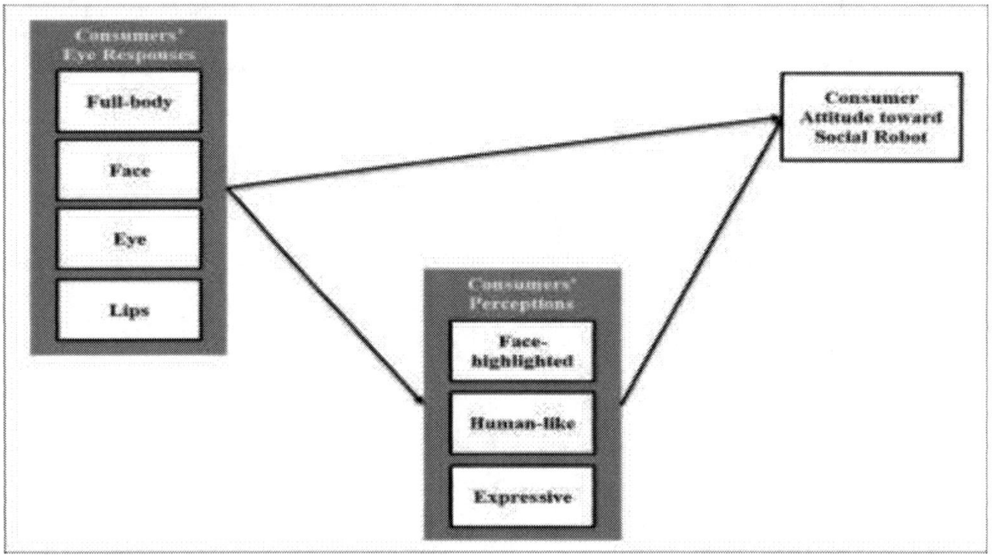

Fig. 1. Research Model Design
Source: Own processing.

3 Research process

3.1 Experiment Participant

This study was conducted with undergraduate and graduate students participating in UROP (Undergraduate Research Opportunities Program) classes at K University in Seoul. A total of 32 participants were recruited and respectively consisted of 17 males (53.1%) and 15 females (46.6%), ensuring a relatively equal sex ratio between men and women. The ages of the participants respectively ranged from their 20s to 40s, with 94% in their 20s, 3% in their 30s, and 3% in their 40s, with an average age of 25.78.

3.2 Measurement Tool

The eye tracking measurement tool was measured using Gazepoint's 'GP3' device. The device recognized the pupil shape of the subject through the infrared sensor of the camera and recorded the central position of the pupil as a 60Hz gaze response per second through image analysis software. The GP3 device was placed under the monitor of the computer where the subject was outputting the stimulus, and the subject's gaze was measured.

Fig. 2. Eye Tracker (Gazepoint's 'GP3')
Source: Gazepoint official website

3.3 Experimental Stimuli

The experimental stimuli used in this study are social robots reminiscent of human appearance and characteristics, and these robots are currently on the market or are known to the public. In order to select a robot to be used in this study among various robots, a pilot experiment was performed. In the pilot experiment, various social robots were presented to 8 college students, and among them, the robot that showed the best emotional expression was selected. However, in this study, factors related to the function of the robot were not considered. Therefore, as a result of pilot experiment focusing on the design of the robot, four types of robots were selected: 'Zenbo', 'Nao', 'Mero-S', and 'Mero-3'.

Fig. 3. Social Robot Stimuli
Source: The official website of Zenbo(1st), Nao(2nd), Mero-S(3rd), and Mero-3(Final Picture)

3.4 Experimental Process and Stimulus Presentation

This study was informed to the participants before the experiment that it was an experiment related to the design evaluation of the social robot so that the experiment participants could focus on the design of the social robot. Therefore, information such as social robot function or technical level was not provided in the experimental stimulus. The specific experimental process is as follows.

The monitor on which the stimulus is presented is 1920 X 1080 size and conforms to the 24-inch standard. The distance between the participant and the monitor was about 50cm to 6cm, and the stimuli were presented to the participants for total 10 minutes. At this time, 'Calibration' was performed to adjust the gaze of the participants before the stimulus was presented. And before the stimuli were presented, in order to control the effect of the participant's gaze on the experiment, a pre-prepared gaze point was first presented, and the starting point of the gaze was fixed. In addition, in order to control the background other than the robot presented, all backgrounds were erased and treated as white.

On the monitor, the stimulus was placed in the center so that the upper and lower and left and right intervals were the same. However, since the size of the actual social robot is different, the image size of the stimulus was adjusted to be the same. As a result, each robot occupies 25% of the area on the monitor and has a size of width(12.5cm) X length(22.5cm). Therefore, participants look at 4 robots on one screen for 30 seconds.

3.5 Measurement

The gaze responses of the participants collected through the experiment of this study were Fixation, First Visit, Total Viewed, and Revisits. Specifically, it is as follows. First, Fixation is a response to how long the participant stayed focused on a specific area. In this study, it was judged that fixation appeared when the gaze stayed for more than 0.1 seconds. Second, First Visit refers to the time it takes for the participant to first look at a specific area. Therefore, First Visit can be used as an indicator of which part of the stimulus the participant wants to see first [27]. Third, Total Viewed is a response to how long the participant's gaze stayed on a specific area. Total Viewed is known as a measurement that can grasp the overall interest of the participants and can discover key features in design [25]. Finally, Revisits refers to the number of times the gaze deviated from a specific area and then returned again. Therefore, it is possible to check the gaze path and attention level of the participant through Revisits [28].

In this study, the face, eyes, and mouth, which are external designs related to the emotional expression of social robots, were designated as main Areas of Interest (AOI). Furthermore, in order to evaluate the overall design of the social robot, the whole body of the robot was also designated as a separate AOI.

In this study, four variables were collected as variables that reflect consumer beliefs about social robots: 'attitude toward robots', 'emotional expression, 'facial prominence', and 'anthropomorphism'. Specifically, it is as follows. First, the attitude toward the robot was measured with the question 'I have a good feeling for the robot' [17]. And, as for the emotional expression, it was checked whether the participant could recognize the robot's expression with the question 'The robot's expression shows well' [29]. Facial prominence was measured as 'the robot's face stands out' [30], and anthropomorphism is a question of 'the presented robot feels like a human', which attempts to reflect the degree of anthropomorphism of the robot [31].

4 Expected conclusions

This study approached the appearance of the robot, one of the various design research flows of social robots, using eye tracking. To verify this, a research model for design evaluation of social robots was constructed by organically linking participants' gaze tracking indicators measured from various areas of interest of social robots, consumer beliefs about social robots, and consumer attitudes toward social robots. Specifically, we proposed and validated a mechanism that a user's attitude toward social robots can be formed through consumer gaze response (direct path) and consumer beliefs (indirect path). This is expected to provide specific insights related to social robot design.

The theoretical and practical implications expected through this study are as follows. As a theoretical implication, first, it can be suggested that an individual's beliefs about social robots are related to gaze response. Second, the target of the consumer's gaze response was extended to the robot's face, subdivided elements, and the entire body. Through this, more detailed and wide range of gaze response data can be reflected in the research model. Finally, the path through which consumers' attitudes toward social robots are formed can be divided into a direct path through gaze response and an indirect path through consumer beliefs.

As a practical implication, first, this study emphasizes the importance of the face, especially the eyes and lips, in the design of social robots. People's evaluation of social robots is similar to evaluating real people. Therefore, it is necessary to include factors that are considered important while evaluating the appearance of a real person in the design of the social robot appearance. Second, the attitude of consumers toward social robots is not formed only by evaluation of one factor, but is formed through consumer beliefs considering various factors in a complex way. Therefore, in the design of social robots, it is necessary to clearly define the value that the robot wants to deliver to consumers, and to derive factors suitable for the value so that the optimal design can be achieved. Lastly, it suggests the possibility that scientific experimental tools such as eye tracking can be utilized in design. Although only eye tracking was used in this study, if various measurement tools such as EEG analysis or skin conduction response are used later, it will be able to contribute to a more scientific and objective design.

Acknowledgements

This research was supported by the BK21 FOUR (Fostering Outstanding Universities for Research) funded by the Ministry of Education (MOE, Korea) and National Research Foundation of Korea (NRF).

References

1. C. Breazeal, Toward sociable robots. *Robotics and autonomous systems*, **42**, 3-4, 167-175 (2003)
2. M. Mori, Bukimi no tani [the uncanny valley]. *Energy*, **7**, 33-35 (1970)
3. T. Fong, I. Nourbakhsh, K. Dautenhahn, A survey of socially interactive robots. *Robotics and autonomous systems*, **42**, (3-4), 143-166 (2003)
4. F. Hegel, C. Muhl, B. Wrede, M. Hielscher-Fastabend, G. Sagerer, Understanding social robots. *IEEE Trans.* In *2009 Second International Conferences on Advances in Computer-Human Interactions*, pp. 169-174 (2009)

5. C. Bartneck, J. Forlizzi, A design-centred framework for social human-robot interaction. In *RO-MAN 2004. 13th IEEE international workshop on robot and human interactive communication*, pp. 591-594 (2004)
6. K. Kim, D. Park, Design Evaluation Mode Based on Consumer Valuesl: Three-Step Approach from Product Attributes. *Perceived Attributes, to Consumer Value," Intelligence Information Research*, **23**(4), 57-76 (2017)
7. M. A. Salichs, Á. Castro-González, E. Salichs, E. Fernández-Rodicio, M. Maroto-Gómez, J. J. Gamboa-Montero, M. Malfaz, Mini: A New Social Robot for the Elderly. *International Journal of Social Robotics*, 1-19 (2020)
8. A. A. Ramirez-Duque, T. Bastos, M. Munera, C. A. Cifuentes, A. Frizera-Neto, Robot-Assisted Intervention for children with special needs: A comparative assessment for autism screening. *Robotics and Autonomous Systems*, **127**, 103484 (2020)
9. H. Chen, H. W. Park, C. Breazeal, Teaching and learning with children: Impact of reciprocal peer learning with a social robot on children's learning and emotive engagement. *Computers and Education*, **150**, 103836 (2020)
10. J. Wirtz, P. G. Patterson, W. H. Kunz, T. Gruber, V. N. Lu, S. Paluch, A. Martins, Brave new world: service robots in the frontline. *Journal of Service Management*, **29**(5) (2018)
11. G. McLean, K. Osei-Frimpong, Hey Alexa.. examine the variables influencing the use of artificial intelligent in-home voice assistants. *Computers in Human Behavior*, **99**, 28-37 (2019)
12. J. Borenstein, R. Arkin, Robotic nudges: the ethics of engineering a more socially just human being. *Science and engineering ethics*, **22**(1), 31-46 (2016)
13. J. Borenstein, R. C. Arkin, Nudging for good: robots and the ethical appropriateness of nurturing empathy and charitable behavior. *Ai and Society*, **32**(4), 499-507 (2017)
14. B. R. Duffy, Anthropomorphism and the social robot. *Robotics and autonomous systems*, **42**(3-4), 177-190 (2003)
15. M. L. Walters, D. S. Syrdal, K. Dautenhahn, R. Te Boekhorst, K. L. Koay, Avoiding the uncanny valley: robot appearance, personality and consistency of behavior in an attention-seeking home scenario for a robot companion. *Autonomous Robots*, **24**(2), 159-178 (2008)
16. M. Blow, K. Dautenhahn, A. Appleby, C. L. Nehaniv, D. C. Lee, Perception of robot smiles and dimensions for human-robot interaction design. *IEEE Trans*. In *ROMAN 2006-The 15th IEEE International Symposium on Robot and Human Interactive Communication*, 469-474 (2006)
17. J. M. Park, K. S. Koo, S. S. Hong, The Study on the Proportional Preference for Humanoid Robot. *Journal of Digital Design*, **14**(2), 725-734 (2014)
18. J. Fink, Anthropomorphism and human likeness in the design of robots and human-robot interaction. In *International Conference on Social Robotics*, pp. 199-208. Springer, Berlin, Heidelberg (2012)
19. B. S. Kim, S. I. Kim, User Preference for the Personification of Public Service Robot. *Journal of Digital Convergence*, **18**(2), 361-366 (2020)
20. C. Breazeal, Emotion and sociable humanoid robots. *International journal of human-computer studies*, **59**(1-2), 119-155 (2003)
21. S. Kim, D. Oh, The Kinesics for emotional expression in the household social robot - Focus on the head movement of the robot. *The Korean Society of Science & Art*, **35**, 69-82 (2018)

22. D. S. Park, Y. H. Pan, A Study on User's Perception and Preference Based on Emotions Expressed through The Eyes of a Social Robot. *Journal of Communication Design*, **67**, 24-35 (2019)
23. S. Jang, H. K. Cho, Analysis of User's Eye Gaze Distribution while Interacting with a Robotic Character. *Journal of Korea Robotics Society*, **14**(1), 74-79 (2019)
24. Y. Liu, F. Li, L. H. Tang, Z. Lan, J. Cui, O. Sourina, C. H. Chen, Detection of humanoid robot design preferences using EEG and eye tracker. *IEEE Trans*. In *2019 International Conference on Cyberworlds*, 219-224 (2019)
25. M. Y. Choi, A Study on the Method of Eye Tracking Analysis Based on the Properties in Visual Perception of User -With Emphasis on the Development of Analysis-Framework for Product Design. *Journal of Korean Society of design science*, **16**(4), 197-206 (2003)
26. E. S. Seo, The Suggestion for the Design of Eye Tracker to Promote the Study on the Gaze Tracking Interface. *Journal of Cultural Product and Design*, **50**, 145-152 (2017)
27. R. J. Jacob, and K. S. Karn, Eye tracking in human-computer interaction and usability research: Ready to deliver the promises. In *The mind's eye*, 573-605 (2003)
28. J. H. Kim, J. Y, Kim, Characteristics of Visual Perception for the Bluetooth Headset Image Based on Eye-Tracking Experiment. *Journal of The Korean Society of Living Environmental System*, **24**(3), 433-445 (2017)
29. S. W. Jeong, Designing Facial Expressions of an Educational Assistant Robot by Contextual Methods. *Archives of Design Research*, **26**(2), 409-435 (2013)
30. Y. J. Oh, Y. S. Shin, J. H. Lee, J. W. Kim, A Study on Interaction Design of Companion Robots Based on Emotional State. *Journal of Digital Contents Society*, **18**(7), 1293-1301 (2017)
31. D. Kim, S. Choi,The Effect of Expressions of Self-consciousness in Social Robots during Social Attraction. *Journal of KIISE*, **46**(7), 653-663 (2019)

Never have a crisis like COVID-19

Yongseok Jang[1], and *Woojin* Lee[2*]

1 California State University San Bernardino, 5500 University Parkway, San Bernardino 92407
2 Kookmin Universsity, 77, Jeongneung-ro, Seongbuk-gu, Seoul, South Korea

Abstract. Current circumstances caused by COVID-19 pandemic pose a unique challenge to businesses. How could businesses survive? What strategic options do businesses have when they effectively engage such a challenge? Delving into the inquiry, we explore and investigate different strategic approaches to measure their effectiveness. For the study's explorative and investigative purposes, our investigation involves a set of tests at both business and corporate levels. Using 224 responses from the leaders and key individuals of Korean startups, we tested the effectiveness of strategic options using the notion of ambidexterity and alliances, and mergers and acquisitions. Using the ambidexterity lens, we found exploitation and exploration indirectly affect firm performances and are still effective under COVID-19. We also found "reinforcing the relationship with existing alliances," a corporate-level approach, to be effective.
Keywords: Ambidexterity, COVID-19, crisis management, corporate strategy

1 Introduction

In times of external shock, firms face different types of crises, such as the reduction of consumer surplus, mostly stem from a combination of loss of customer and reflected as a loss of revenue, or reduction of producers' surplus, which is derived by a few factors such as loss of suppliers and loss of key employees, etc. Current circumstance caused by COVID-19 pandemic poses a unique challenge to businesses. First, it is a combination of natural and economic crisis, compared to 2008 financial crisis which was caused by a massive disruption of financial sector. Second, for the first reason, policy measures countering health risks, such as banning public events, lockdown, and shutdowns, followed and consequently put business at risks of losing customers and supply/distribution channels. In this case, businesses face a dual crisis. First, the crisis strikes internal stakeholders-the health risk of owners and employee. Second, externally, businesses get to struggle by the reduced foot traffic and business infrastructure due to travel restriction. Third, on top of the magnitude of the crisis, the crisis comes with a wide scope and the impact of which is omnipresence, thus making relocation or changing the target markets irrelevant and ineffective options [1].

How could businesses survive? What strategic options do businesses have when they effectively engage such a challenge? In a recent virtual special issue published in Strategic Management Society, strategic options are categorized into four types, retrenchment,

[*] Corresponding author: drlee@kookmin.ac.kr

preserving, innovating, and exit. *Exit* means the discontinuation of the business before the situation goes out of control, and some suggest as a best way to cope with a crisis [2]. Exit may appear a forced outcome, but help businesses save resources for a new project they might develop in the future [3]. Decision to discontinue their business is also up to the context. Based on given findings that businesses are less likely to exit when they face natural disasters [4], it is questionable whether exit is still a valuable option for businesses facing challenges by COVID-19. While exit is in fact widely practiced, we decide not to consider it due to limited accessibility to the business that exited. Instead, we replace exit with another approach that firms can take to respond to the crisis. This approach is corporate-level strategy, such as making new alliances, increasing the number of alliances, or mergers and acquisitions (M&A).

Delving into the inquiry, in this study, we explore and investigate different strategic approaches to measure their effectiveness. For both explorative and investigative purposes of the study, our investigation involves a set of tests at both business level and corporate level. Literature has tested strategic ambidexterity as a key to understanding the role of dynamic capability in times of crisis and found that exploration is effective in adaptation, not exploitation [5, 6]. Given proven relevance, ambidexterity framework is therefore employed as a proper lens to assess business-level strategy. Corporate strategies are considered in multiple angels, but mostly focusing on alliance, M&A. The aim of the study, in this sense, is to present a few pragmatic implications that would help business in responding to a crisis like COVID-19 pandemic.

We find research gap in a multiple dimension, which further signifies the study. First, we see a lack of understanding of how entrepreneurs use multiple strategic tools effectively to address challenges imposed by the external crisis, like COVID-19. Second, regarding the literature of ambidexterity, heterogeneous characteristics of crises are not well accounted for. Despite the difference in origins and natures, external shocks are treated as a backdrop or a treatment that has no effect on firms.

Using 224 responses collected from the leaders and key individuals of Korean startups, we tested the effectiveness of strategic options using the notion of ambidexterity and alliances, and M&A. In the following section, we develop testable model based on our strategic perspective and present what we found from empirical investigation. A brief discussion will conclude our paper with a few potentials for further development.

2 Conceptual model

Focusing on the effect of strategic option, we believe firm performance is function of a combination of operational efficiency and level of innovation. Strategic responses are chosen and employed when firms face external shock and the effectiveness of which is tested regarding the firm's financial performance recorded in the following fiscal year. As briefly explained, borrowing the framework of ambidexterity, the construct of "exploitation" is used to measure the effectiveness of "retrenchment." Likewise, the construct, exploration is used for "innovating." We measure the change of both strategies to measure "preserving." Finally, corporate level strategies are measured at three dimensions, which are reinforcing existing alliances, establishing and increasing new alliances, and finally M&A. We consider business-level strategies as mediators as firms tend to be forced to choose them given path-dependency. In other words, firms with a high level of operating efficiency find "exploitation" easy to pursue and the same for the other. We consider corporate strategies as moderators as decisions made at corporate are relevantly free from the firm characteristics rather it is driven by fit of resource and capability between the partners. In this model, both business and corporate level strategies are considered treatment, thus used observation of the year 2020, causing the outcome of the following year. The baseline model considers the direct relationship between firm characteristics measured by two dimensions, operating efficiency and innovativeness, and financial performance of 2021, one year after the strategic measures were treated. Figure 1 shows our conceptual model that incorporates both ambidexterity and corporate level strategy.

3 Theory and Hypothesis

3.1 Effectiveness of business level strategies

While our focus is on the treatment effect of the focused strategies, we believe the firm performance is largely predicted by two widely accepted two strategic determinants [7], firm innovativeness, and operational efficiency. Using the two pillars that play a pivotal role for firms to create a core capability, we believe testing the effectiveness of the two constructs to find out if they still work during a time of crisis is relevant. As for the first construct, innovativeness is widely known as a good predictor of firm performance [8, 9], which refers to firm wide efforts directed toward business renovation, including the development of new products/services and innovating the existing business model with an aim to attain a sustainable competitive advantage [7]. Operating efficiency is also widely known as a good predictor of firm performance [10], which refers to the firm's wide capability to improve the productivity of standard operating procedures to achieve maximum productivity and thus gain and sustain a competitive advantage [7]. The opposite is still true. Firms' inefficiencies are attributed to suboptimal use of resources, such as overpaying for inputs [11]. Operational efficiency benefits firms, particularly when firms get involved with risk management [12]. We call a combination of these two constructs, hereafter, the firm capabilities as they are generally endogenous factors constant to the firm regardless of a specific external context such as a crisis. We hypothesize therefore that these two firm capabilities predict financial performance even during the time of COVID-19.

Hypothesis 1. Firm capabilities (innovativeness and operating efficiencies) have direct effect on the firm performance.

Focusing on the interaction of the strategies, we move on our focus on the first strategic options firm can take in response to a crisis, retrenchment. *Retrenchment* refers to a strategic response to an externa shock, where business attempt to focus on managing efficiencies when they must deal with a net loss [13, 14]. Aligning to the strategic management literature, this study determines that "retrenchment" corresponds to exploitation, a strategy provided by ambidexterity frameworks. We hypothesize exploitation strategy moderate the effect of firm capabilities of firms.

Hypothesis 1a. Exploitation strategy would mediate the relationship between firm characteristics and firm performance.

The second option we discussed earlier was preserving. *Preserving* refers simply maintaining the status quo and mitigate the impact of the crisis [15], which implies that firms undermining significance of path-dependency might experience hard time in a time of crisis [16]. According to strategies preserving is the safest option especially for firms facing great level of uncertainty. We hypothesize that maintaining the same strategy would moderate the effect of firm capabilities of firms.

Hypothesis 1b. Preserving strategy would mediate the relationship between firm characteristics and firm performance.

Another strategy to be tested is undertaking strategic renewal by *innovating* the way they used to. Innovating business involves widening the business scope by exploring uncharted sectors or new ways of doing business, therefore allow firms to try an alternative market and business opportunities and help firms respond to a crisis [17]. Research found, however, innovation is effective for firms can utilize lack resource and complementary assets [18]. Note that a precondition of "innovating" is extra resource, which is rare to startups, leaving innovating a valuable option for young and small organizations [19]. Again, we found "innovating" corresponds to the other strategic option, exploration. We hypothesize exploring strategy moderate the effect of firm capabilities of firms.

Hypothesis 1c. Exploration strategy would mediate the relationship between firm characteristics and firm performance.

3.2 Effectiveness of corporate level strategies

Partnership and alliances are different from exit as it is a voluntary action that involves "exchange, coordination, sharing, and/or joint development of resources and capabilities" [20]. Partnership and alliances appear to be useful when firms face challenges as they are found to be effective in providing incremental resources [21], and in helping firms diversify [22] as crisis mostly comes with disrupted environmental munificence (abundance of resource available to firms) and disappearance of existing markets [23]. We hypothesize alliance strategy would help firm address uncertain environment. We measure the effect in two dimensions.

First, we contend reinforcing the relations with existing alliances moderate the effect of firm capabilities of firms.

Hypothesis 2a. Reinforcing the relations with existing alliances would moderate the relationship between firm characteristics and firm performance.

Second, we further hypothesize exploring and establishing new alliances moderate the effect of firm capabilities of firms.

Hypothesis 2b. Exploring and establishing new alliance would moderate the relationship between firm characteristics and firm performance.

M&A largely refer corporate level strategies which involves higher level of risk because it requires major resource commitments [24]. Responding to crises, firms may pursue M&A with the aim to achieve decreased competitors to ease industry-level degree of rivalry. Further, the acquisition allows firms to exploit new opportunities presented by environmental disruption [25]. We hypothesize M&A would moderate the effect of firm capabilities of firms.

Hypothesis 2c. Mergers and acquisitions would moderate the relationship between firm characteristics and firm performance.

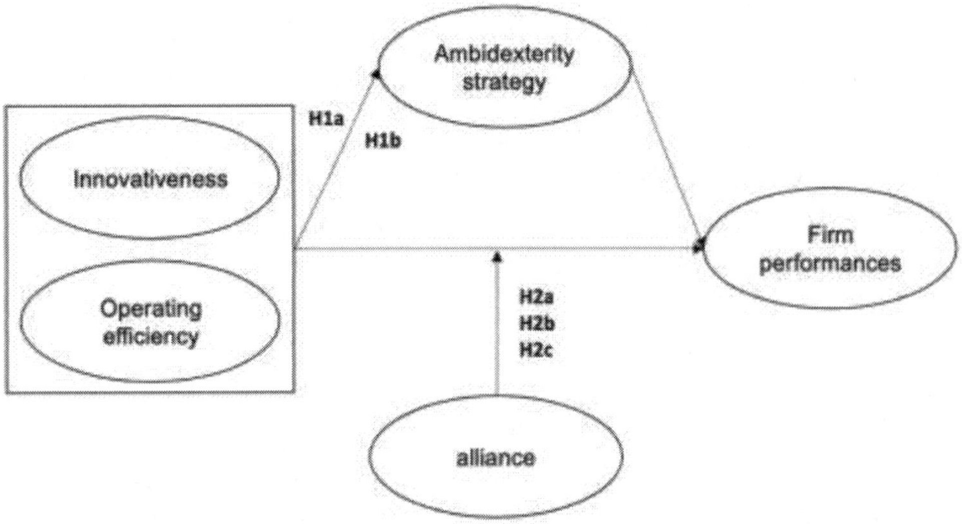

Fig 1. The proposed model
Source: Own processing

4 Method

4.1 Data Collection

For the purpose of the study, we conducted an online survey and collected data from the leaders and key individuals of startups registered of South Korean the Korean Startup Forum and Youth Startup Academy in 2021 (April to Jun) . Korea Startup Forum (KSF) is the nation's largest startup membership organization with active members of more than 1,500 startups and partners. Youth Startup Academy is a public organization supported by Korean SME agency (KOSME), which is dedicated to discover early-stage startups with innovative technologies to support them by providing total package programs to grow them into successful startup companies. In total, 800 companies were invited to participate in this study. The survey questionnaires were distributed to each startup's CEO and/or founder, or team leader, who has a high level of understanding about firm's strategy and circumstances. After cleaning incomplete responses, overall, we collected 224 usable responses.

4.2 Measures

Participants were asked to rate several aspects of strategic orientation and firm characteristics they know of the firm using the following scales. If not indicated otherwise, all scales used a five-point Likert scale ranging from 1 (strongly disagree) to 5 (strongly agree). Table 1 provides overall information about our measures.

4.2.1 Dependent variable

We measured firm performance based on the respondents' perceptional evaluations on their financial performances for a set of three financial indicators, such as annual revenue growth, operating profit, and return on investment. The respondents were asked to provide relevant information of pre-pandemic period, specifically 2021. It our attention to use year 2021's financial performance as an outcome of strategic responses they took in 2020.

4.2.2 Independent variable

Innovativeness is measured a set of eight items to collect information about strategic intensity of firm to engage in R&D. Using the same approach, *Operating efficiency* is measured by eight items scale. After conducting principal component analysis, two items were dropped for cross loading. The respondents were asked to provide relevant information of pre-pandemic period, specifically 2019. It our attention to use pre-pandemic strategic capability of firms as a controlled context.

4.2.3 Focused Mediators

The focused mediators are both exploration and exploitation. We asked the respondents to reflect their strategic intensity on activities to explore and exploit during 2020, they took as strategic responses to respond to the crisis using four item scale to measure both constructs. As a proxy of "preserving," we created a variable to measure how much increase or decrease they made to the strategic intensity of both exploration and exploitation. We find very marginal differences ranging from -.3 to .3. The respondents were asked to provide relevant information of 2012. It our attention to consider them as strategic responses to counter the situation caused by COVID-19.

4.2.4 Focused Moderators

To measure moderating effect of corporate level strategy, respondents rated on three items; a) if they have reinforced relationship they have with existing alliances and partners, b) if they have established new alliances and c) if they have undertaken M&A.

Table 1. Variable Description

	Operational Definition	**Measures**
Control variables	Age – a period in which startup was established Revenue – Annual sales before COVID-19	
Independent variables	Innovativeness – degree to develop the product process and business model Operating efficiency – capacity to improve productivity of standard procedure, technology skills, quality of product/service	5-points Likert
Mediators	Exploration strategy- degree to discover, explore and develop new things Exploitation strategy – degree to maximize efficiency by improving existing technologies and capabilities Preserving – degree to maintaining strategy	5-points Likert
Moderators	Existing alliance – degree of strengthening partnerships with existing companies New alliance - degree of strengthening partnerships with new companies Mergers and acquisition – degree to which mergers and acquisition are attempted with competitors and affiliates	5-points Likert
Dependent variables	Performance – An increase in total revenue, such as total sales, ROS, investment profit of startup	○,120%decreased ~ ○,5 20%increased

Source: Own processing

https://doi.org/10.1051/shsconf/202213201018

5 Analysis and Results

5.1 Hypothesis Testing

Table 2 presents the means, standard deviations, and inter-correlations among the variables under study. Operational efficiency is positively correlated with firm performance (r = .619, p < .01). There was no significant correlation relationship between innovativeness and firm performance (r = .097, ns). Operating capabilities positively correlated with exploration (r = .419, p < .01) and exploitation (r = .437, p < .01) as a mediator, but not with preserving (r = .033, ns).

Table 2. Descriptive and Correlations of Variables from Probationary Participants

Variable	*M*	*SD*	1	2	3	4	5
1. Age	4.05	1.99	1.00				
2. Revenue	1.76	1.10	0.306***				
3. Innovativeness	3.76	0.761	0.042	-0.047			
4. Operating capabilities	3.81	0.664	0.025**	-0.064	0.619***		
5. Exploration	7.07	2.03	-0.006	-0.135*	0.365***	0.419***	
6. Exploitation	0.00	0.09	0.121	-0.097	0.377***	0.437***	0.764***
7. Preserving	0.492	0.559	0.026	0.043	0.054	0.033	-0.206**
8. Alliance 1	3.41	1.30	0.006	-0.186**	0.255***	0.348***	0.689***
9. Alliance 2	3.40	1.26	0.03	-0.122*	0.288***	0.377***	0.648***
10. Alliance 3	2.29	1.38	-0.004	0.064	0.21**	0.023	0.322***
11. Firm performance	3.02	1.16	-0.161**	0.041	0.097	0.222***	-0.062

	6	7	8	9	10
7. Preserving	0.256***				
8. Alliance 1	0.618***	-0.06			
9. Alliance 2	0.66***	0.043	0.802***		
10. alliance 3	0.255***	-0.079	0.339***	0.415***	
11. Firm performance	-0.061	0.078	-0.078	-0.08	-0.103

Chybí source

We began our analysis by examining the means, standard deviations (SD), and correlations for our variables. Table 2 provides descriptive statistics and Pearson correlation of the variables. Hypothesis 1 proposed that ambidexterity (exploitation/exploration/preserving) strategy might mediate the relationship between firm characteristics and firm performance. Moderation tests showed that after controlling for company age, company revenue, exploitation strategy was negatively mediated between operational characteristics and firm performance ($\beta = -.16$, $p < .05$). Thus, Hypothesis 1a was supported. Exploration strategy was also negatively mediated to firm performance ($\beta = -.198$, $p < .05$). Hypothesis 1c was supported. However, preserving strategy was not significantly related to operation capabilities and firm performance. Hypothesis 1b was not supported (Table 3).

Table 3. Regression Results for Mediation Effect of ambidexterity strategy on Operational efficiency and Firm performances

	Dependent Variables	Independent Variables	ß	Std. Error	t value	P	R^2	adj. R^2	F-statistics
Step 1	Firm performance	(intercept)							
		Innovativeness	-0.086	0.125	-0.685	0.494	0.091	0.075	5.503
		Operating efficiency	0.471	0.144	3.278**	0.001			
Step 2	Exploration	Operating efficiency	0.649	0.096	6.745	0.000	0.223	0.213	21.11
	Preserving	Operating efficiency	0.005	0.009	0.486	0.627	0.003	-0.01	0.224
	Exploitation	Operating efficiency	0.713	0.100	7.094	0.000	0.214	0.2035	19.99
Step 3	Firm performance	Operating efficiency + exploration	-0.198	0.078	-2.540	0.012	0.116	0.100	7.217
		Operating efficiency + Preserving	0.913	0.809	1.128	0.260	0.094	0.07811	5.723
		Operating efficiency + exploitation	-0.169	0.075	-2.257	0.025	0.110	0.093	6.773

Source: Authors' current study.

We discussed the moderating effects of existing alliance, new alliance, and the M&A separately. Table 4 shows the results of the moderation. Existing alliance was found to moderate the effect of operational efficiency on firm performance (B = .236, SE = .082, t = 2.885, p < .05). Moderation test with autonomy showed that the interaction between operating capabilities and new alliance did not significantly affect firm perform ace (B = .13, SE = .084, t = 1.539, p = .125). Hypothesis 2b did not receive support. Moderation test with the M&A showed that M&A did not affect the relationship between operating capabilities and firm performance (B = .035, SE = .082, t = .429, p = .668). Thus, hypothesis 2c did receive support.

Table 4. Regression Results for Moderation Effect of Alliance on Operational capabilities and Firm performances

	β	*SE*	*t value*	*P*	supported
Operation capabilities	-0.23368	0.28117	-0.831	0.40684	
Existing alliance	-1.06767	0.32608	-3.274	0.00123	
OC × existing allice	0.23582	0.08173	2.885**	0.0043	Supported
Model R^2	0.143				
adj. R^2	0.1243				
F-statistics	7.33				
Operation capabilities	0.11008	0.2937	0.375	0.70816	
New alliance	-0.66673	0.33479	-1.991	0.04768	
OC × existing allice	0.12973	0.0843	1.539	0.12526	Not Supported
Model R^2	0.124				
adj. R^2	0.105				
F-statistics	6.215				
Operation capabilities	0.33423	0.22139	1.51	0.1326	
Merger & Acquisition	-0.23929	0.33294	-0.719	0.4731	

OC × existing allice	0.0354	0.08249	0.429	0.6683	Not Supported
Model R^2	0.103				
adj. R^2	0.083				
F-statistics	5.049				

Source: Authors' current study.

6 Conclusion and Discussion

In this study, we have explored multiple strategic responses available startups to counter the current pandemic and investigated their effectiveness. We explored three strategic options available to startups, retrenchment, preserving, and innovation. Using ambidexterity lens, exploitation and exploration were used as a set of relevant constructs for retrenchment and innovation, respectively. Consistent with previous literature, we found business response to changing the situation at business level still effective, except preserving. Given long lasting COVID-19, our finding seems to be consistent with what we already know about retrenchment [26, 27].

Not all corporate level approach turned out to be effective, but "reinforcing the relationship with existing alliances." It implies that under the circumstance of COVID19, operational efficiency may benefit from existing partnerships. Making new alliances, on the other hand, may only increase the level of uncertainty that is already high. The effectiveness of M&A may be explained by the relatively scarce resource slack available to startups [22].

This study makes a few meaningful contributions. First, we presented the validity of ambidexterity lens to the context of young startups. Ambidexterity has been studied largely for established business, a relevance of which has been barely applied to the context of entrepreneurial venture. Partly, it was because resource constraints of young startups would drive them into a focused approach, instead of trying two approaches. Our study, we believe, opened a potential possibility of ambidexterity approach for young startups. Second, our study extended the applicability of the ambidexterity lens by examining its validity in the context of the current pandemic. As stated earlier, pandemic, COVID-19 imposed a new set of uncertainty and constraints, we never experienced in a recent history, where businesses are left unguided with which strategic options, they can take addressing the challenge. Presenting the effectiveness of exploitation and exploration, we are confident to suggest business could consider them to try, a managerial implication we make. Second, our study calls for attention to the corporate level strategy, especially the significance of exiting alliance during a time of crisis, to scholars of entrepreneurship research. Corporate-level strategy is largely considered strategic options applicable for large established ones, but our finding says otherwise. Maintaining a close relationship with existing alliances would benefit young startups. Third, during a time of crisis, our findings show that firms should benefit from both business level and corporate level strategies as they are not mutually exclusive to each other. Finally, but not least, our study adds to the literature of crisis management and among a few that tested the effectiveness of strategic responses in the time of COVID-19, the pandemic, a type of crisis humanity never experienced in a recent modern history. We hope that we paved a way for future research stream in this venue. As not all studies are perfect, our studies left room for future development. First, while our study shed light on the role of alliances, we admit that our findings only scratch the surface. We are limited in presenting the detailed story behind it, including the purpose of reinforcing relationships with existing partners, whether it could be to control uncertainty or enter into some sort of smaller scale clan. Future research might make use of secondary data that provide industry restructuring during the pandemic may help us unfold what truly happened during the time. Second, as this study relied heavily on subjective evaluations of the respondents, we ran a risk of forgoing accuracy of the measurement. In our defense, our focus is more on strategic effectiveness, the evaluations of which is often made by the CEO and individuals involved in strategic decisions. Besides, we believe our approach is effective enough in laying out a foundation for further investigation.

References

1. M. Wenzel, H. Krämer, J. Koch, A. Reckwitz, Future and organization studies: On the rediscovery of a problematic temporal category in organizations. *Organization Studies*, **41**(10), 1441-1455 (2020)
2. L. Dai, L. Eden, P. Bemish, Caught in the crossfire: Dimensions of vulnerability and foreign multinationals' exit from war-afflicted countries. *Strategic Management Journal*, **38**(7), 1478–1498 (2017)
3. S. Carnahan, Blocked but not tackled: Who founds new firms when rivals dissolve? *Strategic Management Journal*, **38**(11), 2189–2212 (2017)
4. C. Oh, J. Oetzel, Multinationals' response to major disasters: how does subsidiary investment vary in response to the type of disaster and the quality of country governance? *Strategic Management Journal*, **32**(6), 658681 (2011)
5. A. Gupta, K. Smith, C. Shalley, The interplay between exploration and exploitation. *Academy of Management Journal*, **49**(4), 693–706 (2006)
6. O. Osiyevskyy, G. Shirokova, G, P. Ritala, Exploration and exploitation in crisis environment: Implications for level and variability of firm performance. *Journal of Business Research*, **114**, 227-239 (2020)
7. J. Lee, H. Kwon, N. Pati, Exploring the relative impact of R&D and operational efficiency on performance: A sequential regression-neural network approach. *Expert Systems with Applications*, **137**, 420-431 (2019)
8. M. Benner, M. Tushman, Exploitation, exploration, and process management: The productivity dilemma revisited. *Academy of management review*, **28**(2), 238-256 (2003)
9. T. Liao, W. Lin, W. C, Corporate governance, product market competition, and the wealth effect of R&D spending changes. *Financial Management*, **46**(3), 717-742 (2017)
10. B. Baik, J. Chae, J, S. Choi, D. Farber, Changes in operational efficiency and firm performance: A frontier analysis approach. *Contemporary Accounting Research*, **30**(3), 996-1026 (2013)
11. W. Greene, D. Segal, Profitability and efficiency in the US life insurance industry. *Journal of Productivity Analysis*, **21**(3), 229-247 (2004)
12. Y. Shou, W. Hu, M. Kang, Y. Li, Y. Park, Risk management and firm performance: the moderating role of supplier integration. *Industrial Management & Data Systems* (2018)
13. G. Bruton, D. Ahlstrom, J. Wan, Turnaround in East Asian firms: Evidence from ethnic overseas Chinese communities. *Strategic Management Journal*, **24**(6), 519-540 (2003)
14. R. de Figueiredo Jr, E. Feldman, E. Rawley, The costs of refocusing: Evidence from hedge fund closures during the financial crisis. *Strategic Management Journal*, **40**(8), 1268-1290 (2019)
15. M. Wenzel, Path dependence and the stabilization of strategi premises: How the funeral industry buries itself. *Business Research*, **8**(2), 265–299 (2015)

16. A. Chakrabarti, Organizational adaptation in an economic shock: The role of growth reconfiguration. *Strategic Management Journal*, **36**(11), 1717-1738 (2015)
17. I. Reymen, P. Andries, H. Berends, R. Mauer, U. Stephan, E. Van Burg, Understanding dynamics of strategic decision making in venture creation: a process study of effectuation and causation. *Strategic entrepreneurship journal*, **9**(4), 351-379 (2015)
18. C. Helfat, Know-how and asset complementarity and dynamic capability accumulation: the case of R&D. *Strategic management journal*, **18**(5), 339-360 (1997)
19. J. Brinckmann, J. Villanueva, D. Grichnik, L. Singh, L Sources of strategic flexibility in new ventures: an analysis of the role of resource leveraging practices. *Strategic Entrepreneurship Journal*, **13**(2), 154-178 (2019)
20. N. Park, J. Mezias, Before and after the technology sector crash: The effect of environmental munificence on stock market response to alliances of e-commerce firms. *Strategic Management Journal*, **26**(11), 987-1007 (2005)
21. J. Dyer, H. Singh, The relational view: Cooperative strategy and sources of interorganizational competitive advantage. *Academy of management review*, **23**(4), 660-679 (1998)
22. B. Koka, J. Prescott, Strategic alliances as social capital: A multidimensional view. *Strategic management journal*, **23**(9), 795-816 (2002)
23. W. Wan, D. Yiu, From crisis to opportunity: Environmental jolt, corporate acquisitions, and firm performance. *Strategic Management Journal*, **30**(7), 791-801(2009)
24. A. Pablo, S. Sitkin, D. Jemison, Acquisition decision-making processes: The central role of risk. *Journal of management,* **22**(5), 723-746 (1996)
25. P. Chattopadhyay, W. Glick, G. Huber, Organizational actions in response to threats and opportunities. *Academy of Management Journal*, **44**(5), 937-955 (2001)
26. C. Chadwick, L. Hunter, S. Walston, Effects of downsizing practices on the performance of hospitals. *Strategic Management Journal*, **25**(5), 405-427 (2004)
27. H. Ndofor, J. Vanevenhoven, V. Barker III, Software firm turnarounds in the 1990s: An analysis of reversing decline in a growing, dynamic industry. *Strategic Management Journal*, **34**(9), 1123-1133 (2013)

The Effect of Program Genre on Modes of Media Multitasking and Ad Processing

Hyejin Bang[1]

College of Business Administration, Kookmin University, Seoul, Republic of Korea

Abstract. Despite many studies on the effect of MM on advertising, previous studies on MM considered MM a homogenous phenomenon. Further, whether and how media-related factors predict different modes of MM and how this behavior affects ad processing has been unknown. To fill this gap, the purpose of this study was to examine (a) the effect of program-genre on the occurrence of different modes of media multitasking (MM; utilitarian MM vs. hedonic MM) and (b) how different modes of MM influence the way viewers process ads on the primary screen (i.e., computer screen). A lab-based experiment yielded data for testing the hypotheses. The findings suggest that findings suggest that MM can be classified into two distinct modes: utilitarian MM and hedonic MM. Further, the findings show that participants who watched the sitcom tended to engage in a higher amount of utilitarian MM than those who watched the suspenseful drama; however, both groups engaged in a comparable amount of hedonic MM. The findings also indicate that participants who watched the sitcom demonstrated a lower level of ad memory than those who watched the suspenseful drama. The current study provides meaningful theoretical implications. Further, this study provides useful implications for advertising practitioners and marketers.
Keywords: Media multitasking, Context effect, Ad processing, Limited Cognitive Capacity

1 Introduction

Consider the following scenario: Sarah and her brother Mike are sitting together on the couch watching a popular show, *Saturday Night Live*. As always, both of them are holding their smartphones, and their laptops and iPads are also within reach. While watching the show, Sarah receives a notification from Instagram that her best friend, who is now on vacation in Italy, uploaded new pictures on her feed. To check her status and look at the pictures, Sarah opens the Instagram app on her smartphone and browses the feed while trying to follow the television show. At the same time, Mike suddenly remembers that he has to book a flight to

[1] Corresponding author:: hjbang@kookmin.ac.kr

New York for his friend's upcoming wedding. To check for a good deal, he grabs his laptop and searches Expedia and Orbitz while watching the show.

This scenario represents a normal pattern of media consumption: *media multitasking* (MM), the simultaneous use of two or more media devices at a time [1-3]. Responding to the pervasiveness of this phenomenon, scholars have investigated several questions associated with information processing and media effects related to MM. For instance, prior studies have found that MM resulted in diminished attention, lower quality of exposure, shallow processing, and decreased memory of messages (e.g., ads) in the media [4–6]. On the other hand, MM might also inhibit counter argumentation and increase enjoyment and acceptance of messages [7].

Upon closer examination, the MM behavior of Sarah and Mike indicates two different types of MM. Sarah's second screening can be categorized as more hedonic or entertaining MM, while Mike's can be regarded as more goal-oriented, utilitarian, or information-seeking MM. Previous studies on MM, though, have neglected the different motivations for second screening. These motivations might impact the way people divide their attention between a television show and a second screen and the manner in which they interact with the two screens, and they might ultimately affect the viewing experience and how consumers process the ads delivered to them. Despite the distinctions that might exist between different kinds of MM, scholars have not examined (a) engagement in two different modes of MM or (b) how these two different modes affect ad processing.

To increase understanding of this issue, the current study introduces two distinctive types of MM: *utilitarian* MM and *hedonic* MM, and attempts to examine the situational (i.e., when) predictors of two mode of MM (i.e., hedonic MM and utilitarian MM). In particular, this study proposes program genre as situational factors that affect the types of secondary tasks people engage in while attending to primary media contents. Second, this study aims to investigate how the different modes of MM influences the amount of attention audiences pay to ads appeared on the primary screen and their memory of ads.

2 Literature review

2.1 The Prevalence of Media Multitasking

MM can be generally defined as "performing two or more tasks simultaneously, one of which involves media use" [8, p. 100]. However, two different views of multitasking have emerged: sequential and concurrent. The sequential multitasking perspective assumes that people cannot encode and process more than one stimulus at a time. From this point of view, multiple tasks are sequentially performed, one at a time, through rapid attention switching between multiple stimuli [9]. Other scholars see multitasking as the result of divided attention, assuming that people can process multiple stimuli concurrently instead of switching between tasks [10]. Considering these two distinct definitions, the current study adopted the sequential multitasking perspective, defining MM as "sequential engagement in multiple media activities through rapid attention switching."

MM can be categorized into three types: (a) multitasking between a medium and a non-media activity (e.g., television and conversation), (b) multitasking between two or more media (e.g., mobile device and television), and (c) multitasking between two different tasks within a single medium (e.g., two tasks opened on a computer. The concept of engaging in other activities while consuming media (e.g., watching TV) is not new. Through in-home observation, it was found that audiences engaged in many complementary (e.g., eating, drinking, conversation about the program) and competitive (e.g., chores and hobbies) activities while watching TV [11]. Although most of the multitasking behavior at this time was non-media activity [12], the number of people who concurrently consume multiple media (e.g., the second and third types of MM) is on the rise [2, 13]. This "new normal" media consumption pattern has emerged primarily because digitalization and increased portability of media devices have allowed people to possess a variety of screens and because the widespread availability of high-speed information has fueled the expectation that human cognitive systems can simultaneously process and respond to multiple sources of information.

While overall media use among U.S adults has increased by 20% over the past decade, the amount of time people spend MM has increased more than 119% over the same period [14]. A recent study conducted by IAB indicated that about 80% of U.S. adults used another device (e.g., smartphone, laptop, tablet) while watching TV, indicating that MM has become the default mode for media consumers. Another recent study found that combining multiple new media (i.e., simultaneous use of laptop and mobile phone) was more common than combining two traditional media (i.e., simultaneous use of television and magazines) or one traditional medium and one new medium (i.e., simultaneous use of television and a mobile phone) [13].

2.2 The Effect of MM on Ad Processing

Previous studies on MM have consistently found the detrimental effect of MM on message processing (i.e., decreased attention, low message exposure, and low memory) based on limited cognitive capacity framework [4, 6].

Limited cognitive capacity theory suggests three different levels of information processing: (a) encoding, (b) storage, and (c) retrieval. In the mass communication field, many scholars have operationally equated the "encoding" process with exposure. However, to successfully encode a mediated message, people need to have their sensory receptors engaged (e.g., eyes, ears, skin; Eysenck, 1993) and select which bits of information held in sensory storage will be transformed into mental representations in working memory. That is, without sufficient cognitive resources, such encoding of information cannot occur, suggesting that mere exposure is different from "attention" or "encoding." Because people have a limited pool of attention and mental resources available for cognitive processing [15], MM inevitably interrupts people to encode information. For instance, if an individual searches for an airline ticket on a mobile device during a commercial break for *Saturday Night Live*, he either divides his attention or rapidly switches his attention between the airline site and the ads on the TV screen [6, 10].

This competition for limited cognitive resources is likely to lead to diminished attention to an ad, encoding of sensory input, and ad message recall [5, 10]. In line with this concept, it was found that people who watched TV programs while reading an article showed significantly lower levels of recall and recognition of the article than those who did not engage in multitasking [4]. The effect of MM has also been found to be detrimental to advertising effectiveness, particularly ad memory [1, 16 -18]. Based on these findings, deficits in ad attention and ad memory are likely to occur when program viewing is combined with other media activities.

H1: Ad attention will be lower when people watch a program while MM (vs. engagement with a single media device).

H2: Brand memory will be lower when people watch a program while MM (vs. engagement with a single media device).

2.3 Two Modes of Media Multitasking: Utilitarian vs. Hedonic

Although MM has often been considered a homogenous phenomenon, the combination of tasks or media can vary, and the effects of different forms of MM might differ [13, 19]. According to a recent report from IAB (2015), the top multi-screening activities in which people engage while watching TV include browsing the Internet, checking email, paying bills, catching up on news, social networking, watching another fun video clips, gaming, and shopping. Among those top multi-screening activities, the first four (i.e., checking email, browsing the Internet, paying bills, and catching up on news) might be defined as more goal-directed, instrumental, and utilitarian media consumption, tasks that people are more strongly motivated to accomplish. On the other hand, the last four (i.e., social networking, watching another fun video clips, gaming, and shopping) might be categorized as more non-directed, navigational, and hedonic media consumption, tasks that are primarily for enjoyment. Paying attention to these differences, the current study introduces two different modes of MM, *utilitarian* MM and *hedonic* MM.

In the context of MM, one study identified two motivations for second screening: information seeking and discussion [20]. Also, although the study was limited to online media consumption, media consumption behavior cab be classified into two distinct types: goal-directed and non-goal directed [21]. According to them, the different types of media consumption affect media users' motivation, information processing, and decision-making. In the case of a utilitarian/goal-directed second-screen activity, such as booking a flight ticket or paying a bill, a stronger motivational activation for the task might exist, requiring people to invest a significant amount of their mental resources [22]. On the other hand, in case of a hedonic MM, such as scrolling through posts on Facebook, which usually satisfies media users' hedonic desire or maximizes the enjoyment of media consumption, the level of cognitive effort required is relatively low. Despite such differences between utilitarian MM and hedonic MM, the nature of MM has not been taken into consideration in previous studies. Thus, we examine how media context (i.e., different program genres) affects the occurrence of utilitarian and hedonic MM and how these distinct modes of MM influence ad processing.

2.4 The Effect of Program Genre on Two Different Modes of MM

To uncover when people shift their attention from a primary screen to media-related or non-media related secondary activities, numerous scholars have explored how media content affects audience attention [23-25). Findings from one observational study indicate that entertaining programs (e.g., comedy) more easily distract audience attention than informational programs (e.g., documentary) [25]. However, findings from a field study conducted by Neilson (2015) indicated the opposite pattern. Although results have been controversial, the effect of program genre on the *frequency* of MM has received scholarly attention. However, the effect of program genre on the *pattern* of MM in which people engage remains unexplored.

People tend to save cognitive capacity by avoiding instances involving a high level of cognitive demands [26]. Supporting this perspective, Wang et al. (2015) argued that some MM behaviors can be more cognitively taxing than others and found that people tend to choose MM combinations that are less cognitively demanding in mediated environment [19]. For instance, people tend to choose a secondary task that has a high level of relevance with a primary task, or a MM behavior that has a low level of shared sensory modalities with a primary task (e.g., listening to radio + Facebook) to reduce resource demand. Such findings suggest that the types of MM in which people engage could be influenced by the types of programs being viewed (i.e., primary task). That is, a program that demands high mental resources might increase the extent of hedonic MM, while a program that requires low mental load might enable people to engage in both hedonic and utilitarian MM due to surplus of cognitive resources.

In general, programs that fit into the suspense, drama, and thriller genres are cognitively more engaging due to their narrative structure and identification with characters [25]. On the other hand, programs such as sitcoms and reality shows requires less cognitive demand due to the more compartmentalized structure of the show [25]. Since people tend to choose MM behaviors that are less likely to tax their cognitive resources, programs such as suspenseful shows or drama, might lead to fewer cognitively demanding off-task MM behaviors, such as utilitarian MM (e.g., booking a flight ticket, catching up on news). On the other hand, both hedonic MM (e.g., texting to friends) and utilitarian MM are likely to be selected during programs that are cognitively less engaging, such as comedy shows, because people are likely to have enough cognitive resources leftover for secondary tasks. Therefore, the following research question and hypotheses were proposed.

H3: When watching a suspenseful drama, compared to a sitcom, people will engage in a lower level of utilitarian MM during the show (H3a) and commercial break (H3b).

H4: When watching a suspenseful drama or a sitcom, people will engage in the same level of hedonic MM during the show (H4a) and commercial break (H4b).

Furthermore, building on the limited cognitive capacity framework, we predict that people will be more likely to engage in hedonic MM rather than utilitarian MM, regardless of program genre, because people generally choose the combination of tasks that take up a less amount of their cognitive resources to avoid cognitive fatigue [19]. Thus, the following hypothesis is proposed:

H5: In the context of MM, people are more likely to engage in hedonic MM than utilitarian MM regardless of program genre.

2.5 The Effect of Different Modes of MM on Ad Processing

According to reversal theory [27], the motivational state of an individual can be one of two types: telic (serious-minded) or paratelic (playful-minded). People in the telic state are likely to be motivated by a goal while people in the paratelic state tend to seek enjoyment and have low goal orientation. Consistent with this concept and the classification of media consumption behavior by Hoffman and Novak (1996) [21], utilitarian MM (e.g., information search, paying a bill) is likely to be more goal-directed and demand a higher level of processing, while hedonic MM (e.g., texting, SNS usage) imposes a relatively small cognitive load.

Because utilitarian MM is likely to be more cognitively taxing than hedonic MM, in turn influencing information loss and communication outcomes [19, 28], we predicted that the content people watched on their primary screen would determine their cognitive ad processing. In particular, we predict that people who watch a sitcom might have difficulty processing commercials because the types of the MM in which they engage (i.e., both utilitarian and hedonic) impose a higher cognitive cost. On the other hand, during a suspenseful drama, the detrimental effect of MM on cognitive ad evaluation should lessen because viewers are more likely to engage in hedonic MM than utilitarian MM due to lower available cognitive resources. Thus, we proposed the following hypothesis:

H6: When watching a suspenseful drama, compared to a sitcom, people will show a higher level of ad memory.

3 Methodology

3.1 Overview of the Research Design

A lab-based experiment was conducted to examine 1) how program genre affects modes of MM, and b) how different modes of MM worsen or attenuate the detrimental effect of MM. Dependent variables were ad attention, the types of MM engaged and ad memory, which we measured using self-report. The experiment used a 2 (Tasking Type: Single tasking vs. Multitasking) × 2 (Program genre: Suspenseful drama vs. Sitcom) between-subject experimental design.

3.2 Pretest and Development of Stimulus Material

Prior to a main experiment, two pretest were conducted (a) to classify different types of MM activities into two categories (hedonic MM vs. utilitarian MM), and (b) to choose commercials embedded in the middle of selected programs. In the first pretest, a total of 108 college students participated in the first pretest for extra course credit (M_{age} = 20.90 years (*SD* = 1.03), 79.6% female). To categorize different types of MM activities into two categories (i.e., hedonic MM vs. utilitarian MM), participants were asked to indicate their primary motivations for the top eleven multi-screening activities reported by IAB (2015) on a 7-point scale (1: information seeking, 7: entertainment seeking). The one-sample *t*-test, with a test value of 4, indicates that information seeking activities included checking email, catching up on the news, paying bills searching information about a show and entertainment seeking activities included texting, browsing the Internet, social networking, watching other fun video clips, gaming, online shopping, tweeting/posting about the show on social networks (See Table 1). We used this classification in the analysis stage to measure the amount of hedonic and utilitarian MM.

Table 1. Classification of Media Multitasking Activities

Types of Media Multitasking	Mean	SD	*t*	*p*
Checking Emails	2.28	1.60	-11.19	<.05
Catching up on News	2.88	1.45	-8.02	<.05
Paying Bills	1.87	1.18	-18.68	<.05
Searching Information about A Show	3.29	1.74	-4.20	<.05
Texting	5.36	1.31	10.76	<.05
Browsing the Internet	5.04	1.45	7.50	<.05
Social Networking	5.92	1.22	16.38	<.05
Watching Other Fun Video Clips	5.81	1.46	12.89	<.05
Gaming	5.27	1.89	6.70	<.05
Online Shopping	5.09	1.69	6.70	<.05
Tweeting/Posting About The Show On Social Networks	4.71	1.46	5.08	<.05

Note: Motivations of Media Multitasking Activities were measured on a 7-point scale (1: Information Seeking, 7: Entertainment seeking).
Source: Own processing.

Next, the second pretest was conducted with a total of with 115 college students (M_{age} = 23.50 years (*SD* = 1.12), 60.6% female). To select commercials to embed in the commercial break, we considered a total of fifteen ads across eight different product categories. To eliminate any confounding effects, commercials for unfamiliar brands and commercials that were consistent in overall ad characteristics (e.g., attractiveness, credilbity, overall quality of the ad) were selected. Based on these criteria, commercials for Dilmah ($M_{familiarity}$=1.48, SD=1.00), NOCS ($M_{familiarity}$=2.32, SD=2.06), Jacobs ($M_{familiarity}$=2.55, SD=2.21), and GoMacro ($M_{familiarity}$=1.83, SD=1.50 were chosen.

Based on the pretests, we developed stimulus materials. For each 10-minute clip of the suspenseful dramas (*Blacklist* and *Luther*) and sitcoms (*Hot in Cleveland* and *2 Broke Girls*), we embedded the four selected ads as a commercial break.

3.3 Manipulation

Tasking Type. To manipulate tasking type, we assigned participants one of two tasking conditions: (a) watch a TV program without access to another media device and (b) watch a TV program and freely use a mobile phone (except to make phone calls).
Program Genre. To manipulate program genre, those assigned to the suspenseful drama condition watched *Blacklist* or *Luther* while those assigned to the sitcom condition watched *Hot in Cleveland or 2 Broke Girls.*

3.3.1 Dependent Measures

We measured a total of three dependent variables to assess the effect of program genre on MM and the effect of MM on ad evaluations: (a) amount of attention paid to commercial, (b) types of MM, and (c) ad memory. Ad attention was measured using two seven-point Likert items adopted from Chaffee and Schleuder (1986) [29]. The items included "How much attention did you pay to the commercials while watching the video clip?" and "How much did you focus on the commercials embedded in the video clip?" The types of MM activities in which the participants engaged were measured by asking them to list all of the mobile phone tasks they performed while watching the program and commercials, respectively. To increase the accuracy of their memory, a list of eleven prominent MM activities reported by IAB were given as examples, and they were allowed to look through the browser history on their mobile phones. Then, based on the classification of two types of MM (see Table 1), we generated a composite score for hedonic MM and utilitarian MM, respectively, by adding the number of activities participants reported. Lastly, we measured ad memory using two indicators: ad recall and ad recognition. For ad recall, participants wrote down any brands they could remember seeing during the commercial break [1]. For ad recognition, we provided a list of ten brands (i.e., four target brands and six filler brands); participants then chose as many brands as they could remember seeing during the commercial break [18]. We derived the ad recognition score by subtracting the number of incorrectly chosen brands from the number of correctly chosen brands.

3.3.2 Participants and Procedure

A total of 165 undergraduate courses enrolled at a large southeastern public university in the United States took part in the experiment (82.4% female; 19 to 30 years old [M = 21.54, SD = 2.20]). When the participants signed up, they were asked to bring their mobile phone the day of the experiment. Each student participated in a 60-minute lab session, which took place in a research lab equipped with a computer station. When participants arrived in the lab, they were seated in front of a 27-inch iMac computer and asked to choose one of the four programs available. Then, we randomly assigned participants to either the single-tasking or multitasking condition. To help participants naturally engage in MM behavior, the researcher left the lab once the participants began watching the selected program on their computer screen. Among those in MM condition, 91.4% of them reported that they used an Apple iPhone (5, 5s, 6, 6+, or 7); the rest reported using a Samsung Galaxy (6 or S7).

4 Results

First, to test the effect of media multitasking on ad attention and ad memory, we conducted independent sample *t*-tests. Confirming previous findings and our predictions, people who engaged in MM not only paid less attention to the program but also showed a lower level of ad memory compared to those who watched a TV program without access to a secondary device (See Table 2). Thus, H1 was supported, while H2 was partially supported.

Table 2. Media Multitasking Effect

	Single tasking	Multitasking		
	Mean (SD)	Mean (SD)	*t*	*p*
Ad attention	4.59(1.39)	2.82(.68)	6.33	<.001
Ad recall	.53(.90)	.82(.07)	1.58	.12
Ad recognition	2.56(1.03)	1.70(1.19)	4.18	<.001

Source: Own processing.

Next, to test whether different program genre influenced the type of MM activities in which participants engaged, independent sample *t*-tests were conducted. This study posited that the degree to which people engage in utilitarian MM would vary with program-genre, for the amount of cognitive resources leftover for secondary activities differ. As expected, participants who watched a sitcom (i.e., *Hot in Cleveland or 2 Broke Girls*) tended to engage in a higher level of utilitarian MM (e.g., checking email, catching up on the news) than those who watched a suspenseful drama (i.e., *The Blacklist or Luther*) not only during the program (M_{drama} = .38, SD = .59 vs. M_{sitcom} = 66, SD = .75, t = -2.23, p < .01) but also during the commercial break (M_{drama} = .39, SD = .53 vs. M_{sitcom} = .79, SD = .78, t = -3.19, p < .01). However, the amount of hedonic MM was not significantly different between the two programs (during the program: M_{drama} = 1.28, SD = .87 vs. M_{sitcom} = 1.59, SD = .90, t = -1.93, p > .05; during the commercial break: M_{drama} = 1.83, SD = .98 vs. M_{sitcom} = 1.90, SD = .91, t = -.43, p > .05). Thus, H3 and H4 were supported (Fig. 1.).

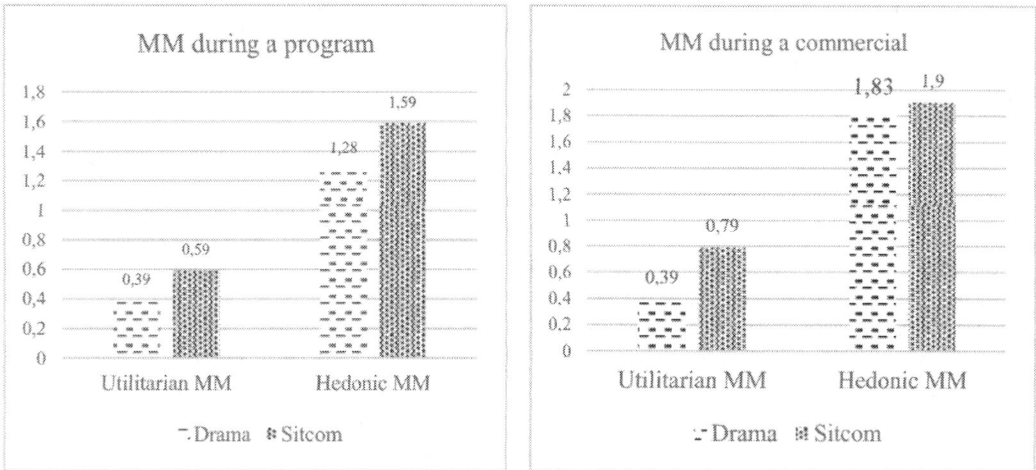

Fig. 1. The Effect of Program Genre on Modes of Media Multitasking
Source: Own processing.

To test H5, which predicted the relative amount of hedonic vs. utilitarian MM for each program genre, we conducted paired sample *t*-tests. Findings suggest that people were more likely to engage in hedonic or entertaining MM (e.g., texting, SNS, gaming) than utilitarian or information-seeking MM (e.g., catching up on the news, searching information) both when watching a suspenseful drama (during program: $M_{hedonic}$= 1.28, SD = .87 vs. $M_{utilitarian}$ = .38, SD = 59, t = 12.97, p < .001; during commercial break: $M_{hedonic}$ = 1.83, SD = .98 vs. $M_{utilitarian}$ = .40, SD = 53, t = 8.22, p < .001) and when watching a sitcom (during program: $M_{hedonic}$ = 1.59, SD = .90 vs. $M_{utilitarian}$ = .66, SD = .75, t = 7.86, p < .001; during commercial break: $M_{hedonic}$ = 1.90, SD = .91 vs. $M_{utilitarian}$ = .79, SD = .78, t = 7.63, p < .001). Thus, H5 was supported.

Lastly, in MM context, we predicted that people who watched a drama would show a higher level of ad memory as they mainly engage in hedonic MM, while those who watch a sitcom engage in both hedonic MM and utilitarian MM at a greater extent, which is likely to diminish one's cognitive capacity. To test the hypothesis, independent t-tests were performed. Among those who freely used their smartphones while watching a TV program on a computer screen, those who watched a suspenseful drama showed a significantly higher level of ad recall (M_{drama} = .48, SD = .86 vs. M_{sitcom} = .13, SD = .74, t=2.40 p<.05) and ad recognition (M_{drama} = 2.04, SD = 1.25 vs. M_{sitcom} = 1.40, SD = 1.07, t=2.85 p<.01), supporting H6. It should be also noted that these differences in ad memory was not found in single-tasking group (Table 3).

Table 3. The Moderating Effect of Program Genre on Ad Memory in the Context of MM

	Drama	Sitcom	*t*	*p*
	Mean (SD)	Mean (SD)		
Ad recall	.48(.86)	.13(.74)	2.40	<.05
Ad recognition	2.04(1.25)	1.40(1.07)	2.85	<.01

Source: Own processing.

5 Discussion

The purpose of this study was to understand a) distinct types of MM, b) the effect of program-genre on the occurrence of different types of MM, and c) its subsequent effect on ad processing. Supporting previous findings [4-6], the current study confirmed the negative effect of MM on attention and memory. When the computer and mobile device were simultaneously accessible, participants generally showed a lower level of ad attention and ad recognition than participants in the single-task group (i.e., computer only). The current study also found that the content people watch on their primary screen determines the modes of MM in which they engage, in turn influencing their cognitive ability to process brand-related information. Supporting the limited cognitive capacity framework [15], our findings indicate that those who watched a cognitively demanding suspenseful drama tended to engage in a lower amount of utilitarian MM than those who watched a sitcom, while both groups engaged in a similar amount of hedonic MM. According to Wang et al. (2015) [15], the cognitive demand of MM depends on task combination, in turn influencing a media users' ability to process information on a primary and secondary screen. Wise et al. (2009) found that a utilitarian or goal-oriented task (e.g., searching information) activated strong motivation than a hedonic or low-purpose media task (e.g., texting friends) [22]. In line with this idea, we found that participants who watched a suspenseful drama showed a higher level of ad recall and ad recognition than those who watched a sitcom, for the former tended to engage in a lower amount of utilitarian MM, which is cognitively more taxing.

The contribution of this study is twofold. First, the previous studies, which examined the media multitasking phenomenon, have neglected audiences' different motivations for second screening activities. To fill this gap, this study introduced two distinctive nature of MM: *utilitarian MM vs. hedonic MM,* and identified the predictors and consequences of different modes of MM. Second, we increased the validity of our findings by creating a more natural media consumption environment. Although previous studies have examined MM in many different forms (e.g., TV screen + Tablet: [18]; Multitasking on a single computer: [1,17]), few studies have investigated this issue in light of the most predominant pattern of MM among college students. To fill this gap, the current study investigated the simultaneous use of computers and mobile devices in a realistic setting.

Although the current study has meaningful implications, the results should be carefully interpreted due to several limitations. First, although we increased the ecological validity of the findings by allowing participants to engage in MM using their own smartphones, this approach might have created confounding effects. For instance, we have found that participants engaged in different amounts and types of secondary activities, potentially affecting their ad processing to varying degrees. Second, we assumed that those who watched a suspenseful drama showed a higher level of ad memory than those who watched a sitcom due to different amounts of utilitarian MM in which they engaged; however, we did not directly test this underlying mechanism in the current study. Thus, based on our findings, future researchers should consider investigating the effect of different types of MM on ad processing in a more controlled setting.

Lastly, the findings are limited to college students aged 20-30 years due to the samples used in the study. Although computers and mobile devices were reported as the most predominant MM combination among college students, other age groups might exhibit different patterns of MM (e.g., TV + mobile device). Although the student sample used in the current study is justifiable because this age group tends to engage in MM more than other age groups, future studies might consider the behavior of other age groups that constitute important targets of advertising. In addition, although individual differences were not considered in the current study, personal characteristics (e.g., attachment style or need for closure) might affect the type of MM behavior in which individuals engage.

6 Conclusion

In conclusion, although many scholars have considered MM a homogenous phenomenon, our experimental findings suggest that MM can be classified into two distinct modes: *utilitarian MM* and *hedonic MM.* Further, it offers the empirical evidence that modes of MM consumers engage in determine consumers' cognitive ability to process brand-related information on the primary screen. Similar to previously identified factors that moderated the detrimental effect of MM on ad processing and memory (e.g., ad appeal [17], processing style [1], congruence between primary and secondary screen activity [30]), this study suggests that program genre determines the types of MM consumers engage in, and eventually have the power to attenuate or worsen the negative effect of MM on ad memory. We hope these findings provide useful guideline on media planning for advertisers, and help motivate future researchers to build on the current research and continue to deepen our understanding in the area of MM.

References

1. B. R. L. Duff, S. Sar, Seeing the big picture: Multitasking and perceptual processing influences on ad recognition. *Journal of Advertising*, **44**(3), 173-184 (2015)
2. E. Ophir, C. Nass, A. D. Wagner, Cognitive control in media multitaskers. *Proceedings of the National Academy of Sciences*, **106**(37), 15583-15587 (2009)
3. Z. Wang, J. M. Tchernev, The "myth" of media multitasking: Reciprocal dynamics of media multitasking, personal needs, and gratifications. *Journal of Communication*, **62**(3), 493-513 (2012)
4. G. B. Armstrong, L. Chung, Background television and reading memory in context: Assessing TV interference and facilitative context effects on encoding versus retrieval processes. *Communication Research*, **27**(3), 327-352 (2000)
5. M. L. Courage, A. Bakhtiar, C. Fitzpatrick, S. Kenny, K. Brandeau, Growing up multitasking: The costs and benefits for cognitive development. *Developmental Review*, **35**, 5-41 (2015)
6. W. Zhang, S. H. Jeong, M. Fishbein, Effects of multitasking and arousal on television content recall and secondary task performance. *Journal of Media Psychology*, **22**(1), 2-13 (2010)
7. S. Chinchanachokchai, B. R. Duff, S. Sar, The effect of multitasking on time perception, enjoyment, and ad evaluation. *Computers in Human Behavior*, **45**, 185-191 (2015)
8. A. Lang, J. Chrzan, Media multitasking: Good, bad, or ugly? *Annals of the International Communication Association*, **39**(1), 99-128 (2015)
9. E. Wood, L. Zivcakova, P. Gentile, K. Archer, D. De Pasquale, A. Nosko, Examining the impact of off-task multi-tasking with technology on real-time classroom learning. *Computers & Education*, **58**(1), 365-374 (2012)
10. B. R. Duff, C. M. Segijn, Advertising in a media multitasking era: Considerations and future directions. Journal of Advertising, **48**(1), 27-37 (2019)
11. D. M. Krugman, G. T. Cameron, C. M. White, Visual attention to programming and commercials: The use of in-home observations. *Journal of Advertising*, **24**(1), 1-12 (1995)
12. S. H. Jeong, M. Fishbein, Predictors of multitasking with media: Media factors and audience factors. *Media Psychology*, **10**(3), 364-384 (2007)
13. H. A. Voorveld, C. M. Segijn, P. E. Ketelaar, E. G. Smit, Investigating the prevalence and predictors of media multitasking across countries. International Journal of Communication, **8**, 2755-2777 (2014)
14. M. W. Becker, R. Alzahabi, C. J. Hopwood, Media multitasking is associated with symptoms of depression and social anxiety. *Cyberpsychology, Behavior, and Social Networking*, **16**(2), 132-135 (2013)
15. A. Lang, The limited capacity model of mediated message processing. *Journal of communication*, **50**(1), 46-70 (2000)
16. P. D. Bolls, D. D. Muehling, The effects of dual-task processing on consumers' responses to high-and low-imagery radio advertisements. *Journal of Advertising*, **36**(4), 35-47 (2007)
17. S. Kazakova, V. Cauberghe, L. Hudders, C. Labyt, The impact of media multitasking on the cognitive and attitudinal responses to television commercials: The moderating role of type of advertising appeal. *Journal of Advertising*, **45**(4), 403-416 (2016)

18. C. M. Segijn, H. A. Voorveld, E. G. Smit, The underlying mechanisms of multiscreening effects. *Journal of Advertising*, **45**(4), 391-402 (2016)
19. Z. Wang, M. Irwin, C. Cooper, J. Srivastava, Multidimensions of media multitasking and adaptive media selection. *Human Communication Research*, **41**(1), 102-127 (2015)
20. H. Gil de Zúñiga, V. Garcia-Perdomo, S. C. McGregor, What is second screening? Exploring motivations of second screen use and its effect on online political participation. *Journal of communication*, **65**(5), 793-815 (2015)
21. D. L. Hoffman, T. P. Novak, Marketing in hypermedia computer-mediated environments: Conceptual foundations. *Journal of marketing*, **60**(3), 50-68 (1996)
22. K. Wise, H. J. Kim, J. Kim, The effect of searching versus surfing on cognitive and emotional responses to online news. *Journal of Media Psychology*, **21**(2), 49-59 (2009)
23. C. G. Christensen, D. Bickham, C. S. Ross, M. Rich, Multitasking with television among adolescents. Journal of broadcasting & electronic media, **59**(1), 130-148 (2015)
24. R. P. Hawkins, S. Pingree, J. Hitchon, B. Radler, B. W. Gorham, L. Kahlor, G. H. Kolbeins, What produces television attention and attention style? Genre, situation, and individual differences as predictors. *Human communication research*, **31**(1), 162-187 (2005)
25. H. A. Voorveld, V. Viswanathan, An observational study on how situational factors influence media multitasking with TV: The role of genres, dayparts, and social viewing. *Media Psychology*, **18**(4), 499-526 (2015)
26. W. Kool, J. T. McGuire, Z. B. Rosen, M. M. Botvinick, Decision making and the avoidance of cognitive demand. *Journal of Experimental Psychology: General*, **139**(4), 665 (2010)
27. M. J. Apter, Reversal theory and personality: A review. *Journal of Research in Personality*, **18**(3), 265-288 (1984)
28. H. Bang, K. King, The effect of media multitasking on ad memory: the moderating role of program-induced engagement and brand familiarity. *International Journal of Advertising*, 1-30 (2020)
29. S. H. Chaffee, J. Schleuder, Measurement and effects of attention to media news. *Human communication research*, **13**(1), 76-107 (1986)
30. R. Angell, M. Gorton, J. Sauer, P. Bottomley, J. White, Don't distract me when I'm media multitasking: Toward a theory for raising advertising recall and recognition. *Journal of Advertising*, **45**(2), 198-210 (2016)

New goods demand patterns in digital economy environment: The case of electric vehicles in Russia

Svetlana Bozhuk[1], and *Nataliia* Krasnostavskaia[1,*]

[1]Peter the Great St. Petersburg Polytechnic University, Saint Petersburg, Russia

Abstract. The trend of using electric vehicles is changing the automotive industry. Electric cars are becoming the most environmentally friendly replacement for combustion vehicles. Knowing the preferences of potential consumers will allow developing effective solutions to create demand for this product. Generating demand should be based on estimating its potential and shaping the consumer profile of this type of transport for market of each country. New goods need special methods to generate demand, since their potential buyers have difficulties in purchase decision making. This paper presents results of a study on prospects in Russia for such new goods as electric vehicles. The study identified factors that ultimately determine the interest of those Russian consumers who have the financial ability to purchase electric vehicles in the near future in electric vehicles. The study demonstrates that consumer prejudices are still there against difficulties in operating electric vehicles. The study confirmed that a number of factors affect the purchase of an electric car in Russia. Expanding the presence of electric vehicles in carsharing companies will significantly improve experience in using this type of transport by potential users. Generating the demand for electric vehicles by applying influence marketing tools is the one of the best solutions.
Keywords: electric vehicles, electric cars, digital marketing, demand, consumer behavior

1 Introduction

The problem of energy saving is becoming more and more urgent in the world [1, 2]. Electric car technology is gaining momentum thanks to the rapid depletion of fossil fuels, as well as environmental concerns. Government programs significantly motivate the development of the electric vehicles market in different countries. Lobbying for conversion of intra-city commercial and urban public transport to electric drive might provide a considerable growth in demand for electric vehicles. However, in Russia, the vector of government support for eco-friendly transport is aimed at financing the gas-fuelled automobile industry. In such difficult market conditions, the main demand for electric vehicles is generated by individual

* Corresponding author: krasnost_nv@spbstu.ru

consumers. Therefore, generating such demand should be based on estimating its potential and shaping the consumer profile of this type of transport for the Russian market.

Automobile manufacturers aim to bring the electric vehicles price offer to the price level of vehicles with conventional internal-combustion engines. Prospects for the Russian electric vehicles market depend substantially on the awareness of representatives of the younger generation, those to choose their own automobiles in the near future, of the advantages of these vehicles.

The digital economy is developing and consumer behaviour is changing [3, 4]. The authors in their scientific papers addressed the issues of consumer preferences and the choice of electric cars. Knowing consumer preferences and electric vehicle needs is key to developing user-friendly solutions for today's sustainable automotive industry.

The role of customer experience was explored and it was demonstrated how this experience affects the adoption willingness of battery electric vehicles. The mean values of the subjective norm, perceived behavioural control, attitudes, and adoption willingness of experienced consumers are significantly higher than those of inexperienced consumers [5].

The preference for battery electric vehicles as opposed to hybrid electric vehicles or normal combustible vehicles was analysed. Strong evidence that previous experience of driving a battery electric vehicles and car sharing are significant factors to state a preference for electric cars. Other factors such as driving range, purchase price, gender, ecological awareness and incentives such as tax exemptions also influence the choice [6].

Customer purchase intention toward new energy passenger vehicles had a significant, positive relationship with price value, environmental consciousness, quality value, and emotional value. On the contrary, customer purchase intention does not show a significantly positive relationship with social value and government subsidy policy [7].

The Thai car buyers pay attention to the performance factor of electric vehicles (driving range, speed, safety) whilst not being concerned much with the infrastructure (charging facilities) and the financial factors (purchase price, operation and maintenance costs, resale price) of electric vehicles [8].

Potential customers have high expectations regarding the comfort and range of an electric vehicle. One of the most important barriers and obstacles for a broad market uptake are the currently high purchase prices for this new technology [9].

The models are dominated by household characteristics including household income, household size, household vehicle ownership, number of licensed drivers and duration of car ownership. Only two city-level transportation variables have an effect - higher taxi density and higher bus density reducing car purchase. Cold weather, population density gross domestic product per capita positively influence car purchase, while urbanization rate reduces car purchase [10].

Electric car rental may be an effective mode to popularize the electric car, before the cost of the electric car can be reduced substantially [11].

A positive attitude to the choice of an electric car is affected by the use of the different promotion tools especially a test drive.

As electric vehicles become more readily available, sales will depend on consumers' interest and understanding. Electric vehicles rollout has focused too much on technology, and not enough on consumers [12].

The effects of a first-time experience on the evaluation of battery electric vehicles by potential consumers were explored and assessed. The experiment showed that short test drives with battery electric vehicles are a suitable means to support the widespread promotion of electric cars [13].

Vehicle manufacturers should provide customers with more chances to test drive the vehicles, which can give customers direct and true emotional feelings about driving new energy passenger vehicles [14].

The authors also investigated the barriers to the purchase of electric cars.
Information on the orientations and perceptions of passenger car enthusiasts and identifies potential socio-technical-economic barriers to the adoption of electric vehicles was provided [15].

It was reviewed how perceived risk factor influences purchase intention referring to electric vehicles. It was examined how perceived risk influences the purchasing mechanism of electric vehicles whilst buying intention is referred to as a main notion within [16].

The major barriers to a more widespread introduction of battery electric vehicles beyond early adopters are the limited range, charging limitations, and costly batteries. An optimization model to estimate the potential for battery electric vehicles to replace one of the conventional cars in two-car households and to viably contribute to the households' driving demand was developed [17].

It is necessary to take into account the fact that electric cars are a relatively new category of products, and most car users have little experience in using them. Also, consumers often have a misconception about electric cars, do not trust them and avoid buying an electric car.

Longitudinal post-adoption data from early adopters of an electric car model was used and showed that consumers' perceptions of innovative aesthetic value buffer the effect of product-related hedonic experience on attitudes towards the product. Innovative consumers are especially sensitive to innovative aesthetic value, so in their co-creation efforts, managers should seek out these customers, because doing so can increase early adopters' long-term product satisfaction and word of mouth, both of which expand the breadth of diffusion for the innovations [18].

It was investigated how Product-Service Systems influence customer acceptance of innovations, since customer acceptance is responsible for the success of innovative products. The main gaps found with electric vehicles are their limited range and high purchase costs. Customers perceive those gaps as less considerable if they rent electric vehicles in a carpool [19].

There is still a mismatch between supply and demand in electric vehicles. Information acceleration in a test studio can improve the traditional approaches of linear acquisition of information about customers and therefore estimation of adequate pricing and commercialisation in an early decision-making phase and can have a positive influence on decision-making behaviour under uncertainty [20].

The industrial dynamics of electric vehicles using product life cycle and eco-innovation concepts were analysed. Large incumbents and start-up firms have targeted different consumer markets with their electric vehicle models. Start-up firms developed electric vehicle models for niche markets (sports cars and low speed vehicles) while large incumbents generally developed electric vehicle models that are more in line with current customer demands [21].

The development of electric vehicle sales depends on the coordinated work of all stakeholders besides direct consumers: the state, car manufacturers and energy suppliers.

Governmental subsidies, environmental policies and effective marketing strategies are needed to support the electric cars market along technological advancements such as improvement in the electric car's battery life [22].

In markets where introduction of modern clean vehicles had been met with supportive policies, their demand has been relatively strong. Factors stated by respondents to influence their opinions of these vehicles included the policy environment, social networks, media influence, and the car-buying experience [23].

The purpose of the study is to determine the portrait of the target audience of electric vehicles in Russia, to identify the factors influencing the purchase of an electric vehicle.

2 Methodology

The study involved 450 respondents over 30 years old. A previous study that looked at potential electric vehicle buyers looked at young people aged 17–30. However, this study focused on those potential buyers who have the necessary financial capabilities to buy a car in the near future.

The study included features that may be divided into feature-factors (10) and resulting features (15).

Lack of experience in using the car creates the consumer's uncertainty when buying it [5, 6]. If a family owns a car, it creates a better consumer understanding of the conditions and requisites for operating a vehicle. In addition, availability of a car is an indirect characteristic of the family income level.

Frequency of the car use is a feature that directly characterizes consumer experience. An unquestionably best solution would be the first-hand experience from driving an electric vehicle. The respondents have no experience of driving electric vehicles in Russia. Trip distance characterizes the need to move. Considering the limited battery mileage of electric vehicles, this feature is an important characteristic. Since the limited trip distance is a barrier to buying an electric vehicle, it is necessary to assess the impact of that feature on consumer intentions. A car being a major purchase, it is rarely made without somebody's recommendation, so was decided to include that feature in the study.

Factors also included the respondent's opinion on reasonable consumption and environmental friendliness of the car: whether it makes sense to have two cars in the family, whether the car is a strong source of pollution. The number of drivers in the household affects the family's striving for diversity of vehicles [10]. In Russia, an electric vehicle is frequently the second or even third car in the family. The authors considered it important to measure this feature during the survey. The consumer's awareness of the environmental impact of the vehicle is significant for displaying interest in purchasing an electric vehicle [6, 14]. So, this feature was included in the list of features to be measured.

Resultant features (dependent features) reflected the respondents' opinions on the interest, reasons and problems of using an electric vehicle, as well as information on their intention to purchase an electric vehicle.

Respondents' opinions on causes for popularity of electric cars reflect their understanding of the advantages of such vehicles, and their key benefits for the respondent. Consumers' opinions on the potential difficulties of operating electric car provide information on the barriers to such a purchase. It can be assumed that such barriers are: the user's financial limitations (cost of the vehicle), operating conditions (shortage of charging stations and service, weather, state of the roads), electric car performance characteristics (speed and roominess).

One of the reasons for failure of programs promoting electric cars is insufficient information given to customers, so this feature was also included in the study as a purchase barrier [24].

The initial set of features-factors characterizing potential consumers of electric vehicles is presented in Table 1.

Table 1. Features-factors characterizing potential consumers of electric vehicles

Description of the feature-factor	Measurement scale	Number of categories/values	Range of values
Gender	Nominal	2	Male/female
Age	Ratio	-	[over 30 years: 31, 32, …]
Household income	Interval	11	1 – less than \$500/month 2 – \$501-\$1000/month 3 – \$1001-\$1500/month 4 – \$1501-\$2000/month 5 – \$2001-\$2500/month 6 – \$2501-\$3000/month 7 – \$3001-\$3500/month 8 – \$3501-\$4000/month 9 – \$4001-\$4500/month 10 – \$4501-\$5000/month 11 – more than \$5001/month
The presence of a car in the family	Nominal	2	Yes/no
Number of cars in the family	Ratio	-	[1, 2, …]
Who influenced the choice of the car when buying	Nominal	6	Father/mother/ husband/wife/self/other
Frequency of car use in the family	Ordinal	3	1 – every day (frequent) 2 – once or twice a week (average) 3 – several times a month (rare)
Distance of trips in the family	Ordinal	3	1 – to other cities of Russia, to other countries (long trips) 2 – no further than to the countryside (average) 3 – within the city limits only (short)
Whether it makes sense to have two cars in the family	Ordinal	3	1 – feasible in any case 2 – feasible under certain conditions 3 – unfeasible
Whether the car is a strong source of pollution	Ordinal	5	1 – no, absolutely not 2 – no 3 – I can't answer 4 – yes 5 – yes, absolutely

Source: Own processing.

The initial set of resultant features characterizing potential consumers of electric vehicles is presented in Table 2.

Table 2. Resultant features characterizing potential consumers of electric vehicles

Description of the resultant feature	Measurement scale	Number of categories/values	Range of values
Problems of purchasing electric vehicle in Russia:			
Underdeveloped infrastructure	Nominal	2	Yes/no
Severe weather conditions	Nominal	2	Yes/no
Poor awareness of offer in electric vehicles	Nominal	2	Yes/no
Prohibitive pricing	Nominal	2	Yes/no
Limited battery mileage	Nominal	2	Yes/no
Complexity of operation and repair	Nominal	2	Yes/no
Limited roominess	Nominal	2	Yes/no
Low speed	Nominal	2	Yes/no
Important factors when purchasing electric vehicle:			
Developed infrastructure	Nominal	2	Yes/no
Special incentives from government	Nominal	2	Yes/no
Available service	Nominal	2	Yes/no
Available price	Nominal	2	Yes/no
Long range with the battery	Nominal	2	Yes/no
Attractive design	Nominal	2	Yes/no
Personal interest in purchasing electric vehicle in the near future	Interval (Thurstone scale of equidistant intervals)	11	0 – no chance 1 – almost no chance 2 – very weak probability 3 – some probability 4 – sufficient probability 5 – quite good probability 6 – high probability 7 – likely 8 – highly likely 9 – almost sure 10 – no doubt

Source: Own processing.

Attributing a feature to a particular type of scale is determined primarily by its content and is of fundamental significance, since the possibility of using a particular data analysis tool depends on it.

The data was analysed using SPSS software.

3 Results

At the first stage of the study, descriptive statistical analysis of each feature was performed. Among the respondents' answers to the question "Personal interest in purchasing electric vehicle in the near future", two options prevailed corresponding to positions on the Thurstone scale of equidistant intervals: 2–3 and 5–7.

Feature distribution at the 0.01 significance level differs from the normal one, since p-values corresponding to the Shapiro-Wilkes test (W statistics) and D'Agnostino test (K2 statistics) substantially lower than the value pre-set (0.01).

Non-parametric closeness indicators were also calculated: Spearman and Kendall rank correlation coefficients (for ordinal factors), Chuprov and Kramer contingency coefficients (for nominal factors), plus the significance of each factor influence on the result, which was evaluated using the Mann-Whitney test (for alternative factors) and Kruskal-Wallis test (for factors taking more than 2 values).

For nominal features, such as "The presence of a car in the family", "Who influenced the choice of the car when buying", the most frequent values were determined. The vast majority of respondents (74%) have a car in their family, and husband, wife are the persons who influenced the choice of the car most frequently (65%).

The distribution analysis for the group of alternative features "Problems of purchasing electric vehicle in Russia", which includes 8 alternative features, corresponding to specific barriers to use, identified three main reasons (they were marked by more than 50% of respondents):

- underdeveloped infrastructure (72%),
- prohibitive pricing (67%),
- severe weather conditions (54%).

The distribution analysis for the group of alternative features "Important factors when purchasing electric vehicle", which includes 6 alternative features, corresponding to specific barriers to use, identified three main reasons (they were marked by more than 50% of respondents):

- long range with the battery (68%),
- available service (61%),
- special incentives from government (59%).

The results demonstrate proper assessment of conditions and rather good awareness of key problems in operating electric vehicles in Russia. The study confirmed the presence of standard concerns about electric vehicles among the Russian consumers.

Generalizing the results on the paired relations, we may conclude that males develop a stronger intention to purchase an electric vehicle with age 45-50. As for females, the strongest intention to purchase an electric vehicle is at the age 38-43. In general, females are less inclined to purchase an electric vehicle than males.

Those concerned with such problems of using electric vehicles in Russia as "Limited battery mileage", "Limited roominess", "Low speed", are less likely to purchase an electric vehicle among the respondents. On the contrary, those who believe that the key problem with electric vehicle penetration in Russia is "Underdeveloped infrastructure", were more willing to purchase an electric vehicle in the near future.

Summarizing results of the comprehensive statistical data analysis, we can offer the following characteristics of the potential Russian consumer of electric vehicles in Table 3.

Table 3. Characteristics of the potential Russian consumer of electric vehicles

Consumer characteristic	Typical value
Gender	Male
Age	45-50
Household income	More than $3001/month
The presence of a car in the family	Yes
Number of cars in the family	1-2
Who influenced the choice of the car when buying	Member of the family
Frequency of car use in the family	Every day
Whether it makes sense to have two cars in the family	Feasible in any case
Whether the car is a strong source of pollution	Yes
Problems of purchasing electric vehicle in Russia	Underdeveloped infrastructure
Important factors when purchasing electric vehicle	Long range with the battery
Personal interest in purchasing electric vehicle in the near future	High probability

Source: Own processing.

It should be noted that the results of the multiple correspondence analysis corelate with those from the paired association analysis and the ordinal regression estimation. This makes our conclusions more confident.

4 Discussion

In this study, the object of the study was people over 30 years old. People of this age most often have a formed opinion about environmental problem. More than 40% of the respondents also showed a formed opinion about the problems of transport and issues related to the possible use of electric vehicles.

Electric cars in Russia are perceived as innovative products. Potential consumers are characterized by low awareness of the characteristics of electric vehicles. It should also be noted that at present, programs to raise awareness of potential consumers about the characteristics of electric vehicles are not widely used. Those technological advances that allowed car manufacturers to improve consumer properties of electric vehicles significantly are not known to the lay public.

Given a level playing field, such as vehicle class and price, it can be expected that choosing an electric vehicle is still unlikely. Consumers cite factors as the main problems: underdeveloped infrastructure, prohibitive pricing and severe weather conditions. The provision of electric vehicles as car-sharing vehicles, as well as the expanding use of another electric vehicle, electric scooters, can increase not only the awareness of potential consumers, but also allow them to have their first experience of driving electric vehicles.

The revealed portrait of potential consumers of electric vehicles shows that these people are ready to consider the electric car as a second car, they definitely declare the presence of environmental problems, are rational and expect that the use of the electric car will be beneficial.

The study confirmed that a number of factors affect the purchase of an electric car, for example, the presence of a second car in the family.

Characteristics describing the value orientation of consumers showed a stronger influence than socio-demographic characteristics. Supporting responsible consumption and protecting the environment matters for potential consumers of electric vehicles.

5 Conclusion

Social media promotion of the idea of electric vehicles should be used to raise awareness and engagement in the purchase. The best solution consists in generating demand for electric vehicles via the influence marketing tools. First, bloggers, as opinion leaders, contribute to raising the audience awareness of offer in electric vehicle and developing its interest in innovation. Second, the product recommendation comes not from the company to the consumer, but from consumer to consumer in that case. It increases significantly consumers' confidence in the information received.

The purpose of the study was to determine the portrait of the target audience of electric vehicles in Russia, to identify the factors influencing the purchase of an electric vehicle. The research objectives for the potential target audience have been achieved. A portrait of the target audience was identified, characteristics that would facilitate the purchase of an electric vehicle in Russia in the future while contributing to the solution of the problems named by the respondents were identified.

This research work was supported by the Academic Excellence Project 5-100 proposed by Peter the Great St. Petersburg Polytechnic University.

References

1. S. Bozhuk, N. Krasnostavskaia, T. Maslova, N. Pletneva, The problems of innovative merchandise trade in the context of digital environment. *IOP Conference Series: Materials Science and Engineering*, **497**, p. 012115 (2019)
2. S. Bozhuk, T. Maslova, N. Kozlova, N. Krasnostavskaia, Transformation of mechanism of sales and services promotion in digital environment. *IOP Conference Series: Materials Science and Engineering*, **497**, p. 012114 (2019)
3. N. Krasnostavskaia, N. Pletneva, M. Kupriyanova, S. Golovkina, The level of involvement and the nature of the stimulus as factors in the decision-making process on the purchase of handmade goods on the Internet. *IOP Conference Series: Materials Science and Engineering*, **940**, p. 012069 (2020)
4. N. Krasnostavskaia, T. Maslova, L. Ruglova, M. Chigir, Problems of forming marketing competencies in the digital economy. *IOP Conference Series: Materials Science and Engineering*, **940**, p. 012066 (2020)
5. R. Liu, Z. Ding, X. Jiang, J. Sun, Y. Jiang, W. Qiang, How does experience impact the adoption willingness of battery electric vehicles? The role of psychological factors. *Environmental Science and Pollution Research*, **27**(20), 25230–25247 (2020)
6. A. S. Darup, M. Guillen, X. Piulachs, Consumer preferences for electric vehicles in Germany. *International journal of transport economics*, **45**(1), 97–122 (2018)
7. A. Mangmeechai, A. Yuan, Factors Influencing Customer Purchase Intentions towards New Energy Vehicles in China. *The International Journal of Interdisciplinary Environmental Studies*, **13**(1), 1–15 (2018)

8. T. Thananusak, S. Rakthin, T. Tavewatanaphan, P. Punnakitikashem, Factors affecting the intention to buy electric vehicles: empirical evidence from Thailand. *International Journal of Electric and Hybrid Vehicles*, **9**(4), 361 (2017)
9. J.-A. Bühne, D. Gruschwitz, J. Hölscher, M. Klötzke, U. Kugler, C. Schimeczek, How to promote electromobility for European car drivers? Obstacles to overcome for a broad market penetration. *European Transport Research Review*, **7**(3), 30 (2015)
10. Z. Ling, C. R. Cherry, H. Yang, L. R. Jones, From e-bike to car: A study on factors influencing motorization of e-bike users across China. *Transportation Research Part D: Transport and Environment*, **41**, 50–63 (2015)
11. R. Miao, J. Cao, K. Zhang, B. Chen, Z. Jiang, L. Wang, Value-added path of service-oriented manufacturing based on structural equation model: the case of electric car rental for instance. *International Journal of Production Research*, **52**(18), 5502–5513 (2014)
12. P. D. Larson, J. Viáfara, R. V. Parsons, A. Elias, Consumer attitudes about electric cars: Pricing analysis and policy implications. *Transportation Research Part A: Policy and Practice*, **69**, 299–314 (2014)
13. F. Hinnüber, M. Szarucki, K. Szopik-Depczyńska, The Effects of a First-Time Experience on the Evaluation of Battery Electric Vehicles by Potential Consumers. *Sustainability*, **11**(24), 7034 (2019)
14. A. Mangmeechai, A. Yuan, Factors Influencing Customer Purchase Intentions towards New Energy Vehicles in China. *The International Journal of Interdisciplinary Environmental Studies*, **13**(1), 1–15 (2018)
15. F. E. Ciarapica, D. T. Matt, M. Rossini, P. R. Spena, Quality, environmental and economic factors influencing electric vehicles penetration in the Italian market. *XVIII Summer School "Francesco Turco" 11-13-September-2013*, pp. 358–363 (2013)
16. D. Thilina, N. Gunawardane, The effect of perceived risk on the purchase intention of electric vehicles: an extension to the technology acceptance model. *International Journal of Electric and Hybrid Vehicles*, **11**(1), 73 (2019)
17. S. Karlsson, What are the value and implications of two-car households for the electric car? *Transportation Research Part C: Emerging Technologies*, **81**, 1–17 (2017)
18. C. Nagel, J. H. Schumann, Post-adoption buffering effects of innovative product aesthetics. *Creativity and Innovation Management*, caim.12363 (2020)
19. D. M. Schmidt, F. Braun, S. A. Schenkl, M. Mörtl, Interview study: How can Product-Service Systems increase customer acceptance of innovations? *CIRP Journal of Manufacturing Science and Technology*, **15**, 82–93 (2016)
20. H. Proff, T. M. Fojcik, Pricing and commercialisation of electric mobility - dealing with high market uncertainty. *International Journal of Automotive Technology and Management*, **16**(1), 30 (2016)
21. W. Sierzchula, S. Bakker, K. Maat, B. van Wee, The competitive environment of electric vehicles: An analysis of prototype and production models. *Environmental Innovation and Societal Transitions*, **2**, 49–65 (2012)
22. S. W. Lim, K. F. Chen, E. H. Yap, System Dynamics of Electric Cars (EC) Usage and Support Infrastructure in Malaysia. *Applied Mechanics and Materials*, **627**, 342–346 (2014)
23. A. Dini, S. Washington, G. Hawkins, Understanding barriers to consumer demand of plug-in vehicles in Australia. *O'Keeffe, B (Ed.) Australasian Transport Research Forum 2013 Proceedings. Australasian Transport Research Forum, Australia*, pp. 1-13 (2013)
24. K. Dudenhöffer, Why electric vehicles failed. *Journal of Management Control*, **24**(2), 95–124 (2013)

Does COVID-19 really make people risk aversion in investment decision-making?

Muhammad Shadab Iqbal[1], and *Lin* Li[2,*]

[1]Graduate School of Venture, Hoseo University, Seoul, Korea
[2]College of Business Administration, Kookmin University, Seoul, Korea

Abstract. The economic fallout from COVID-19 pandemic changes individuals' investment perceptions and behaviors in a tremendous way. Consequently, investment decision-making has been affected as people have to adjust to the new environment. This study aims to study whether COVID-19 really make people risk aversion due to the economic slowdown. Our empirical results are analyzed from household finance data in U.S in July 2021. It is found that COVID-19 proximity, income, and occupation are positively associate with risking taking in investment decision-making, while age and family size are not. This study contributes to the newly emerged body of knowledge on post pandemic investment decision-making and risk behavior analysis and provide implications for financial investment institutions.
Keywords: Risk aversion, Risk taking, Investment decision-making, COVID-19 pandemic

1 Introduction

Coronavirus (COVID-19) pandemic has swept the globe and affected nearly all aspects of life, from economic to politics, from workplace to everyday activities. For example, the unemployment rates are on the rise in many counties and most of countries are in recession. The International Monetary Fund (IMF) estimated that the world economy declined by 4.4% in 2020, which is the worst since the Great Depression in the 1930s [1].

The economic fallout from COVID-19 changes individuals' investment perceptions and behaviors in a tremendous way. Consequently, investment decision-making has been affected as people have to adjust to the new environment. Risk perception shows how risks are interpreted from different thoughts and reality [2]. Risk perception plays a vital role in influencing household behavior, especially in investment decision making among uncertain environment [3]. It is important to study the influence of a negative shock (COVID-19 in this case) on households, which are essential to the economic system in the micro level. Individuals in households' response to the pandemic both in perceptions and behaviors, which demand more research to obtain a better understanding of the household financial investment decision-making response to the negative shock.

* Corresponding author: lilin@kookmin.ac.kr

The research gap is that limited number of related studies have mixed results on risk aversion or risk taking during COVID-19 pandemic. For example, it is found that households tend to become risk aversion amidst the COVID-19 [4]. However, it is discovered that individuals' risk tolerance increase for the early stages of COVID-19 by using an incentivized experiment [5], and one study found that there is no change in risk aversion or risk taking in a risk elicitation task during COVID-19 [6].

The mixed results intrigue this study to study whether COVID-19 really make people risk aversion in the investment decision-making due to the economic slowdown. This study contributes to the newly emerged body of knowledge on post pandemic investment decision-making and risk behavior analysis and provide implications for financial investment institutions.

2 Methodology

2.1 Data collection

The data is collected from 200 household finance data across the U.S. through Amazon Mechanical Turk (MTurk) service in July 2021. MTurk is a gig economy platform and serves as a crowdsource website and a digital labor platform [7]. Survey data from MTurk is found to have no significant difference in terms of reliability when compared to other survey methods [8].

The potential problem of using MTurk is that participants receive payments may influence the objectivity of the data. However, it is argued that respondent's extrinsic motivation with the payment is not a major concern if a reasonable payment is given for the time spending on answering the questions [9]. In this study, USD 50 cents were paid to the participants for using 4 to 5 minutes of their time, which indicate that the representativeness of the data sample is not a major concern.

2.2 Empirical model

This study follows the extant study [4] and uses logistic regression analysis model for analysis. The model for the empirical analysis is shown as below:

$$Y_i = \alpha + \text{income}_i + \text{family size}_i + \text{age}_i + \text{occupation}_i + \text{covid proximity}_i + z_i + \varepsilon_i \quad (1)$$

Where Y_i is the dependent variable (i.e., risk aversion as 0 and risk taking as 1), Zi is a set of control variables, and ε_i is an idiosyncratic random error for a household i.

2.3 Measurement of variables

For independent variables, income is measured by 9 groups, family size is measured by how many members are in a household, and age is measured by 7 age groups. Occupation is measured by 7 groups, the bigger the number is, the closer the job is related to white-collar, service, and financial related jobs. The COVID proximity is measured by the number of COVID-19 cases per 100,000 people in the neighbourhood based on home postal code in the U.S., where 1 indicates low proximity and 7 indicates high proximity.

For the control variables, education is measured in 7 groups, and gender is measured by 2 groups, 1 indicates females and 2 indicates males.

As to the dependent variable, investing in safe-haven and traditional assets (such as gold) for more than 50% in the household investment portfolio is considered as risk aversion and it is measured as 0. On the other hand, investing more than 50% in emerging and risker assets (such as cryptocurrency) is regarded as risk taking in the household investment decision-making and it is measured as 1.

3 Results and Discussion

3.1 Logistic regression analysis

Logistic regression analysis model is used in this study. Since the dependent variable was coded as 0 (risk aversion) or 1 (risk taking), logistic regression analysis was chosen because the outcome variable is based on logit equivalent [10]. In addition, the maximum likelihood estimation (MLE) was used to estimate the path coefficients. The model is shown as below:

$$\ln\left[\frac{P(y_i = 1)}{P(y_i = 0)}\right] = b_0 + b_1x_1 + b_2x_2 + \cdots + b_nx_n \quad (2)$$

Where x_1 to x_n are the n predictors, y_i is the outcome.

3.2 Data analysis results

Summary statistics are shown in Table 1 as below. Number of objects, mean value, standard deviation, maximum value, and minimum value are included in the summary statistics.

Table 1. Summary statistics

Variables	Obs	Means	Std.	Min.	Max.
Income	200	3.74	0.132	1	9
Family Size	200	4.98	0.112	1	7
Age	200	5.54	0.080	1	7
Occupation	200	2.50	0.102	1	7
COVID	200	4.91	0.115	1	7
Education	200	3.36	0.106	1	7
Gender	200	1.50	0.035	1	2

Source: Own processing.

Before testing the hypotheses, correlation analysis and a multicollinearity test were conducted, and the results are shown in Table 2. If the correlation between two variables is above 0.80, it is problematic [11]. The correlation matrix indicates that the highest value of the correlation coefficient was 0.359, showing acceptable correlations among variables. In addition, the variance inflation factor (VIF) values were checked for multicollinearity. The VIF values of this study varied from 1.047 to 1.241, which is less than the threshold value of 5.0 [12], meaning no multicollinearity issue.

Table 2. Correlations and results of multicollinearity analysis

	(1)	(2)	(3)	(4)	(5)	(6)	(7)	VIF
Income (1)	1							1.241
Family Size (2)	0.142*	1						1.092
Age (3)	0.171*	0.187**	1					1.088

Occupation (4)	0.225**	-0.041	0.136	1				1.081
COVID (5)	0.359**	0.214**	0.145	0.144**	1			1.221
Education (6)	0.180*	0.046	-0.052	0.055	0.113	1		1.047
Gender (7)	0.079	0.029	-0.005	0.017	0.149*	0.001	1	1.025

*$p<0.05$; **$p<0.01$

Source: Own processing.

The logistic analysis results are shown in Table 3. Model 1 has only the control variables, and both control variables were significantly related to risk taking: education (β=0.234, $p<0.005$), gender (β=0.705, $p<0.005$). The results mean that the higher the education level is, the higher chance that risk will be taken in the investment decision-making. Moreover, males are more likely to take risk than females in the investment decision-making.

Model 2 includes the independent variables, and the Cox & Snell R^2 was increased from 0.056 to 0.629, and Nagelkerke R^2 was increased from 0.075 to 0.849, indicating the strength of connections between the independent variables and the dependent variables increase significantly. Income (β=2.219, $p<0.001$), occupation (β=0.770, $p<0.001$), and COVID proximity (β=1.025, $p<0.01$) were significantly related to risk taking. While family size (β=0.049, $p>0.05$) and age (β=-0.549, $p>0.05$) were not significantly related to risk taking.

Table 3. Results of logistic regression analysis

Dependent variable: Risk Aversion (0) or Risk Taking (1)	**Model 1 (Control variables)**			**Model 2 (Model 1 + Independent variables)**		
	B	Standard Error (SE)	*p*-value	B	SE	*p*-value
Control variables						
Education	0.234	0.099	0.018*	0.353	0.221	0.110
Gender	0.705	0.292	0.016*	1.815	0.682	0.008
Independent variables						
Income				2.219	0.366	0.000***
Family Size				0.049	0.224	0.827
Age				-0.549	0.296	0.063
Occupation				0.770	0.240	0.001***
COVID Proximity				1.025	0.253	0.000***
Constant	1.805	0.582	0.002	15.935	3.192	0.000***
R^2 (Nagelkerke)	0.075			0.849		
R^2 (Cos & Snell)	0.056			0.629		
d.f.	2			2		
P	0.003			0.003		
-2 Log Likelihood Ratio	265.650			78.797		

Notes: *$p<0.05$; **$p<0.01$; ***$p<0.001$

Source: Own processing.

3.3 Discussion of results

The interpretations of our findings are described as below. First, it is interesting to find out that the higher proximity of COVID-19 is in terms of COVID-19 cases in the neighbourhood per 100,000 people, the higher chances are that people are more risk taking in their investment decision-making. The possible explanation is that households may become risk aversion amidst the COVID-19 [4], or no change in risk aversion or risk taking during COVID-19 [6] in general as extant studies found out. But we believe that when there are actually more COVID-19 cases in the neighborhood, people might consider the COVID-19 pandemic as a more concrete event due to a short spatial distance [13], thus process and analyze the investment plans more carefully, resulting in associating risks as low-level means [13], which indicates increasing the risk tolerance [5] and taking more risks in their financial decision-making. On the contrary, if there are less cases in the neighborhood, the COVID-19 pandemic could be processed as a dreadful event that happens far away, which results in associating risks as high-level means [13], indicating less mental process is used in the financial decision-making process and resulting in less changes in the risk tolerance.

Second, income and occupation are found to be positively associated with risk taking in the investment decision-making during the COVID-19 pandemic. The higher the income is, the higher chance individuals take more risks in investment decision-making, the possible explanation is that individuals with more disposable income could afford to take bigger risks with their investments. As to the occupation, we found that when the jobs are more white-collar, service, and financial related, more risks could be accepted in the investment decision-making, the reason could be that more financial knowledge makes it comfortable to accept volatility in their investment portfolios and make risker decisions in their investments.

Third, family size and age are found to be not significantly associated with risk taking in the investment decision-making during the COVID-19 pandemic. The findings are consistent with [14] that the number of children nor the number of dependents influence the level of risk tolerance. As to the age, there is a tendency that the older the individuals are, the less likely they would like to take risks in their investment decision-making, yet the associate is not significant.

4 Conclusion

In conclusion, this study found that COVID-19 proximity, income, and occupation are positively associate with risking taking in investment decision-making, while age and family size are not. COVID-19 pandemic may be considered as a concrete event and raise risk tolerance if there are more cases in the neighborhood despite the economy uncertainty that has caused by the pandemic in general. Therefore, COVID-19 pandemic does not necessarily make people more risk aversion as expected. In addition, whether households take more risk in their investment decision also related positively with income and occupation, but not with age and family size.

This study contributes to the newly emerged body of knowledge on post pandemic investment decision-making and risk behavior analysis and provide implications for financial investment institutions. Even though it is important to invest in safe-haven assets through various times and situations, it cannot be assumed that households naturally become risk aversion due to the COVID-19 pandemic. Therefore, financial investment institutions should not only focus on finding safe-haven assets, but also finding emerging assets that could benefit the returns of individuals in a timely manner.

References

1. L. Jones, D. Palumbo, D. Brown, Coronavirus: How the pandemic has changed the world economy [online], Available at: https://www.bbc.com/news/business-51706225 (2021)
2. A. Nadya Septi Nur, L. Lutfi, The influence of risk perception, risk tolerance, overconfidence, and loss aversion towards investment decision making. *Journal of Economics, Business, & Accountancy Ventura,* **21**(3), 401-413 (2019)
3. D. Forlani, J. W. Mullins, Perceived risks and choices in entrepreneurs' new venture decisions. *Journal of business Venturing,* **15**, 305-322 (2000)
4. P. Yue, A. Gizem Korkmaz, H. Zhou, Household financial decision making amidst the COVID-19 pandemic. *Emerging Markets Finance and Trade,* **56**(10), 2363-2377 (2020)
5. J. Shachat, M. J. Walker, L. Wei, The impact of the Covid-19 pandemic on economic behaviours and preferences: Experimental evidence from Wuhan. *ESI Working Paper,* **20**(33) (2020)
6. M. Angrisani, M. Cipriani, A. Guarino, R. Kendall, J. Ortiz de Zarate, Risk preferences at the time of COVID-19: an experiment with professional traders and students. *FRB of New York Staff Report,* **927** (2020)
7. T. M. Daly, R. Nataraajan, Swapping bricks for clicks: Crowdsourcing longitudinal data on Amazon Turk. *Journal of Business Research,* **68**(12), 2603-2609 (2015)
8. M. Buhrmester, T. Kwang, S. D. Gosling, Amazon's Mechanical Turk: A new source of inexpensive, yet high-quality data? *Perspectives on Psychological Science,* **6**(1), 3-5 (2016)
9. P. B. Lowry, J. D'Arcy, B. Hammer, G. D. Moody, "Cargo Cult" science in traditional organization and information systems survey research: A case for using nontraditional methods of data collection, including Mechanical Turk and online panels. *The Journal of Strategic Information Systems,* **25**(3), 232-240 (2016)
10. S. Banerjee, A. Y. Chua, A theoretical framework to identify authentic online reviews. *Online Information Review,* **38**(5), 634-649 (2014)
11. M. Licht, Multiple regression and correlation.In *L.G. Grimm & P.R. Yarnold (Eds.), Reading and understanding multivariate statistics*, Washington, DC, American Psychological Association, 19-64 (1995)
12. M. A. Schroeder, J. Lander, S. Levine-Silverman, Diagnosing and dealing with multicollinearity. *Western Journal of Nursing Research,* **12**(2), 75-187 (1990)
13. Y. Trope, N. Liberman, Construal-level theory of psychological distance. *Psychological Review,* **117**(2), 440-450 (2010)
14. T. A. Hallahan, R. W. Faff, M. D. McKenzie, An empirical investigation of personal financial risk tolerance. *Financial Services Review-greenwich,* **13**(1), 57-78 (2004)

Valuation of construction work on architectural heritage and its influence on project implementation decision

Lucie Brožová[1], *Iveta Střelcová*[1], *Stanislav Vitásek*[1], *Radan Tomek*[1] and *Tomáš Krulický*[2*]

[1]Czech Technical University in Prague, Faculty of Civil Engineering, Thákurova 7, Prague 6, Czech Republic
[2]Institute of Technology and Business in České Budějovice, School of Expertness and Valuation, Okružní 517/10, 370 01 České Budějovice, Czech Republic

Abstract. The article deals with the description of the methodology for determining the prices of construction work used most often in the reconstruction of historic buildings. The procedure is based on parametric estimating of buildings and on traditional tools and principles of pricing. The historic building is divided into structural parts and technological units. Structural elements are individually valued at the unit price, which was created on the basis of sub-budget. Very strict rules are set for the financing of cultural monuments and the method of valuation is based on the price systems' methodology. The research also included an analysis of market prices and a subsequent comparison with indicative prices. Proper estimate of historic buildings' reconstruction cost is also an important input for the decision-making process regarding the actual implementation of the project, its method of execution and its life cycle cost determination.
Keywords: cultural monument, historical objects, valuation of reconstructions, parametric valuation, price analysis

1 Introduction

Preservation of cultural monuments is one of the important ways to get to know our past. Caring for them is not an easy task, but it is the crucial one for the National Monuments Institute (NPÚ). A significant part of this care is the restoration and maintenance of architectural monuments. It is therefore necessary to pay considerable attention to the economic aspect of this care, specifically in terms of valuation. The uniqueness of every such cultural monument and the emphasis on the delicate way of intervention make it difficult to value such buildings. Each cultural monument requires a different degree of intervention, and each element that appears in the monument should be treated as unique. Valuation of reconstructions of historic buildings is problematic both in the phase of the first cost estimates and in the creation of the itemized budget of the reconstruction. The biggest problem in both cases is the high specificity and uniqueness of the objects. It is

* Corresponding autor: krulicky@mail.vstecb.cz

important to establish methodologies that contribute to full recovery and prevent the use of inappropriate materials and workflows, which is still occasionally happening.

The aim of this article is to outline the method of valuing construction work on historic buildings in the phase of the first cost estimates, as well as in creating an itemized budget for reconstruction. The proposed methodology is based on an analysis of market prices of construction work most often performed during repairs of historic buildings.

Proper estimate of historic buildings' reconstruction cost is also an important input for the decision-making process regarding the actual implementation of the project. It also helps to predict its life cycle cost and to reduce risks during the reconstruction works.

2 Literature Review

The issue of valuing construction work on immovable cultural monuments has not yet been addressed in more depth in the professional literature. There are many publications that deal with the preservation of cultural heritage objects, but they usually do not address construction costs and other economic aspects [1, 2, 3].

Many authors also study 3D modelling of historic buildings [4, 5, 6]. However, the economic aspect and the calculation of construction costs lag behind.

In construction, estimation of costs is important for both the investor who finances the investment and the contractor who has to estimate the costs and achieve a level of profit.

To determine the price of construction works, we have to know the direct and indirect costs, so that the contractors can estimate costs of works based on the unit price. This method requires a lot of experience from the contractor. The unit price must include all of the mentioned cost's elements. Traditional method of cost estimating is based on calculation of separate costs elements: the direct cost (labour, materials, and equipment), indirect cost and profit analyze article[7].

The cost estimating in construction is the main topic of the article by Bernhard Bauer et al. [8]. According to the authors, each cost estimate has peculiarities and has to be adapted for each project. The better the project documentation, the more accurate the cost estimate and price determination.

The article by Radovan Majer et al [9] also deals with the issue of estimating in construction. The article targets the methods of profit calculation, because budgetary factors are frequent cause of business failure in construction.

The link between building life and life cycle costs is analysed by Heralová [10]. Macek [11] in detail analyzes the costs of real estate maintenance. Karásek [12] also dealt with life cycle maintenance cost planning. Heralová et al. [13] deals with the issue of maintenance and renovation of cultural heritage objects. This is a very specialised area of cost optimization.

The reconstruction of the cultural monuments can also be seen as an investment into the historical substance of the region. From the economic point of view, it can provide new benefits as Dlask and Beran [14] claim.

The procedures leading to the creation of sub-budgets (i.e. few aggregated estimates) and budgets for the structural elements of historical buildings were applied by the External Thermal Insulation Composite System (ETICS) by Dlask et al. [15]. For unique historical structures, it is possible to separate selected items and use them in budgeting.

Vrbka et al. [16] focuses on the synthesis of valuation methods that determine the total amount of investment. Hromada et al. [17] deals with the description of the methodology for calculating the reproduction value of buildings.

Macek et al. [18] wrote the article about the issue of maintenance and renovation of cultural heritage objects and about the cost optimization.

3 Valuation of construction work on architectural heritage

Valuation of construction work on immovable cultural monuments methodically on the valuation of construction activities related to historic buildings. The uniqueness of individual historical building's elements causes higher demands on the technologies used in comparison with classic (modern) building structures during their revitalization. Therefore, the valuation proposal itself is based on an analysis of market prices for construction work used in historic buildings in combination with traditional tools and pricing methods for the regular ones.

The valuation methodology is based on seven pillars:

- Price systems for buildings.
- Budgetary principles.
- Individual estimates of items using a standard estimating formula.
- Hourly billing rates.
- Principles of parametric valuation of buildings and division of structural elements respecting the needs of immovable cultural monuments.
- Sub-estimates in relation to defined structural elements.
- Analysis of market prices for construction work used in historic buildings.

3.1 Price systems for buildings

The price system is a comprehensive database on assembly and construction work, building materials and products. It includes information systematically classified into items. These individual items consist of an identifier (code), a description and a unit of measure, supplemented by price and technical conditions intended for the calculation of necessary costs and the calculation of the unit price. The reported values in the price systems are obtained by statistical methods from the data of the respondents (usually real companies' bid prices from recent tenders) in the structure of the estimating formula.

Usually, each country has its own national classification system or at least the usual structure for classifying building structures and works. In the Czech construction environment, it is the Classification of Building Structures and Works (TSKP) on which the price systems issued by private engineering companies are based.

Price systems are primarily intended for the needs of building construction. However, they also have their use for the line constructions, where they are applied for construction projects that are not financed by the State Fund for Transport Infrastructure (SFDI), but by other public contracting authorities (regions, municipalities, etc.). These are usually simplier objects, such as third-class roads, culverts, footbridges, etc. [19]

3.2 Budgetary principles by price system

For the valuation of construction work on immovable cultural monuments, we use the TSKP classification. The TSKP classification is based on construction parts (Main Construction Production (HSV), i.e. underground construction plus framework; Associated Construction Production (PSV), also known as work trades; and assembly price lists (M price lists). Specifically, for construction work related to demolition or repairs in the HSV construction work, separate catalogs (price lists) intended for demolition and repairs and maintenance are used. Conversely, in associated construction output, items for valuing these activities are part of individual catalogs. They are divided into individual parts:

- Part A - Establishment of structures.
- Part B - Demolition (disassembly) of structures.
- Part C - Repair and maintenance of structures.

Demolition prices include costs arising from a specific construction activity, including putting rubble on a pile or loading it onto a truck. It is also necessary to consider the technological handling of rubble at the level of in-site transport of rubble and demolished materials and off the site, i.e. the removal of rubble and demolished materials to the landfill, including the landfill fee.

The prices of intra-site transport of rubble and demolished materials include costs:

- For horizontal transport either by carrying or using a construction wheel or using a conveyor.
- For vertical transport:
    - with the use of mechanization-crane, construction elevator, or even a construction winch
    - with limited mechanization - without the use of a crane, i.e. only with a construction lift or winch
    - hand-carried on stairs.

The method of measurement for individual demolished structures is based on its volume, e.g.: for walls, foundations, columns, pillars, ceilings, the quantity is calculated in m3 of the volume of structures into which the entire structure is decomposed for estimating purposes, e.g.:

- The volume of masonry with insulating cavities is considered including cavities.
- For demolition of half-timbered masonry or masonry into a steel frame, the prices of demolition of the relevant type of masonry can be used and for these construction activities the volume of the wooden or steel structure is not deducted.

On the other hand, there are demolished structures that are not included in the total volume, e.g.:

- volume of window openings,
- volume of door openings,
- volume of niches, if the visible area of individual niches is greater than 0.5 m2 and at the same time the depth is greater than 0.15 m,
- volume of chimneys, ventilation and other vents with their cross-section greater than 0.05 m2,
- volume of protrusions lower than 50 mm [20].

3.3 Individual estimation of items using a standard estimating formula

Individual estimates target direct costs for determining own costs or for creating a price in cost components with respect to specific conditions, technology and organization of production, place, time. The basis for determining the price of a building for most types of company prices is the estimation of costs of individual structures and works. In many cases, before this estimate, the costs of individual cost components must be estimated due to individual conditions (items of the so-called estimating formula – see the picture below). Individual estimate presupposes clarity of technologies, construction conditions, built information system regarding prices and own costs, own normative basis for determining consumption, perfectly performed preparation with regard to construction risks, estimation of business risks and changes in prices and other input conditions over time and finally skilled estimators with a sense of company's capabilities and situation of the market.

The costs for the implementation of a certain construction activity, which are directly related to its implementation, are estimated for the so-called estimating unit. The estimating unit is defined as the output defined by the name, qualitative or delivery conditions and unit of measure. The estimating unit can be a building object or its part, individual construction work or structure, assembly work, man-hour, etc [21].

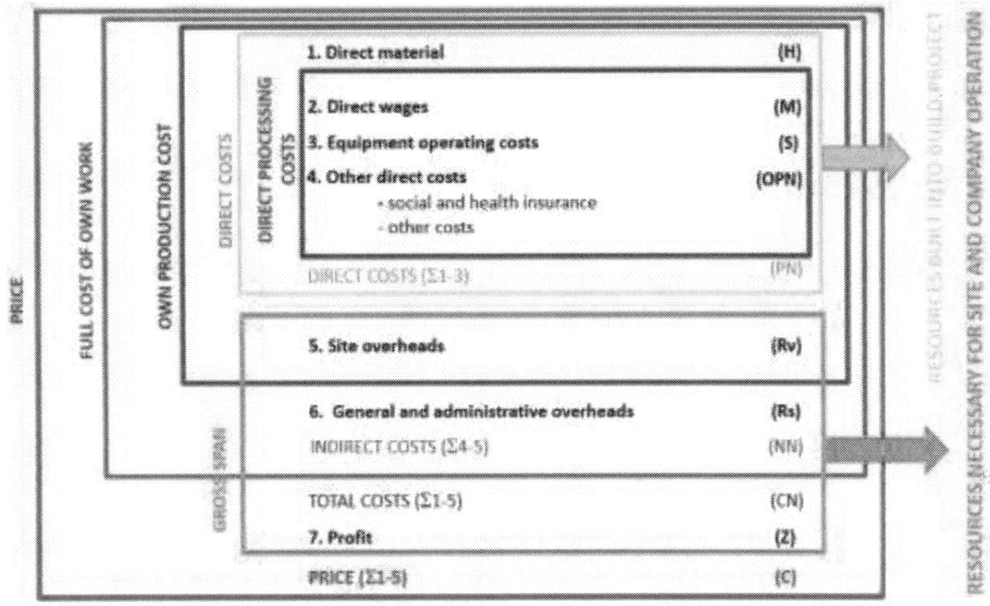

Figure 1. Estimating formula
Source: Own processing.

3.4 Hourly billing rate

We use it where it is difficult to determine the output per unit of production (unit of measure is either m2, m3, ...). Above all, it is a matter of determining the costs of restoration work, for which it is not possible to determine their extent in advance (e.g. work on stucco elements or facade, renaissance cornice or such).
Hourly billing rate includes:

- Labor costs, including levies, costs of small materials, basic equipment for a specific activity, overheads, profit.

Hourly billing rate does not include:

- Costs for restoration materials, for machines, are mainly rental of machines, transport costs, scaffolding, demolition and construction site equipment.

3.5 Principles of parametric valuation of buildings and division of structural elements

Parametric valuation of buildings is a method based on estimating construction costs using selected or input parameters (basic dimensional quantities) of the building. Another fundamental prerequisite is the division of a building into structural and technological units (referred to as structural elements), which are directly linked to selected input parameters (e.g. width, length and height of the building).

Structural elements are individually valued on a defined estimating unit (usually a common unit of measure from budgeting practice). The valuation itself can be done in two ways. It is either the use of available price systems or the application of prices from the market, which are obtained by the company on the basis of the implementation of previous construction projects transformed to the required price level. The application of market prices is a more accurate method of valuation, but it requires the company to have both a uniform way of collecting information on costs in relation to structural elements, and a system for monitoring trends and predicting price developments.

The final estimate of construction costs will be created by the sum of valued individual structural elements that are part of the subjected historic building.

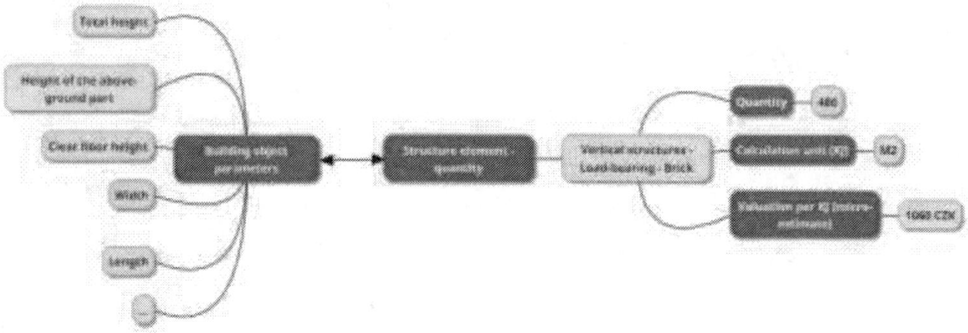

Figure 2. The link between the parameters of the building and the brickwork structural element
Source: Own processing.

The advantage of parametric valuation lies in the rapid determination of costs for construction work based on selected technical information about the building. Nevertheless, this method of valuation is not completely common in the Czech Republic. Price (budget) indicators are closest to this valuation method. However, these indicators do not reflect as many input parameters about the building as the parametric valuation method does, so they are less accurate. An example of a complete classification according to the proposed structure of the division is shown below in the figure for vertical load-bearing masonry structural elements.

 https://doi.org/10.1051/shsconf/202213201022

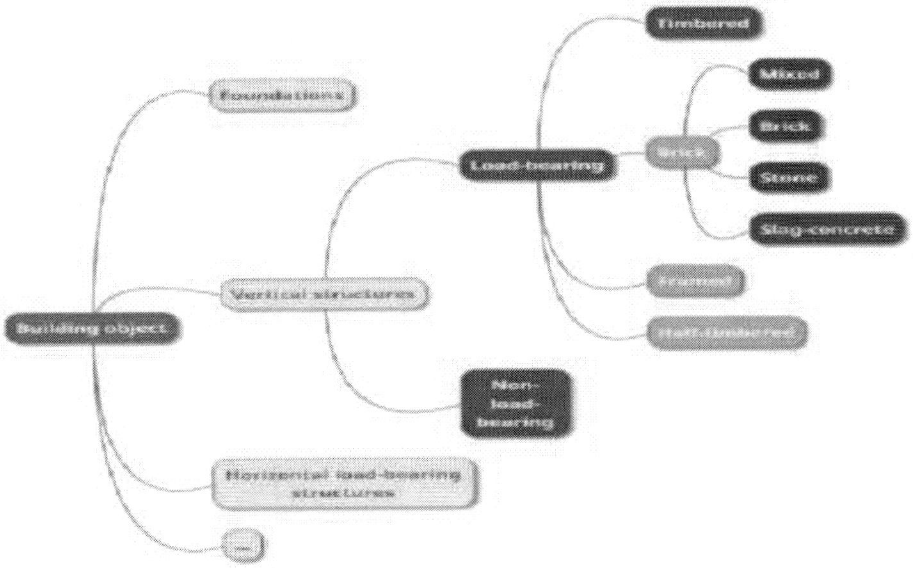

Figure 3. Example of hierarchical division for vertical load-bearing masonry structural elements
Source: Own processing.

3.6 Sub-estimate of structural elements

The unit price, which is assigned to the defined structural elements, is based on the principle of the so-called sub-estimate. It consists of several items of the CS ÚRS price system (currently at price level II / 20), where the unit price of a structural element is created by the sum of partial prices (so-called indicative prices in the structure of the estimating formula) for all sub-estimate's items. The estimating unit of the sub-estimate corresponds to the usual unit of measure of the structural element. In each sub-estimate, there are type items of these activities:

- Demolition of the existing structure
  - demolition of the structure,
  - intra-site relocation of rubble,
  - removal of rubble to the construction dump,
  - rubble deposit fee,
- Creation of a new structure
  - new construction, including surcharges,
- Internal & external scaffolding (according to the position of the structure)
  - auxiliary scaffolding,
  - facade scaffolding,
  - assembly,
  - disassembly,
  - rent (surcharge),
- Material transfer
  - transfer of materials from the site depot to the place of the structure's construction.

The following table shows an example of creating a sub-estimate for demolition and subsequent construction of a masonry partition made of ceramic bricks without surface treatment [22].

Table 1. Sub-estimate of vertical non-load-bearing structure element - brick partitions, ceramic, without surface treatment, unit price of 2 173 CZK / m2

	3	Vertical and complete construction				935,90
1	342241162	Partitions made of bricks full length 290 mm strength P 15 on MC 140 mm thick	m2	1.000	745.00	745.00
2	342291121	Anchoring of partitions to load-bearing structures	m	1.000	108.00	108.00
3	346971132	Insulation under partitions against sound spreading, double, made of sand, cardboard and MC width up to 200 mm	m	1.000	82.90	82.90
	9	Other construction and work, demolition				410.63
4	949101111	Scaffolding of auxiliary height of scaffolding floor up to 1.9 m	m2	1.200	47.80	57.36
5	962031133-1	Cutting of partitions from fired, solid, hollow bricks	m2	1.000	114.00	114.00
6	998017003	Movement of materials with limited mechanization for buildings up to 24 m of height	t	0.235	1 020.00	239.27
	997	Moving rubble				826.60
7	997013155	In-site transport of rubble and demolished materials for buildings up to 18 m of height with limited mechanization	t	0.261	1 630.00	425.43
8	997013501	Transport of rubble and demolished materials to a landfill or intermediate landfill up to 1 km with the composition	t	0.261	233.00	60.81
9	997013509	Surcharge for the removal of rubble and demolished materials to the landfill ZKD 1 km over 1 km	t	2.349	10.45	24.55
		0.2610 * 9 " Recalculated by the quantity coefficient		2.349		
10	997013631	Landfill fee (landfill) for construction waste mixed waste code 170 904	t	0.261	1 210.00	315.81

Source: Own processing, based on CS ÚRS price system.

4 Analysis of market prices for construction work used in historic buildings

Due to the complexity of valuing construction works, repairs and reconstructions of historic buildings, a comparison of indicative and bid prices for completed constructions was performed. This comparison was made on 16 bid budgets for the reconstruction of historic buildings located in several regions of the Czech Republic, specifically the Pardubický, Plzeňský, Královehradecký, Středočeský, Moravskoslezský, Ústecký, Jihomoravský, Jihočeský and Vysočina.

The comparison of indicative and bid prices was performed on a file that contains 464 representative items from 16 bid budgets. These items were defined by the CS ÚRS price system. The comparison included not only individual work items (R-items), but also items that specify the delivery of a material or product. These items are mainly of the list of works of the Associated Construction Production (PSV).

Construction works on historic buildings took place at different times - from the 2nd quarter of 2013 to the 2nd quarter of 2019. In contrast, the indicative prices (IP) of the CS ÚRS price system with which the bid prices (BP) were compared were in the current price level of the 1st quarter of 2021. In order for these bid prices to be comparable with the indicative prices, it was necessary to unify the price level for all prices for the 1st quarter of 2021 using coefficients. Data published by the Czech Statistical Office (ČSÚ) were used to determine these coefficients.

Every quarter, the ČSÚ publishes a publication entitled "Construction Work Price Indices, Construction Work Price Indices and Construction Production Cost Indices". The publication contains several tables in which the percentage development of prices compared to the previous period is recorded.

To determine the development of prices of construction work for individual sections and trades, we use the table Price indices of construction structures and work according to TSKPstat, which has the following breakdown (see Picture 4 below).

TSKP code	name
	Construction and work of HSV
1	**Earthworks**
	of which:
12	Excavations and dig-througs
13	Dug excavations
16	Relocation of the excavation
17	Soil construction
2	**Foundation, strengthening of rocks**
	of which:
27	Basics
3	**Vertical and complete structures**
	of which:
31	Walls of buildings
33	Columns and pillars
34	Walls and partitions
4	**Horizontal structures**
	of which:
41	Ceilings and ceiling structures of buildings
5	**Communication**
	of which:
56	Base layers of roads, airports and areas
57	Covers of roads, airports and areas of aggregate or bitumen
58	Covers of roads, airports and areas of concrete and other materials
59	Covers of roads, airports and paved areas
6	**Surface treatment, flooring and installation of fillings**
	incl.:
61	Treatment of interior surfaces
62	Modification of external surfaces
63	Floors and floor constructions
8	**Long-distance and connecting lines**
	of which:
87	Plastic and glass pipes
9	**Other construction and work, demolition**
	of which:
94	Scaffolding and construction lifts
96	Demolition of structures
97	Punching holes and other demolition work
99	Mass transfer and debris handling

TSKP code	name
7	**Construction and work of PSV**
71	**Insulation**
	incl.:
711	Insulation against water, moisture and gas
712	Insulation - coatings
713	Thermal insulation
72	**Sanitary technical installations**
	incl.:
721	Sanitary technical instalations - sewerage
722	Sanitary technical instalations - water supply
723	Sanitary technical instalations - gas pipeline
725	Sanitary technical instalations - fixtures
73	**Central heating**
	of which:
733	Central heating - distribution pipes
734	Central heating - fittings
735	Central heating - radiators
74	**Electrical equipment**
	incl.:
743	Electrical equipment - rough assembly
744	Electrical equipment - wiring
747	Electrical equipment - assembly of power equipment
75	**Technological equipment**
76	**Structures**
	of which:
762	Carpentry structures
763	Mounted structures
764	Plumbing structures
765	Hard roofing structures
766	Joinery structures
767	Locksmith structures
768	Steel structures
77	**Floors**
	incl.:
771	Tile floors
772	Stone floors
775	Folded floors
776	Coating floors
777	Cast floors
78	**Finishing job**
	incl.:
781	Finishing work - tiling
783	Finishing work - coatings
784	Finishing work - surface painting

Figure 4. Classification of building structures and works according to TSKP
Source: Czech Statistical Office (ČSÚ).

The distribution of individual items into the HSV and PSV sections (according to the division of the TSKP based on the CS ÚRS price system) is evident from the following figures (Figure 5).

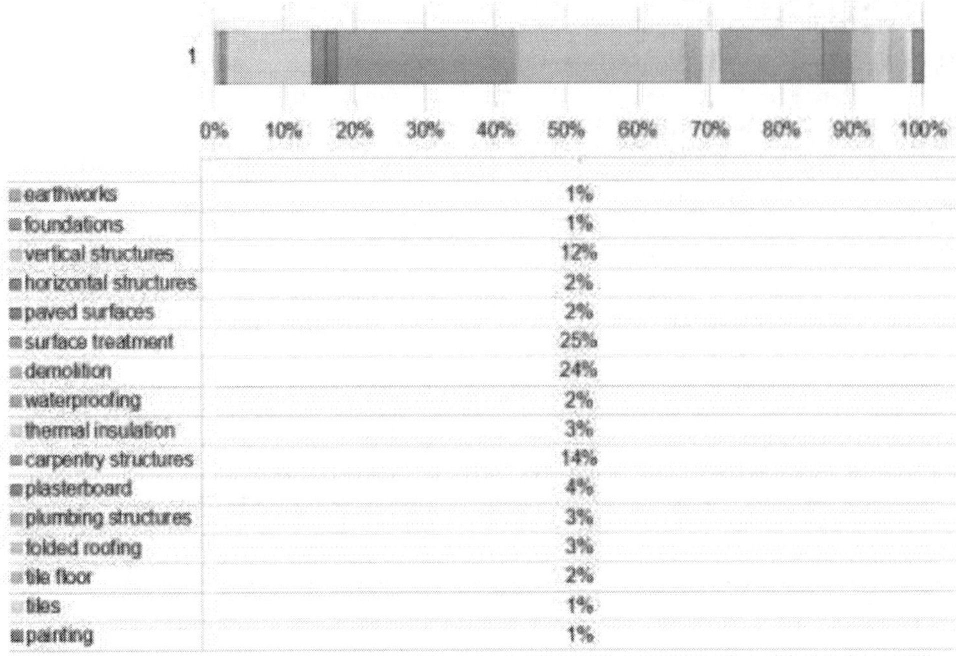

Figure 5. Composition of the compared sample showing the most frequently used items in the reconstruction of historical monuments
Source: Own processing.

Items of joinery and locksmith constructions do not occur in the examined sample. These are mainly construction works, including materials, which relate to the renovation and replacement of window openings, door leaves, gates, staircase structures, grilles, railings, etc. These structural elements are mainly estimated for construction work repairs and reconstructions of historic buildings and in the estimates are recorded as so-called R-items. In the statistical file, bid prices and indicative prices were compared and the percentage difference between them was calculated for all items.

In the statistical file, the items of the HSV and PSV sections of the bid prices were compared with the indicative prices of the CS ÚRS price system. Based on the comparison, the percentage difference between them was calculated and the average determined.

The standard deviation was determined for the whole sample and subsequently also for the individual sections of HSV and PSV, a histogram was created with the frequency of occurrence for the following intervals:

- bid price (BP) is lower by more than 30.00% than indicative price (IP),
- BP is lower by 20.00 - 29.99% than IP,
- BP is 10.00 - 19.99% lower than IP,
- BP is 0.00% - 9.99% lower than IP,
- BP is higher by 10.00% - 0.00% than IP,
- BP is higher by 10.01% - 20.00% than IP,
- BP is higher by 20.01% - 30.00% than IP,
- BP is higher by more than 30.00% than IP

The arithmetic mean is the mean value of the examined sample. This is a statistical quantity of the position. Its definition is such that we add all the values of the statistical unit character and divide them by the number of these units.

$$\bar{x} = \frac{1}{n}\sum_{i=1}^{n} x_i \quad (1)$$

Where:
n... number of values in the statistical file,
x... value of the statistical file,
x̄... average of the values of the statistical file.

The standard deviation determines how much the values are scattered or deviated from the average. It is defined as the square root of the variance. It can only take on positive values and has the same unit as the values in the statistical file. This is a statistical value characterizing variability.

$$\sigma = \sqrt{\sigma^2} = \sqrt{\frac{1}{n}\sum_{i=1}^{n}(x_i - \bar{x})^2} \quad (2)$$

Where:
σ ... direct deviation,
σ^2.. scatter,
n... number of values in the statistical file,
x... value of the statistical file,
x̄... average of the values of the statistical file.

The average difference between the bid prices and the indicative prices of the whole statistical sample is -14.23%. This means that, on average, bid prices are lower than the indicative prices. The standard deviation for the whole file was 15.35%. The following figure shows the above-mentioned statistical quantities as well as a histogram of the frequency of occurrence for defined intervals.

Statistical sample ......464 items
Average .... x̄ = - 14.23%
Standard deviation ..... σ = 15.35%

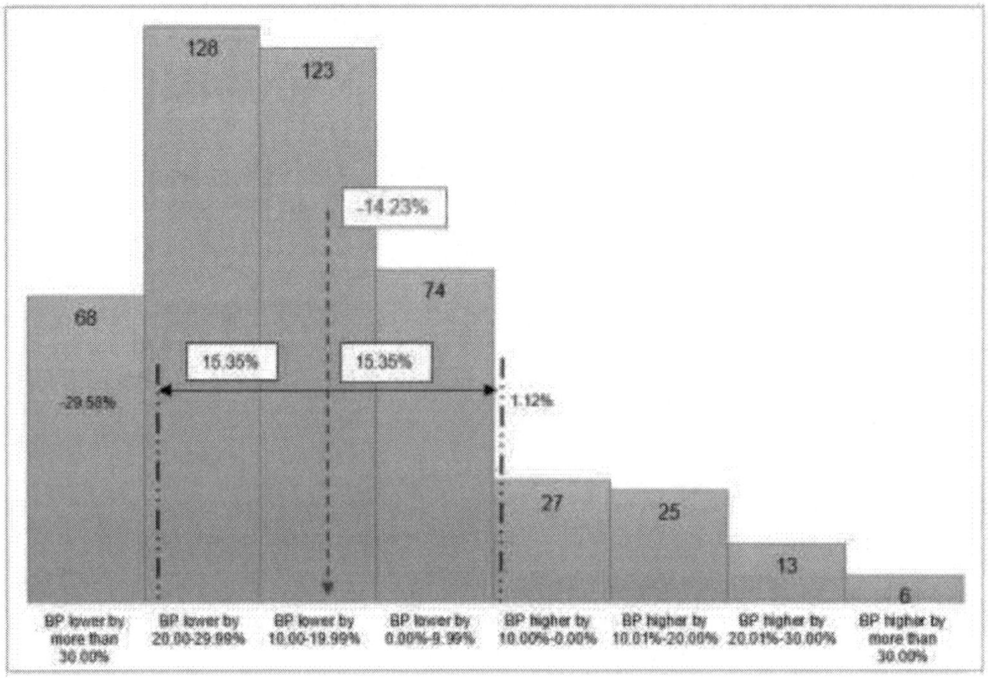

Figure 6. Statistics of the whole analyzed file
Source: Own processing.

Among the examined sections, the most numerous were the vertical structures, surface modifications and demolition of structures, while in PSV (work trades) carpentry structures were the most common.

In the section of vertical construction, load-bearing foundation masonry, columns, walls, partitions, lintels and chimneys are valued. Repairs and maintenance include work related to, for example, walling of holes, replenishment of masonry, grouting of masonry, replenishment of cornice masonry, installation of rolled beams in prepared holes and related demolition work, which involves cutting out pockets for anchoring fasteners, etc. This section belongs to the more numerous ones in terms of the number of items in the examined set.

The average difference between the bid prices and the indicative prices of the statistical sample of vertical structures is -9.61%, i.e. on average, bid prices are lower than the indicative prices. The standard deviation of vertical structures is 15.83%. The following figure shows the above-mentioned statistical quantities as well as a histogram of the frequency of occurrence for defined intervals.

Statistical sample .... 56 items
Average .... $\bar{x} = -9.61\%$
Standard deviation.......... $\sigma = 15.83\%$

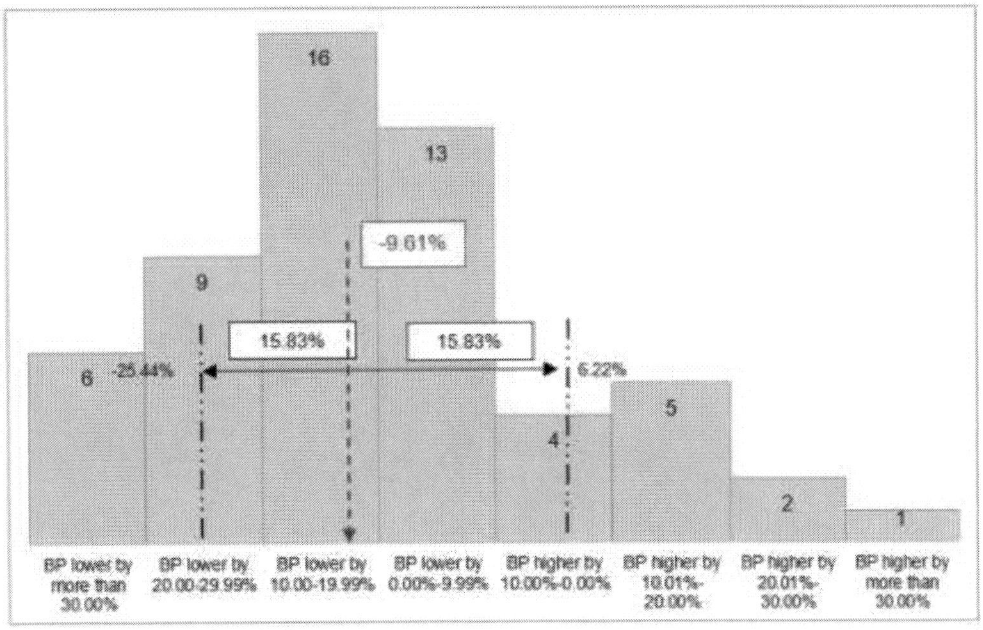

Figure 7. Statistics of vertical structures
Source: Own processing.

In this part, the construction works are divided into three larger groups, these are internal and external surface treatments and floor surface treatments such as floor embankments, screeds, anhydrites and cement screeds. In internal and external surface treatments, we find mainly plasters of walls, ceilings, columns and ceilings, as well as contact thermal insulation systems made of various materials. Among the items valuing repairs to historic buildings, the most common items are items for leveling uneven surfaces, repairing plasters of smaller areas and filling grooves.

The average difference between the bid prices and the indicative prices of the statistical sample of surface treatments is -11.25%. This means that, on average, bid prices are lower than the indicative prices. The standard deviation of surface treatments is 16.55%. The following figure shows the above-mentioned statistical quantities as well as a histogram of the frequency of occurrence for defined intervals.

Statistical sample .... 118 items
Average .... $\bar{x}$ = - 11.25%
Standard deviation........... σ = - 16.55%

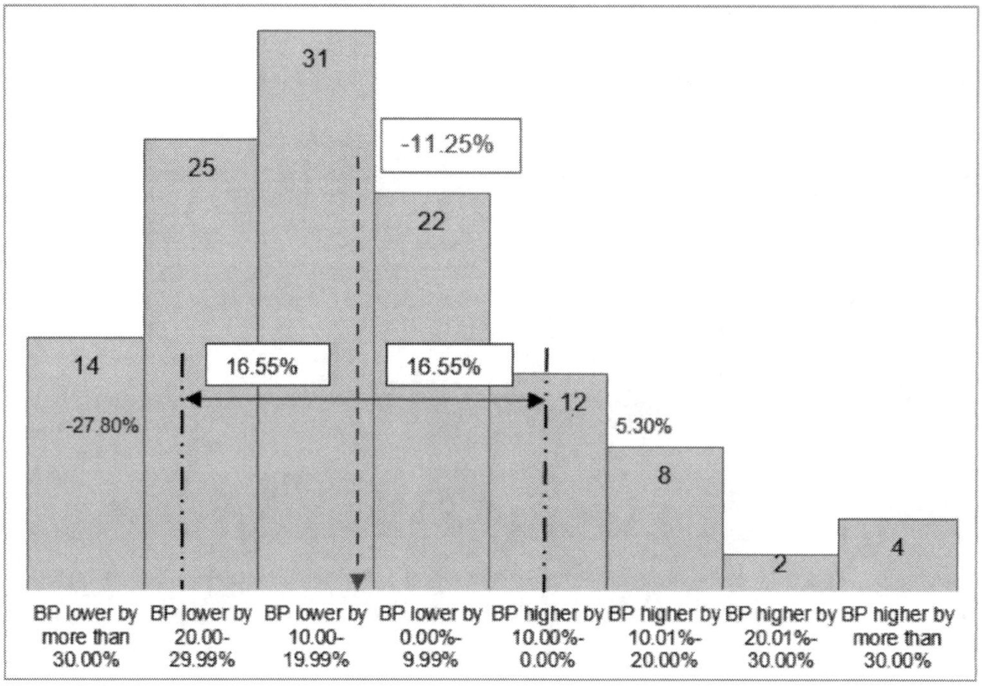

Figure 8. Surface treatment statistics
Source: Own processing.

The next most numerous group of items is the section of structures' demolition. The most common items were the demolition of masonry and partitions, loosening of beam heads, demolition of rolled beams, demolition of floors and subfloors, cutting of masonry and lining, cutting of pockets in masonry, fixing of ceilings, beating of interior and exterior plaster, demolition of window frames and masonry openings. , scraping joints in masonry, cutting tiles. This section also includes items valuing the removal of rubble and demolished materials to landfills.

The average difference between the bid prices and the indicative prices of the statistical sample of structural demolition is -15.56%. This means that, on average, bid prices are lower than the indicative prices. The standard deviation for the demolition of structures is 16.34%. The following figure shows the above-mentioned statistical quantities as well as a histogram of the frequency of occurrence for defined intervals.

Statistical sample ……..111 items
Average ……………… $\bar{x}$ = - 15.56%
Standard deviation........ σ = 16.34%

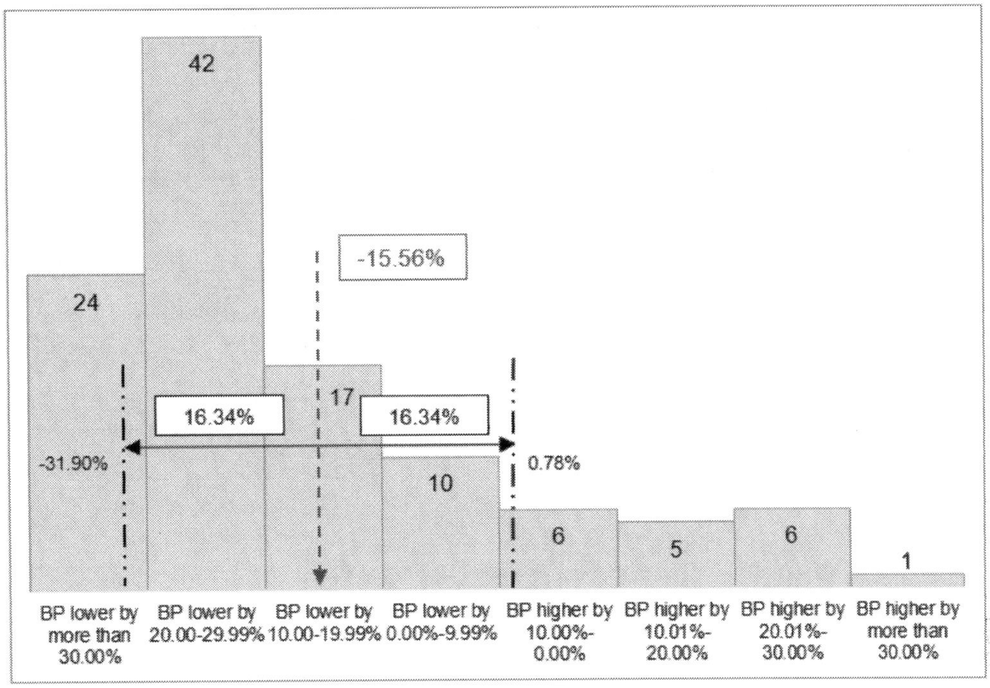

Figure 9. Statistics of demolition of structures
Source: Own processing.

Within carpentry constructions it is especially the assembly and disassembly of bound truss structures, battens and roof formwork which is being valued. Frequent items were carving, cutting out and completing the roof bond. In addition to roof structures, there are also items for dismantling and assembling floors from planks and wall cladding, laying cushions under the floors, cutting and completing ceiling beams.

The average difference between the bid prices and the indicative prices of the statistical sample of carpentry structures is -16.95%. This means that, on average, bid prices are lower than the indicative prices. The standard deviation for carpentry structures is 12.01%. The following figure shows the above-mentioned statistical quantities as well as a histogram of the frequency of occurrence for defined intervals.

Statistical sample ....67 items
Average .... $\bar{x}$ = - 16.95%
Standard deviation...........σ = 12.01%

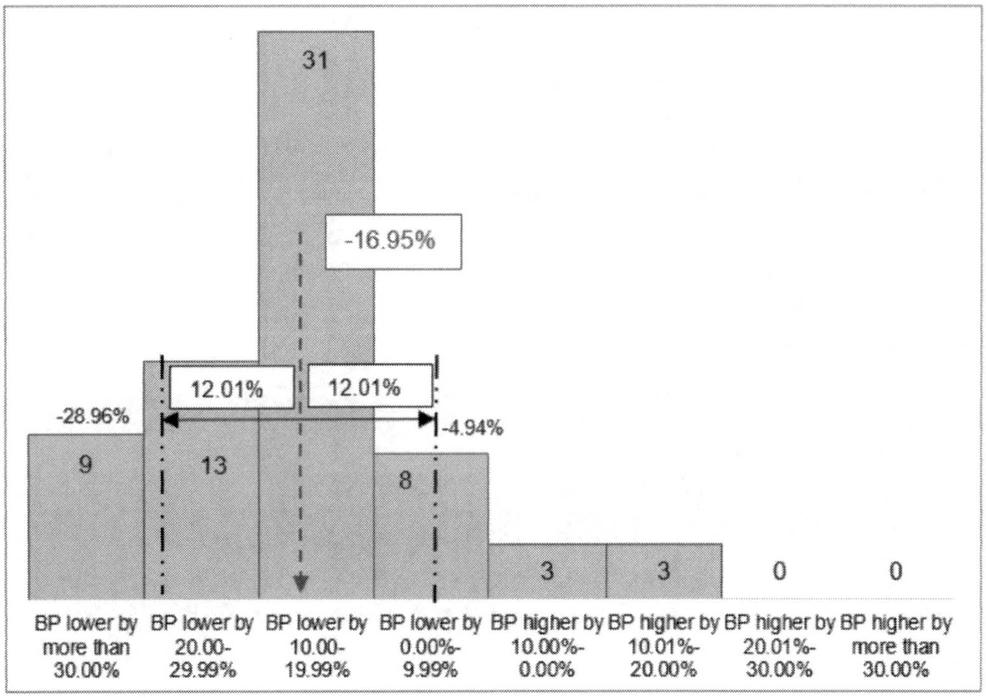

Figure 10. Statistics of carpentry structures
Source: Own processing.

4 Results

Valuation of reconstructions of historical buildings is quite problematic due to the nature of the works and an uncertainty in estimating their number. The aim of the indicative price analysis was to find out whether they actually correspond to prices on the construction market. It is necessary to realize that if the work on the building is subject to the restoration regime and can be performed only by a restorer with a valid permit, then it is necessary to value these works individually, because the indicative prices do not take restoration repairs into account. During reconstructions, you can also find a relatively large number of works that do not appear in the estimating software and are valued individually in the bids. These are mainly carpentry and locksmith constructions and artistic elements.

The price analysis was performed on a statistical sample of 464 items from 16 bid budgets. In comparison with the indicative prices of the CS ÚRS price system in the price level I / 2021, the bid price was lower than the indicative price for 85% of the examined items.

A detailed percentage distribution of the differences between the bid and indicative prices can be seen in the following figure, when the bid price was 20.00-29.99% lower than the indicative price for 28% of the items. The average difference for all items in the sample was -14.23%.

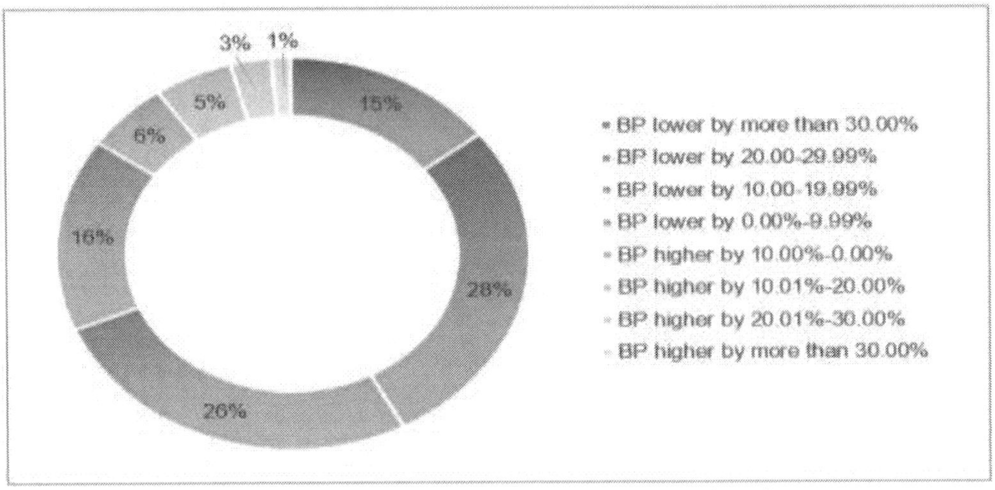

Figure 11. Percentage distribution of differences between bid and indicative prices
Source: Own processing.

Based on the performed analysis, valuation by indicative prices by the CS ÚRS price system for valuing reconstructions of historic buildings seems to be appropriate. When creating a bid price, it is always necessary to respect the specificity of a particular project and consider a possible adjustment of these prices. If the project includes restoration elements that are intended to be repaired by the restorer, these components must be valued separately.

5 Conclusion

The presented paper brings a methodology for the valuation of construction work in the reconstruction of historic buildings. The valuation method is based on an analysis of market prices for construction work used in historic buildings in combination with traditional pricing tools and methods. Based on the performed research, structural elements and technological units were identified for the specific sections. The unit price, which is assigned to these defined units, is based on the principle of sub-estimates and the CS ÚRS price system was used for its creation. The suitability of using this pricing system is based on the performed analysis and comparison of indicative and market prices on a statistical sample of 464 items from 16 bid budgets.

The proposed methodology is used in the follow-up application software MonuRev, where it is used for the valuation of the restoration and maintenance of selected immovable cultural monuments. The outputs of this application software can be used by the public sector in planning the restoration of immovable cultural heritage. The development of the MonuRev application is implemented within the NAKI II project of the Ministry of Culture. Ultimately, the application should be able to determine the cost of renovation and maintenance of historic buildings during the life cycle of buildings and enable investors to make relevant complex decisions and help reduce potential risks of such projects.

References

1. M. Philokyprou, E. Limbouri-Kozakou, An overview of the restoration of monuments and listed buildings in Cyprus from antiquity until the twenty-first century. *Studies in Conservation*, **60**(4), 267-277 (2014)
2. R. H. White, Editorial. *The Historic Environment: Policy & Practice*, **7**(4), 281-282 (2016)
3. B. Espion, R. Engels, M. Provost, Civil engineering heritage: country profile – Belgium. *Proceedings of the Institution of Civil Engineers - Engineering History and Heritage*, **170**(4), 197-205 (2017)
4. I. Sipiran, P. Lazo, C. Lopez, et al., Retrieval of cultural heritage objects. *Computers & Graphics*, **100**, 1-20 (2021)
5. J. Kęsik, M. Milosz, J. Montusiewicz, K. Samarov, Documenting the geometry of large architectural monuments using 3D scanning – the case of the dome of the Golden Mosque of the Tillya-Kori Madrasah in Samarkand. *Digital Applications in Archaeology and Cultural Heritage*, **22** (2021)
6. D. Fritsch, J. F. Wagner, B. Ceranski, S. Simon, M. Niklaus, K. Zhan, G. Mammadov, Making historical gyroscopes alive–2D and 3D preservations by sensor fusion and open data access. Sensors, **21**(3) (2021)
7. E. Plebankiewicz, A. Leśniak, Overhead costs and profit calculation by polish contractors. *Technological and Economic Development of Economy*, **19**(1), 141-16 (2013)
8. B. Bauer, J. Koppelhuber, J. Wall, D. Heck, Impact factors on the cost calculation for building services within the built environment. *3rd International Conference on Sustainable Civil Engineering Structures and Construction Materials - Sustainable Structures for Future Generations (SCESCM)*, **171**, pp. 294-301 (2017)
9. R. Majer, H. Ellingerová, J. Gašparík, Methods for the calculation of the lost profit in construction contracts. *Buildings*, **10**(4) (2020)
10. R. S. Heralová, Life cycle costing as an important contribution to feasibility study in construction projects. *6th Creative Construction Conference (CCC) In Procedia Engineering*, **196**, pp. 565-570 (2017)
11. D. Macek, Building maintenance and renovation. *Conference on Central Europe towards Sustainable Building in Proceedings of the CESB 10: Central Europe towards Sustainable Building—From Theory to Practice*, pp. 669–672 (2010)
12. J. Karásek, J. Pojar, L. Kaločai, R. S. Heralová, Cost optimum calculation of energy efficiency measures in the Czech Republic. *Energy Policy*, **123**, 155-166 (2018)
13. R. S. Heralová, D. Macek, E. Hromada, I. Střelcová, L. Brožová, S. Vitásek, J. Pojar, R. Bouška, Cost optimization for renovation and maintenance of cultural heritage objects. *IOP Conference Series: Earth and Environmental Science In: Central Europe towards Sustainable Building (CESB19)*, **290** (2019)
14. P. Dlask, V. Beran, Long-term infrastructure investment: a new approach to the economics location. *E+M Ekonomie a Management*, **19**(3), 40-56 (2016)
15. D. Čápová, P. Dlask, O. Horák, P. Kalčev, Generace malých rozpočtů [Generations of small budgets]. *Business & IT*, **8**(2), 54-62 (2018)
16. J. Vrbka, P. Junga, T. Krulický, Methodology for determining the rate of return on rental of built-up land. *Ad Alta-Journal of Interdisciplinary Research*, **9**(2), 364-370 (2019)
17. E. Hromada, R. S. Heralová, P. Dlask, T. Krulický, Methodology for calculating the reproduction value of buildings. *EDP Sciences - Web of Conferences In: Innovative Economic Symposium 2020 – Stable Development in Unstable World (IES2020)*, **91** (2020)

18. D. Macek, R. S. Heralová, E. Hromada, I. Střelcová, Cost optimization for renovation and maintenance of cultural heritage objects. *IOP Conference Series: Earth and Environmental* Science, **290** (2019)
19. R. Schneiderová, S. Vitásek, L. Brožová, I. Střelcová, *Cost Estimating*. Prague: Czech Technical University in Prague (2020)
20. Budgeter's handbook: *Cost of construction works*. Prague: ÚRS, Price system ÚRS (2017)
21. S. Vitásek, R. S. Heralová, *Cost Estimating*. Prague: Verlag Dashöfer (2018)
22. Software KROS 4 2021. Company ÚRS CZ

The use of the BIM method for sustainable care of historic buildings: Methodological framework

Stanislav Vitasek*

Czech Technical University in Prague, Faculty of Civil Engineering, Czech Republic

Abstract. The article presents the use of the Building Information Modelling (BIM) method applied to historic buildings (HBIM). This is a very current topic with a direct link to the development of digital innovations. The HBIM approach presented in the article is primarily targeted at the owners/managers of historic buildings where its greatest expected benefit for this type of buildings is, in particular, in Facility Management. The key output of the article is the group of proposed parameters, which represents the requirements for the data on the part of both the facility manager, and the public authorities collecting selected information about historic buildings. As an example, the proposed parameters were applied to the All Saints' Church, where the "BIM technologies" were used for the elaboration of the project documentation.
Keywords: BIM, Digital Innovation Development, Facility Management, HBIM, Historic Building.

1 Introduction

The article presents the use of the Building Information Modelling (BIM) method in the context of sustainable care of immovable heritage buildings. The linking of the BIM approach to historic buildings is referred to as Historic Building Information Modelling (HBIM) in professional sources. The BIM approach, in general, is currently a topic of intensive research across the whole building industry thus representing the digital development of the segment. [1, 2]

The BIM/HBIM (hereinafter referred to as BIM) method presented in the article is primarily targeted at the owners/managers of historic buildings, where its greatest expected benefit for this type of buildings is, in particular, in Facility Management (FM). A well-executed digital model of a building significantly simplifies the building management process, not only due to the standardized information format across the owner/manager portfolio. [3, 4] The objective of the article is to present, at the basic methodological level, how this standardized format (of comprehensive and structured data) can be captured with the use of the key parameters that should be part of the digitization process of the project documentation of individual historic buildings for the purposes of a subsequent database processing. The model application of the method to a historic building is performed on a medieval church, which has recently undergone renovation and selected "BIM (digital) technologies" have been applied to it primarily for the purpose of compiling the existing-

* Corresponding author: stanislav.vitasek@fsv.cvut.cz

conditions, design and as-built project documentation. In the article itself, however, the author focuses on how to fill the digital building model (BIM model; an example of a church in a digital environment - Fig. 1) with data, primarily with non-graphical information, so that the data can be subsequently transferred from the BIM model to meet the data needs on the part of selected public authorities managing statistical and informative (basic) data on historic buildings, and the facility managers themselves.

Fig. 1. Historic building processed with the use of the BIM method

Source: ScienceDirect https://www.sciencedirect.com/science/article/pii/S1296207417300754#fig0030.

2 Literature review

The Building Information Modelling is still perceived as a new concept in the construction sector, representing a significant technological progress in the information technology that has become apparent in recent years. It combines a digital environment with a real one with a great potential to change the traditional ways of work in the pre-project planning, construction and management phases of construction projects. The BIM method has several definitions coined by the leading experts in the construction industry and building associations. The most frequently quoted definition according to the author of the article is. The BIM technology produces one or more accurate, digitally generated virtual models of a building supporting the building's design throughout all phases of its life cycle, enabling more enhanced analyses and checks than manual (traditional) processes. When completed the computer-generated models contain the precise geometry and data needed for the construction, production and the procurement process according to which the building is delivered, including its subsequent implementation. [5] All interpretations connected with the BIM method definition generally have some common signs based on:

- sharing information between individual parties in the planning, construction and subsequent management of a built asset,
- 3D visualization and localization of structures or the entire building,
- faster decision-making associated with the project management and an overall higher economic efficiency of the project.

The issues directly related to the topic of BIM and historic buildings are usually associated with specific buildings in professional publications. In this respect, an interesting article where a digital twin of a historic building was generated with the use of the "BIM technologies" and a scheme was subsequently designed for the integration of non-graphical data from the BIM model into the facility management system presents the St-Pierre-le-Jeune Church in Strasbourg (France). The scheme used there is inspiring for a basic overview of this topic, although some of the conclusions are the product of the time when it was published (2019). However, the mind map used to transfer data between the BIM model and the FM system is still relevant at the general level. [6] A similar article, but from the Italian environment, is associated with the historic structure of the Carlo III Bridge of the Carolino Aqueduct, where the BIM model again served as the source of information not only for the bridge management, but also for its restoration. The basic "BIM technologies" like the laser scanner, photogrammetry and topography were applied there. The article concludes by weighing the benefits of digital technologies against their drawbacks where the benefits of these advanced technologies unambiguously outweigh the drawbacks. [10] A different type of article, from the Spanish environment this time, compares the digital modelling of a historic building (the geometric aspect of the BIM model is primarily considered) of the Pavilion of Carlos V in Seville, which is executed in two ways. The first method represents the classic digitization of the existing project documentation, while the other applies the Point Cloud technology to survey the as-built layout of the building. The article concludes by comparing the different results in the building's geometry, where the authors identify the point cloud method as a more accurate way of generating a BIM model. [7] A similar study was also conducted by the colleagues from the Italian University of Naples Federico II and the Chinese Liming Vocat University, who dealt with the design of a parametric approach to the elaboration of methodologies for the digitization of heritage buildings with the use of the "BIM technologies". The result of their research can be expressed by the proverbial expression "less is more", as they recommend modelling simple geometric shapes of structures and including basic non-graphical information. [11, 13] A specialized study directly related to the design of the BIM model linked to the key execution document, the so-called BEP (BIM execution plan), was applied in the British environment. The authors focus on the essential points of the digital projection, particularly the graphic design, where they recommend the LOG 300 level for the passportization of historic buildings according to the Level of Development methodology. [8] An interesting article addresses the renovation projects of historic buildings, where the authors use selected algorithms and digital "BIM technologies" to design an optimum grid system for the restoration of a traditional Chinese building element. It is an intriguing way of using digital technologies, which can be easily transferred to any other building element. [9] A very practical study was conducted by the colleagues from the Polytechnic University of Valencia, who investigated several levels of the BIM method benefits for historic buildings: from the very design of the renovation project through visitor management and visitor protection to the digital promotion of the monument. The conclusions of their research recommend applying the BIM method on the widest possible scale. [12] However, according to the author of this article, some of the visions were too progressive and may only have a practical impact in a few years from now.

The operation of historic buildings in connection with long-term sustainable development is further addressed in studies [14, 15]. The authors found that the operation and maintenance of historic buildings requires higher costs due to, for example, the need to respect historical technological procedures and historic building materials and take into account the opinion of the office of monument care.

3 Digital model of a building

The digital model of a building serves as the key pillar of the BIM method. The BIM model can represent a database of information from the pre-project planning, construction, operational and maintenance phase, i.e. covering the entire life cycle of a construction work. In the context of historic buildings, the BIM models will most likely include primarily the information associated with different facility management and statistical requirements. The BIM model itself, therefore, represents a significant part of the project documentation in the format of graphical and non-graphical information (for more detail see the subchapters below). The key agent involved in the standardization of this information is referred to as the Data Standard for Buildings (DSB). Based on DSB, a BIM model is generated.

Based on the type of building structure and trades (structural system, sanitary engineering, electrical installations, etc.), the digital model of a building is usually divided into individual partial digital models of the building. The data from all partial BIM models are integrated in the so-called BIM coordination model [4]. Within the thematic scope of the article, the author did not focus on the overall concept of building information management, which is a more complex framework for the BIM method capturing the respective processes, communication, data sharing, etc. Our primary topic is the information related to the digital model of a building, which is of the greatest benefit for the management of historic buildings.

4 HBIM – basic requirements for information

In the perspective of the owner/manager of a historic building, as has been mentioned repeatedly in the text, the BIM method is beneficial primarily for the facility management of the building. The key phase before generating the BIM model itself is the identification of the data (depending on the subsequent use of the data), which the owner/manager needs to "collect" for FM and, based on these requirements, the compilation of a "data cookbook" for the designer (DSB). In other words, DSB incorporates both the graphical requirements for the virtual form of the structures/technologies in the digital environment (so-called elements), and the requirements for the non-graphical information (e.g. for individual items of the passport system). This means that the resulting BIM model frequently also contains, in addition to the graphical representation of the building, the more substantial non-graphical information assigned to each element representing the structure/technology.

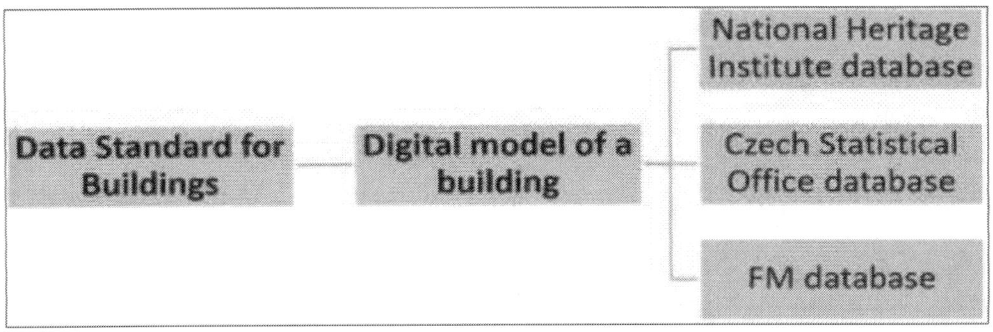

Fig. 2. Links between the DBS, digital model of a building and individual databases
Source: Own processing.

The non-graphical information can be divided into two basic types. Firstly, there are complex (common) parameters containing information about the building as a whole, and, secondly, data assigning specific (partial) properties (parameters) to individual structures/technologies in the building. Within the scope of the article, the author addresses the first group of common parameters in more detail, i.e. the situation where the same set of information is assigned to all the elements in the BIM model (Chapter 4.3). Explicitly, in the case of the renovation of the All Saints' Church, the data needed by selected public authorities managing the statistical and informative (basic) data on historic buildings and by the facility manager are presented (an illustrative scheme of data linkage is part of the scheme above - Fig. 2). On the contrary, the group of sub-parameters unique to each type of structure/technology is approached more generally in the article (Chapter 4.4).

4.1 Laser scanning and photogrammetry

The key "BIM technology" in relation to the generation of BIM models of not only historic buildings is laser scanning. It is a process in which we obtain accurate information about the scanned object by means of determining its spatial coordinates. Individual points carry information based mostly on the principle of the spatial polar method. Hence, each point is identified by the oblique length, the horizontal angle, the zenith angle, and it can also carry the colour shade value. The result is a point cloud that is transformed into a BIM model (sample output in the Fig. 3).

Fig. 3. Example of a laser scanning result – point cloud
Source: Autodesk http://docs.autodesk.com/3DSMAX/16/ENU/3ds-Max-Help/images/GUID-B3A18050-A741-4CA6-85F4-9683C8CBB4B-low.png.

Another technology beneficial for mapping historic buildings is photogrammetry, whose surveys are most often performed with the help of drones and airplanes. It is a measurement method used for contactless collection of data on the Earth's surface, i.e. the process of remote sensing and measurement, mostly during a flight. After the information is collected in the field, a point cloud and, subsequently, a BIM model of the object are created from the photographs and other data.

4.2 Graphical information

Graphical information specifies the type of drawing, graphics, location in space, scale, shape, etc. In the case of requirements for graphical data, due to the absence of national standards, most creators of "BIM data" usually refer to the international method of marking the graphical levels according to the Level of Geometry (LOG) method. LOG determines the geometric level of detail of the BIM model or its individual elements (structures/technologies processed in the virtual environment) for various phases of the project documentation, and thus defines which parts of structures/technologies should be drawn in the model and which should not. The scale is usually from LOG 100 to LOG 400 (for an example of a model structure see the figure below - Fig. 4), where the detail of the documentation for the building permit is recommended at the LOG 300 level, while for the documentation for the construction process and building passports, the recommended version of detail is LOG 350.

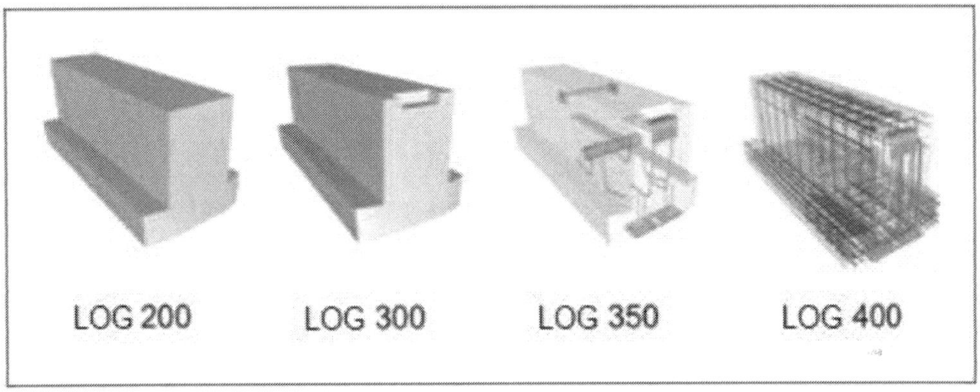

Fig. 4 Example of graphical information levels of detail in LOG
Source: Areo blog https://blog.areo.io/content/images/2016/12/LED-BIM-example.png.

Each increase in the geometric level of detail also proportionally increases the complexity and the time needed for the BIM model generation. This is subsequently reflected in the cost of its creation, too. Therefore, it is advisable to consider the required detail in relation to the subsequent use carefully.

4.3 Non-graphical information – common parameters

Non-graphical information generally enables the description of individual characteristics of a building which are usually not graphically expressed. These are, above all, names, dimensions, areas, additional information about the building, etc. In international methodologies, the level of detail of non-graphical information is referred to as the Level of Information (LOI). In principle, it works in the same way as the LOG method in the case of the demands for graphics.

The common parameters represent the properties (parameters) at the identification level of the whole historic building for which the BIM model is being created. Hence, it is a group of parameters that will be part of all elements integrated into the digital model of the building. The identification of the key information for facility management requires some experience in working with this process, therefore, the author of the article drew on the required data on historic buildings collected by the National Heritage Institute (NHI), the Czech Statistical Office (CSO) and the Roman Catholic Church (RCCH). NHI and CSO manage their data in a database, which is fully compatible with the BIM method concept, which itself is based on a database environment. This implies that there may be a smooth (automated) data transfer between the BIM model and the NHI and CSO databases. In principle, the proposed method of work is applicable to any public and private companies working in database environments.

Table 1. Common parameters for the All Saints' Church

Property	Format	Requirement for information	Historic building
Name of the building	Text	RCCH, NHI	All Saints' Church
Origin	Text	RCCH, NHI	13th century
State	Text	RCCH	Czech Republic
Region	Text	RCCH, NHI	Usti nad Labem Region
District	Text	RCCH, NHI	Litomerice
Municipality	Text	RCCH, NHI	Litomerice
Cadastral Territory	Text	RCCH, NHI	Litomerice
Address	Text	RCCH, NHI	Dlouha; Jezuitska
Classification of Types of Constructions (CZ – CC)	Number	CSO	127211
Classification of Buildings (KSO)	Number	CSO	801 47
Catalogue Number I (Building's ID)	Number	RCCH	895484
Catalogue Number II (Building's ID)	Text	NHI	21077/5-1748
Catalogue Number III (Building's ID)	Text	CSO	57481-544-5
Historical (architectural) style	Text	NHI	Gothic; Baroque
Type of heritage care	Text	NHI	Cultural Monument
Subject to heritage care since	NUM	NHI	1958
Type of building	Text	NHI	Church
Number of overground stories	Number	RCCH	1
Number of underground stories	Number	RCCH	0
Enclosed area	Number	RCCH	563
Built-up area	Number	RCCH	125
Floor area	Number	RCCH	110

Source: Own processing.

Table 1 displays the information required by both the public authorities (NHI and CSO) and the facility manager (RCCH). For example, NHI has the need to classify historic buildings based on their purpose, namely in accordance with the national classification system labelled the Classification of Buildings (KSO) used in the conservation practice, and, on the contrary, the needs of CSO are incorporated in the Classification of Types of Constructions (CZ-CC). The information needed by RCCH was identified by the author by selecting it from the FM System's database, which is used by RCCH. It includes the properties related primarily to the location and the dimensional characteristics of buildings.

4.3 Non-graphical information – type (partial) parameters

The second group of properties is associated with the specific requirements of individual structures/technologies for specific information. Typically, this is the type of data related to material parameters, dimensional characteristics, service life, etc. Compared to the first group of *Common Parameters*, this group is significantly larger in volume. This is primarily due to the wide range of the used building materials and the different execution technologies. The interest on the part of individual public authorities (NHI and CSO) in this information also significantly differs. Or, sometimes, it is almost zero. In contrast, the facility manager (RSC) is interested in these data. RCCH can subsequently use the data actively for facility management, the planning of renovations and other activities.

Property	**Format**	**Description**	**Historical object**
Structure's Marking	Text	-	brick wall
Structure's Origin	Number	Presumed year of structure's making	-
Last Modification	Number	Presumed year of renovation/modernization	-
Element Code	Number	Classification of building structures and technologies	35864
Structure's Condition	Text	Current condition of the structure (choices – good, neglected, dilapidated)	good
Structural Function	Text	Indicates the element's structural function (choices – load-bearing, non-load-bearing)	non-load-bearing
Element's Location	Text	Indicates the element's location (choices – inside, outside)	inside
Surfacing – Inside	Text	The type of surface treatment used	lime plaster
Surfacing – Outside	Text	The type of surface treatment used	cement plaster
Thickness	Number	Numerical value of the element's thickness in mm	300
Height	Number	Numerical value of the element's height in mm	2 600

Table 2. Type parameters – a brick wall

Source: Own processing.

The specification of the relevant parameters for individual structures/technologies suitable for the passport of historic buildings (issues related to DSB) is beyond the scope of this article, and, therefore, only one example of potential parameters for a brick wall is included (Table 2). In historic buildings, it is usually difficult to fill in some of the required parameters, which are usually commonly available only for newly built or not very old buildings. These are typically the parameters like the Origin, the Last Modification of the Structure, etc.

5 Conclusion

The article presents, at the basic methodological level, how the BIM model can be data linked with various database tools using a specific historical building (All Saints' Church), which has recently been renovated, as an example. The "BIM technologies" at the point cloud level were applied to compile the project documentation during its renovation. The author of the article has designed a group of non-graphical information (Common Parameters) for the generated BIM model. In the design of specific parameters for the All Saints' Church, the author was inspired by the key authorities that already work with the data on a regular basis and require them from the owners of buildings (i.e. CSO and NHI). These parameters, in turn, further complement the requirements set by the historic building's facility manager (RCCH), who also collects selected data about the owned buildings in a database. The key output of the article is Table No. 1, which displays the proposed Common Parameters.

In general, HBIM represents another specific digitization branch of the building industry (thematically with a direct link to the development of digital innovations), which needs to be conceptually elaborated in adequate detail to achieve the required efficient processing of information needed for sustainable immovable heritage conservation. The work method proposed in the article is principally applicable to any public and private companies that work in database environments. The database-processed data and their automated transfer from one database to another represents a modern way of working advisable primarily for the organizations managing quite a few buildings.

References

1. P. Matejka, Utilization of digitized Building Data and Information Models (BIM) in value estimation of building in rural areas. *18th International Scientific Conference on Engineering for Rural Development (ERD)*, pp. 1693-1698 (2019)
2. G. Rocha, L. Mateus, J. Fernandez, V. Ferreira, A Scan-to-BIM Methodology Applied to Heritage Buildings. *Heritage*, **3**(1), 47-65 (2020)
3. R. Bouska, R. S. Heralova, Opportunities for use of advanced visualization techniques for project coordination. *6th Creative Construction Conference (CCC)*, pp. 1051-1056 (2017)
4. M. Stransky, P. Dlask, Process of matching work items between bim model and cost estimating software. *17th International Scientific Conference on Engineering for Rural Development (ERD)*, pp. 856-864 (2018)
5. CH. M. Eastman, *BIM handbook: a guide to building information modeling for owners, managers, designers, engineers and contractors*. 2nd ed. Hoboken: Wiley (2011)
6. X. Yang, Y. Lu, A. Murtiyoso, M. Koehl, P. Grussenmeyer, HBIM Modeling from the Surface Mesh and Its Extended Capability of Knowledge Representation. *ISPRS International Journal of Geo-Information,* **8**(7), 301 (2019)
7. J. E. Nieto-Julia, D. Anton, J. J. Moyano, Implementation and Management of Structural Deformations into Historic Building Information Models. *International Journal of Architectural Heritage*, **14**(9), 1384-1397 (2020)
8. D. Heesom, P. Boden, A. Hatfield, S. Rooble, K. Andrews, H. Berwari, Developing a collaborative HBIM to integrate tangible and intangible cultural heritage. *International Journal of Building Pathology and Adaptation*, **39**(1), 72-95 (2021)
9. J. Yang, X. Wang, C. Wu, C. Bai, Regularized Reconstruction of Grid System for Traditional Chinese Timber Structure Building in HBIM. *International Archives of the Photogrammetry, Remote Sensing and Spatial Information Sciences*, **42-2**(W15), 1229-1233 (2019)
10. A. Conti, L. Fiorini, R. Massaro, C. Santoni, G. Tucci, HBIM for the preservation of a historic infrastructure: the Carlo III bridge of the Carolino Aqueduct. *Applied Geomatics* (2020)

11. M. Capone, E. Lanzara, Scan – to – BIM vs 3D Ideal Model HBIM: Parametric Tools to Study Domes Geometry. *International Archives of the Photogrammetry Remote Sensing and Spatial Information Sciences*, **42-2**(W9), 219-226 (2019)
12. E. Salvador-Garcia, Integrating HBIM models in the management of the public use of heritage buildings. *Canadian Journal of Civil Engineering*, **47**(2), 228-235 (2020)
13. L. Zhang, Research on Protection and Restoration of Heritage Buildings Based on Heritage Building Information Model (HBIM) Technology and Its Application. *Advances in Social Science Education and Humanities Research*, **417**, 239-245 (2020)
14. E. Hromada, R. S. Heralova, P. Dlask, T. Krulicky, Methodology for Calculating the Reproduction Value of Buildings. *Innovative Economic Symposium 2020 – Stable Development in Unstable World*, **91**, 1034 (2021)
15. D. Macek, R. S. Heralova, E. Hromada, I. Strelcova, L. Brozova, S. Vitasek, J. Pojar, R. Bouska, Cost Optimization for Renovation and Maintenance of Cultural Heritage Objects. *IOP Conference Series: Earth and Environmental Science*, **290**, 12155 (2019)

Evaluation of impact of the Covid-19 pandemic on rental housing prices in the city of Brno between 2020 and 2021

Petr Junga[1,*], *Radka* Smolinská[2], *Tomáš* Krulický[3] and *Veronika* Machová[4]

[1] Mendel university in Brno, Department of agricultural, food and environmental engineering, Zemedelska 1, 613 00 Brno, Czech Republic
[2] Mendel university in Brno, Department of agricultural, food and environmental engineering, Zemedelska 1, 613 00 Brno, Czech Republic
[3] Institute of Technology and Business, Faculty of expertness and valuation, Okruzni 517/10, 37001 Ceske Budejovice, Czech Republic
[4] Institute of Technology and Business, Faculty of expertness and valuation, Okruzni 517/10, 37001 Ceske Budejovice, Czech Republic

Abstract. The aim of the paper is an application of the basic principles in determining rental housing prices and factors that may affect them. In the experimental part, an analysis of rental housing in the city of Brno is performed for the 2020 – 2021 period affected by the covid-19 pandemic. The analysis is processed for individual city districts and divided according to apartment layout. Finally, all outputs are compared and the real estate market development is determined with a focus on the biggest changes and their occurrence.
Keywords: rental housing, apartment, market price, the city of Brno, Covid-19

1 Introduction

At present, there are several basic options on the market for addressing housing needs. One of the most sought alternatives is owner-occupied housing, while the other common one is rental housing, where determining rent prices depend on many factors. The most important factors include apartment size, layout as well as location.

The Covid-19 pandemic emerged very suddenly, and the majority of real estate experts feared a decline in property prices similar to 2007. A study by [1] raises a question of revisions most likely to affect apartment prices. SHAP values were used to apply given models, with the values obtained subsequently coming to conclusion that time is the most dominant and consistent variable for the real estate market, specifically renting [2] claim that the rental housing market span is very small worldwide. As a result, macroeconomic damage to the market stability is reported. The authors simulate a model where macroeconomic stability is represented by reforms strengthening the rental housing market. The model´s first

* Corresponding author: krulicky@mail.vstecb.cz

step is to discuss criteria for choosing housing. The next step is a proof of private rent attenuation. Finally, a DSGE model is introduced satisfying housing needs.

Currently, most people are looking for their new homes online. However, it is possible to come across very little known information that online housing markets provide to future tenants. [3] address the issue in the United States using American Community Survey census statistics and Craigslist geolocated advertising minilones. It was found that future tenants lack essential information about the socio-economic layout of neighbourhoods. The authors focus on describing and identifying of types of information commonly available in statements in various neighbourhoods.

A document created by [4] discusses behaviour of housing companies engaged in long-term rental housing and common rental housing and the impact of competition on these companies with the use of numerical stimulation. Their results show that if companies make strategic adjustments, even though they are only minor adjustments, it is important to maintain the system stability. Companies decentralized with long-term rental of apartments, which have a higher market share, can influence rentals of general rental use.

A similar issue is addressed by [5]. Again, the paper is related to the USA focusing on millions of Craigslist rental statements. It was revealed that differences between tenants are in demographic differences such as age, language, rent or household size. The online environment more frequently offers housing markets in larger cities. Although the Craigslist contains detailed and valuable data, it does not evenly represent all market segments. Technologies such as the Craigslist create new human interactions and have the ability of planners to track.

On the contrary, [6] primarily aims at types of properties and the dynamic link related to price and rent in Singapore. The past analyses do not take into account links between property types and are static. The authors deal with segments of the housing market that may be affected in various market cycles by shocks resulting from macroeconomics. The research found that price is determined by rent and yet the properties are interconnected. Another consequence involves shocks that affect declines in rents for non-residential housing and demand from one property type to another [7] draw their attention to European cities, specifically Krakow, and the impact of the Covid-19 pandemic on rents. Results of the research conducted by the authors showed a decrease of 6-7% in the first wave, which was in contrast with assumptions of real estate agents (about 13%). In the next wave, the subsequent decline was immediately seen (by 6.25%). Owing to the study, however, the decline in rents was precisely determined and forecast. Regardless of how much economy changes next year, rent prices will continue to fall [8] examine dynamic correlations between rent and price in the four-stage real estate markets, i.e. offices, housing, factory and retail. They focus mainly on the Hong Kong area between January 1993 and December 2008, where the Dynamic Conditional Correlation Multivariate Stochastic Volatility (DCC-MSV) model and the Bayesian Markov Chain Monte Carlo (MCMC) model are applied to capture time-varying price and volatility correlations. Throughout the period under review, the results show that rental volatility is significantly lower than price volatility in real estate markets for each type. Regarding the above mentioned stages, it was found that the correlation between rent and price in the housing market differs the most in time compared to the other three markets. In addition, the correlation in the office market is more stable than in the factory market. In 1997 and 2003, they were defined as two very unstable periods.

However, [9] study how the Covid - 19 pandemic in its first months affected the housing market in 49 largest regions of the United States. Using pandemic severity measures and a unique rental list dataset, the authors found that the Covid - 19 pandemic reduced median and average rents only in Latin Quarters. White neighbourhoods, on the other hand, are experiencing an increase in rents.

Other paper compares the static and social consequences of a simple rental housing market model and homeowners in a single framework. To examine dynamics of the comparative model, [10] generates an impulse response of rental housing prices as well as apartment and house occupancy patterns. The model forecasts an external supply of housing as opposed to realism, where the author uses the version of free entry to the supply as an extended direction [11] analyses the occurrence of housing subsidies using a Finnish panel, where the main sample included 500 000 households that received housing allowances in the 2000 – 2008 period. The results show that a significant part of the subsidies goes to rental prices. Considering subsidies, however, the demand for rental housing increased, which is not in line with the objective of subsidies. Results calculated by the author are very similar to those presented in studies that deal with the same research problem. Provided that all calculations from the studies are correct, the author recommends reviewing the housing subsidy system reform [12] looks into artificial neural networks able to create a model that will forecast the ability to survive future financial crisis in a particular company. Out of three selected neural networks, the aim is to find a specific network able to classify the company based on input data.

The main prerequisite for identifying value generators for construction companies is the use of the EVA Equity indicator, as presented by [13]. Data including financial statements of construction companies in the 2006 – 2015 period are used and arranged in a table, and based on the results, adequate methodologies and identification of generators are developed. The following variables are considered to be the most important indicator: equity, bank loans, economic growth for the current period, and others [14] also focus on construction, which is an important part of the national economy. The outcome of their research involved strong impact on construction development with a relatively small group of companies. As regards the results, the authors deduce the future development of the construction industry based on a selection of 59 companies. They also emphasize that it is necessary to verify the representation of companies over time [15] deals with life cycle cost (LCC) analysis. According to the author, the commonly used approach is used in the early phase of projects, since the greatest benefit from LCC is obtained there. In the Czech Republic, the public sector initiates applications for LCC quantification when deciding on construction. The author summarizes her experience from LCC, where benefits (and significant difficulties) of the public sector are identified and summarized [16] are concerned with processing of quantity take-off (QTO) from data generated from the BIM model. A case study and exploratory methods are employed as well as an outdated process used for quotations of 2D documents. Subsequently, the authors deal with construction pricing without lengthy calculations provided from BIM calculations.

The main goal of the paper by [17] is the use of data collection methods, particularly primary data used to determine costs of energy efficiency measures. Furthermore, the authors present results of a model that is very unique, i.e. national cost calculation, and is seen as the optimal calculation method. In selected types of buildings, a method of primary energy consumption is used.

2 Development of rental housing prices on the real estate market around the year 2019

In general, it may be argued that rent prices copy (with a certain time lag) real estate prices. The assumption in 2019 was that rent prices would soon reach their maximum. This was mainly due to the fact that such prices in the city of Brno had been growing at a very fast pace, with rental housing prices rising even more dynamically in 2019 than in 2018, which could therefore be considered a period with optimistic prospects. The economy of the Czech Republic was at its peak in 2019 and can be regarded prosperous. Thus, it was more advantageous for the solvent part of the population to invest money so that inflation does not reduce its value. The largest demand was related to small flats, which is one of the main reasons why the average price per m^2 in Brno was the highest for small apartments (CZK 331 per m^2), as may been in the Table below. In terms of the Czech real estate terminology, the Table also includes specific abbreviations of apartments on the basis of their layout. The smallest apartments (hereafter referred to as "1+kk") consist of a single room with a kitchenette. Larger apartments contain either two rooms witch a kitchenette (hereafter referred to as "2+kk") or three rooms with a kitchenette (hereafter referred to as "3+kk"). There are also apartments including 4 rooms with a kitchenette (hereafter referred to as "4+kk").

Table 1. Average rent prices in Brno (October, 2019)

Apartment Layout	Average Rent [CZK/m^2]	Total Average Rent [CZK/m^2]	Average Deposit for Services
1+kk	331	10950	2200
2+kk	269	13500	2700
3+kk	221	16750	3500
4+kk (and larger)	191	19850	4000

Source: Development of real estate prices in Brno and its surroundings, 4th quarter of 2019, *remaxpro.cz* [online], Ing. Tomáš Fráňa, 01/2020 [cit. 2021-04-02] Available at: www.remaxpro.cz/vyvoj-cen-nemovitosti-v-brne-a-okoli-4-ctvrtleti-2019/.

From a global point of view, however, neither the European nor the world economy gave the impression of promising development during that period. The German economy was on the verge of recession, the United States were about to impose tariffs on European cars, and Britain was about to withdraw from the European Union. Therefore, an economic crisis onset was predicted for a number of reasons. In the Czech Republic, the Czech National Bank (CNB) sought to slow down the overly dynamic real estate market since 2016 by reducing the availability of mortgage loans and raising interest rates. The CNB strategy was functional until 2018, as from 2019 interest rates began to fall again. The downward trend did not stop until January 2020 and interest rates began to rise again. The increasing tendency of mortgage loans led to a decrease in their availability and at the same time a decrease in demand for mortgages. The important fact is that real estate prices were already very high at that time. In combination with the estimated economy development, it was therefore assumed that real estate prices in the next period would rather stagnate or grow slightly. However, the COVID-19 pandemic, which broke out in the Czech Republic in March 2020, became a key factor in the development of prices on the real estate market.

2.1 Real estate market in the city of Brno during the Covid-19 pandemic

As stated by Fráňa (2020), there are two categories of buyers in the case of buying and selling real estate. The first category comprises people who are not worried about their future (provided they have a job, business, sufficient financial background, etc.) and who have a tendency to purchase. The second category includes people who postpone purchasing of real estate, mainly due to the following reasons:

- concerns about Covid-19 (they do not want to view properties),
- uncertain future (fear of losing their job or falling income),
- fear of getting into debt or preferring to keep money in cash,

- expecting a fall in prices (speculation with the influence of pandemic prices, etc.)

Considering rental housing, Brno has a fairly stable base of residents. The number of tenants who have chosen rental housing over owning real estate is not declining, and rather an increase in rental housing preferences may be expected due to the current uncertainty. An integral part of those interested in rental housing involves people living in Brno only temporarily. Such tenants can be divided into several key groups:

The first group consists of **students** as the city of Brno has several high schools and universities. The problem with apartments rented and occupied by students can occur when high schools and universities are closed. The tenants may then tend to ask for a reduction or remission of rent. Other group of people involves tenants **living in Brno for work reasons**. In the pandemic period, the economy is weakened and the question is how long the pandemic measures will last and what the state of the economy will be after this period. So far, redundancies have been selective and only some areas of services are weakened. According to [21], however, terminations of rental housing have been minimal so far. A very specific group involves **foreign students and foreign employees**. Due to quarantine measures and border closures, this group is the only one that the Brno rental market can lose. It remains to be seen when and under what conditions they will be able to return and whether their jobs will still be available.

As reported by [19], the COVID-19 pandemic has disrupted production worldwide and is causing major GDP losses. According to available results, the year 2020 is considered to be one of the worst in economic history. According to Bloomberg, the world economy has suffered one of the worst declines since World War II, due to the pandemic and the response to it.

[20] states that the first and most affected category is related to short-term apartment rentals. A few years ago, the Airbnb phenomenon largely dominated in big cities. Many investors acquired real estate for the purpose of short-term rentals to foreigners. At the same time, however, short-term accommodation and the said platform depend on the number of visitors to the given city, its influx of visitors from abroad and tourism. In this respect, the short-term accommodation market is currently completely suspended and there is hardly any demand for this type of accommodation nowadays. Apartments for short-term rent remain empty, while the solution is to offer apartments for long-term rent. Freeing of such a large capacity of apartments will currently have a major impact on the rental market.

[21] notes that during March 2020, an accelerated change of tenants was recorded in Brno. During March to April 2020, real estate agencies registered a rapid increase in advertisements. The main influence on their rapid growth is mainly due to apartments from short-term housing platforms of the Airbnb type, which are currently without use. In Prague, the impossibility of using apartments for short-term rent poses a huge problem, with a substantial number of apartments being in the Airbnb mode. However, Brno has a significantly lower number of tourists than Prague, and for this reason there are fewer apartments and houses rented in relation to the number of apartments for long-term rent. The vast majority of Airbnb apartments are only in the historical part of Brno, or close to the city centre.

3 Material and methods

The experimental part of this paper is focused on evaluating price development on the apartment rental market in the city of Brno monitored between June 2020 and May 2021 and affected by the Covid-19 pandemic. Input data source subjected to analysing includes advertisements of apartments for rent on real estate websites, namely sreality.cz, bezrealitky.cz, as well as advertisements of individual real estate agencies. Overall, there are

4 362 advertisements used and put into different categories. The advertisements are continuously recorded and data are divided in tables taking account of individual time sequences, i.e. 06/2020, 09/2020, 12/2020, 03/2021, 05/2021. Subsequently, they are separated according to the respective city districts, and are further sub-divided on the basis of a layout into three main categories (the first category: 1+kk and 1+1 – the latter being apartments with a single room and a kitchen; the second category: 2+kk and 2+1 – the latter being apartments with two rooms and a kitchen; the third category: 3+kk and 3+1 – the latter being apartments with 3 rooms and a kitchen). The last assessed factor involves conditions of rental housing units, where apartments are classified into the above-standard category (apartments in new buildings or recently reconstructed apartments), the standard category (apartments in average conditions for regular use) and the below-standard category (apartments of older date, not reconstructed or maintained). Then, prices per m^2 and their development in the period under review in individual locations and categories are determined.

Additionally, it should be noted that the city of Brno is the second largest city in the Czech Republic and is located in the South Moravian Region in the southeast of the Czech Republic. The city consists of 29 city districts, 48 cadastral areas and, according to the Czech Statistical Office, has about 381 000 inhabitants (data from 2020) and its area is 230.18 km^2. The minimum altitude is 190 m above the sea level, while the maximum altitude is 497 m above the sea level.

To achieve a telling result, the first step in evaluating the collected data is to assess the overall situation and eliminate certain extremes (for instance, non-standard apartments with bathrooms in hallways or outside the apartments or, conversely, apartments above standard with saunas, duplex / multi-storey apartments, apartments with oversized terraces, etc.). The elimination is followed by a comparison of values for individual periods. Regarding the available statistical methods, mode, median or arithmetic mean can be used, with mode being the value that has the highest frequency. In this case, however, the data are divided into sub-categories depending on which part of the city they fall under. As some categories do not contain enough data, it is inappropriate to use mode. Median is used as the mean value of sets which are extremely different. Since extremes are excluded here, the use of median values is inappropriate as well. The following is therefore applied:

$$\bar{x} = \frac{1}{n} \cdot \sum_{i=1}^{n} x_i \qquad (1)$$

Where:
n = the number of apartments for rent / advertisements.

4 Results and discussion

Particularly included are results of the Real Estate Market Analysis in the city of Brno carried out on the basis of apartment classification according to several basic characteristics, i. e. city districts, apartment layout and apartment conditions.

4.1 Development of rental housing prices on the local real estate market

Following from the experimentally analysed data, it is possible to determine certain development of the rental housing market in Brno. The first analysis is summarized in Figure 1, where the graph shows price development of individual apartments divided according to their layout in the period under review without distinguishing between city districts or apartment locations.

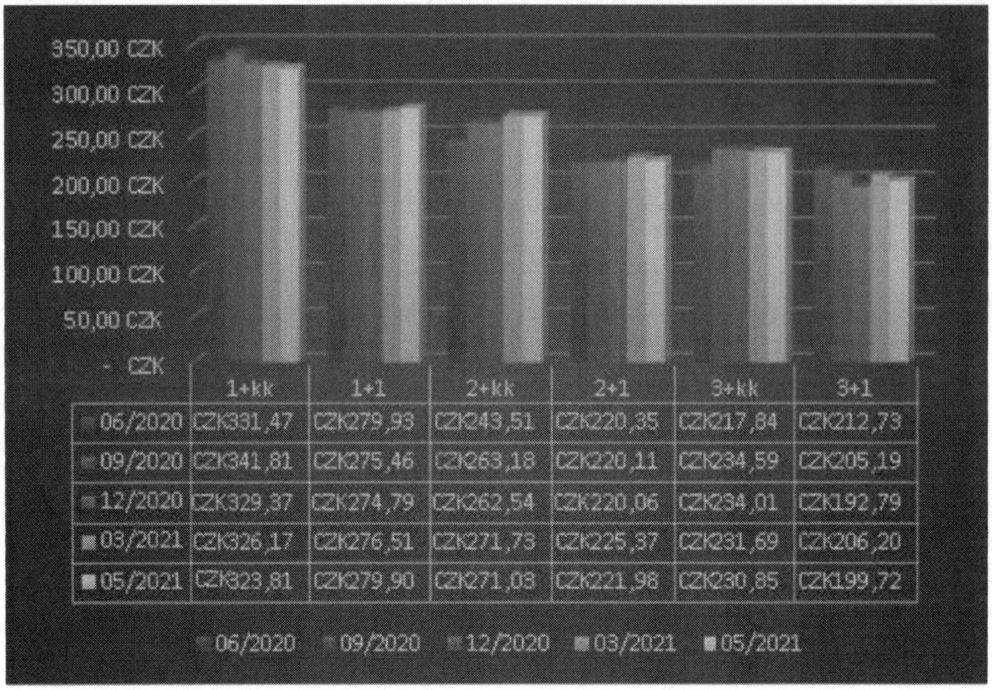

Figure 1. Summary analysis of rental housing prices in Brno according to layout. (CZK/m^2)
Source: Own processing.

It can be stated that the price of an apartment should grow in direct proportion to its increasing layout. At the same time, however, it is observable on the market (and this fact also follows from the analysis performed) that small-size apartments are currently the most sought-after category. Apartments with the 1+kk layout are the most expensive and, as the analysis proves, the average price per m^2 decreases with an increasing layout.

4.2 Development of rental housing prices on the real estate market in relation to city districts of Brno

Based on the number of properties advertised for renting, city districts are divided into three categories. The first category includes the following districts: Bosonohy, Chrlice, Ivanovice, Jehnice, Kníničky, Ořešín, Tuřany, Útěchov and Vinohrady. No apartments for rent with sufficient informative value that could be used in the analysis were recorded here as the districts are located in the suburbs of Brno. The second category consists of the following districts: Bohunice, Černovice, Jih, Jundrov, Kohoutovice, Komín, Líšeň, Maloměřice and Obřany, Medlánky, Nový Lískovec, Slatina, Starý Lískovec and Žebětín. There are more apartments for rent here than in the first category, but the number of their advertisements is not sufficient to determine a genuine informative value. The last category comprises the following city districts with a high number of advertised properties: Bystrc, Královo Pole, Řečkovice and Mokrá Hora, Sever, Střed, Žabovřesky and Židenice. They are either located in the city centre and its immediate surroundings or development projects and intensive construction works are taking place there. Thus, there is a greater interest in them resulting in a larger amount of apartments for rent. An overall comparison of prices in the last category of districts is shown in Figure 2.

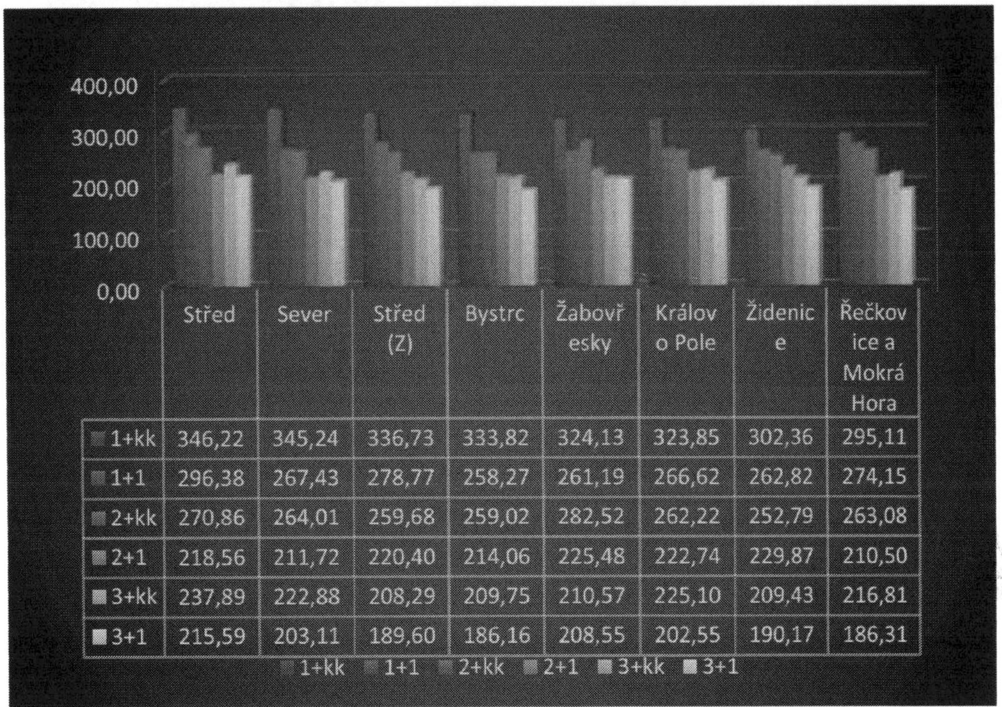

Figure 2. Summary analysis of rental housing prices in Brno based on city districts. (CZK/m^2)
Source: Own processing.

Having categorized the collected data according to the city districts, it can be determined that the most lucrative apartments can be found in the city district of Brno - Střed (without the cadastral district of Zábrdovice). Other lucrative apartments are located in the city district of Brno - Sever. They are followed by apartments in the city district of Brno - Střed (including the cadastral district of Zábrdovice) and Bystrc with their average price being in a similar price range, and also the Žabovřesky and Královo Pole districts with only slight nuances in the average price development. The last city districts whose average price per m^2 is the lowest of the evaluated locations are Židenice and Řečkovice and Mokrá Hora.

4.3 Development of rental housing prices on the real estate market in relation to apartment standards

The last assessed factor affecting prices is related to what impact apartment conditions have on prices. In order to make the analysis and price determination as accurate as possible, the last examined quarter, i.e. 05/2021, was used along with the city district of Brno - Střed, where most advertised apartments are situated. The results are presented in Figure 3.

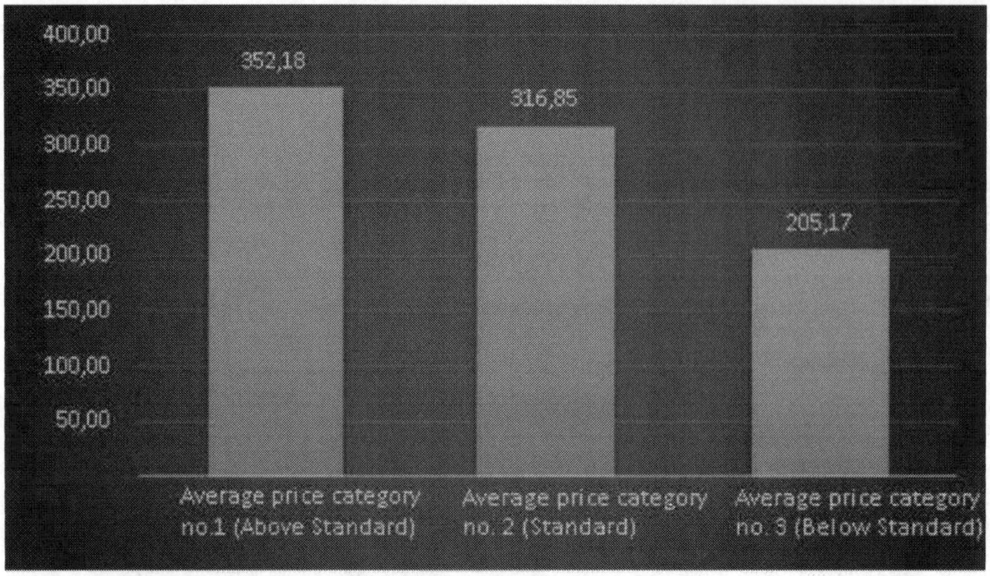

Figure 3. Development of rental housing prices depending on apartment standards and equipment
Source: Own processing. (Value in CZK)

As can be seen from the analysis results, property conditions significantly affect rent prices. It follows that reconstructed or modernized and newly constructed (above-standard) apartments are more expensive at the price of CZK per m^2 (by 11.2%) than standard apartments, whilst the prices of below-standard apartments are more than 35.2% lower than the standard ones.

5 Discussion

Considering the analysis of rental housing development in the city of Brno, it is essential to collect a sufficient amount of data in relation to the period under review to be able to achieve an adequate informative value. The number of advertisements used in the analysis amounted to 4 362, with their given groups shown in Figure 4.

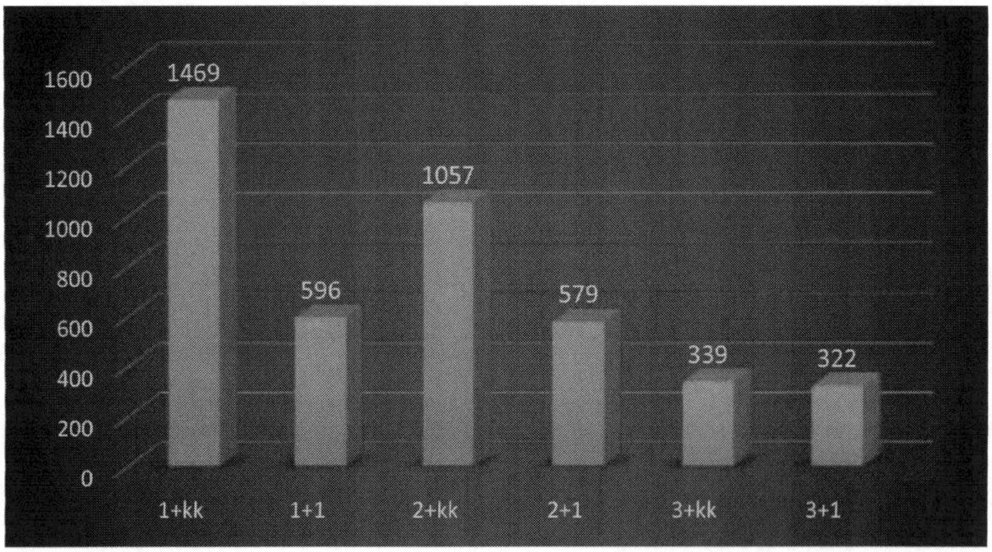

Figure 4. Number of advertisements classified on the basis of apartment layout
Source: Own processing.

Collected data from the advertisements indicate that apartments with the 1+kk layout (small apartments) are the most advertised ones for rent. This is confirmed by a 2019 survey by the Trikaya real estate agency listing purchases and rentals of small apartments as one of the most lucrative.

The lucrativeness of small apartments is also reflected in the price analysis results of rental housing in Brno. According to Trikaya, the average price of a 1+kk apartment in Brno was CZK 331 per m^2 (as of October 2019). In relation to the pandemic situation, the director of the RE/MAX Pro real estate agency, Ing. Tomáš Fráňa, informed about the rising number of apartments advertised on the real estate market in Brno in April 2020. With the pandemic being on a slight decline over the summer months, the situation stabilized and, as based on the analysis of rental housing in 06/2020, the price remained at 331 CZK per m^2 for 1+kk apartments. However, the analysis also shows that consequences of the COVID-19 pandemic become most evident for 1+kk apartments. To clearly illustrate that, below is a graph of rent price development of 1+kk apartments (Figure 5) and a trend curve of the pandemic development (Figure 6).

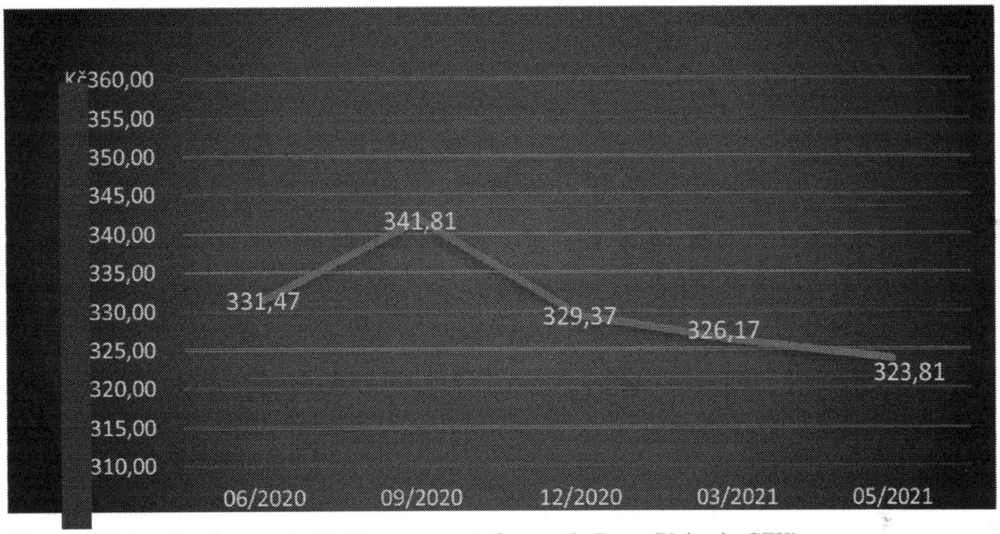

Figure 5. Price development of 1+kk apartments for rent in Brno (Value in CZK)
Source: Own processing [22].

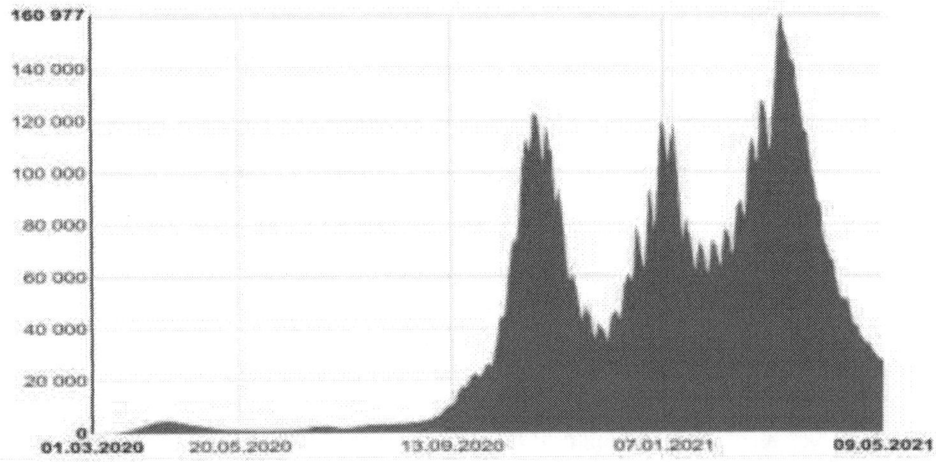

Figure 6. Trend development of the number of people with laboratory-proven COVID-19
Source: The Ministry of Health of the Czech Republic, 2021, available at: onemocneni-aktualne.mzcr.cz/covid-19

[23] claims that Brno has a lower number of tourists than Prague, hence there are fewer apartments rented in the form of Airbnb. It can therefore be assumed that the impact on the real estate market in Brno will not be so noticeable. However, the above analysis shows that at the turn of 09/2020 and 12/2020 there are significantly more advertisements, with the majority of them being unoccupied apartments and "hotel-type" apartments. Thus, the deteriorating COVID-19 situation and the increased number of advertisements can be considered as the cause of price decrease in small apartments between 12/2020 and 05/2021. That also corresponds to the assumption of the Association for the Development of the Real Estate Market stating that prices of properties for sale may fall in the short term, but the decline will not be long-term. This follows from the Trend Report 2020. The situation is different for apartments intended for rent, where the already mentioned short-term rentals pose the main issue.

6 Conclusion

The paper presents an experimental evaluation of the price development on the real estate market during the pandemic Covid 19. The evaluation works with data obtained from the real estate market of the statutory city of Brno in comparison with input data sources obtained from various internet portals. In the evaluation of the price development, we deliberated between the available statistical methods, but in the end, it is not possible to use any of the static methods, so our chosen formula was used.

The results of the experimental evaluation of the price development in the real estate market finally showed that as far as the apartments with 1+1 and 2+1 layout are concerned, the price has remained unchanged with only slight nuances during the period under study. However, in the case of flats with a layout of 2+kk and 3+kk, there was an increase in prices per sqm in the period 09/2020 (despite the fact that the increase in prices for 3+kk flats is not as significant as for 2+kk flats, where the price increased from CZK 243.51 per sqm to CZK 271.03 per sqm). However, according to the available Trikaya statistics, the price of 2+kk flats reached CZK 269 per sqm. Thus, this increase can be considered as a comparison of the price level after the decline that occurred in the first wave of the pandemic. Another factor for the increase may be the lower number of 2+kk and 3+kk apartments advertised and the fact that university students are returning in September to start the new semester. As a result, larger flats, which are often shared by students, will also be used.

References

1. A. Grybauskas, V. Pilinkiene, A. Stundziene, Predictive analytics using Big Data for the real estate market during the COVID-19 pandemic. *Journal of Big Data,* **8**(1), 105 (2021)
2. M. Rubaszek, M. Rubio, Does the rental housing market stabilize the economy? A micro and macro perspective. *Empirical Economics*, **59**(1), 233-257 (2020)
3. M. Besbris, A. Schachter, J. Kuk, The Unequal Availability of Rental Housing Information Across Neighbourhood. *View Web of Science Researcher ID and ORCID*, **58**(4), 1197-122 (2021)
4. L. L. Mu, X. Y. Qin, Y. Y. Wang, Dynamic Analysis of the Rental Prices of Long-Rental Apartments and Ordinary Rental Housing Based on the Impact of Long-Rental Apartment Enterprises' Competitive Behaviors. *Discrete Dynamics In Nature And Society* (2021)

5. G. Boeing, Online rental housing market representation and the digital reproduction of urban inequality. *View Web of Science Researcher ID and ORDCID.* **52**(2), 449-468 (2020)
6. P.I. Ji, S. Bogati Bhandari, The price-rent dynamics and linkage of urban housing evidence from Singapore. *International Journal Of Housing Markets And Analysis,* Available at: https://www.emerald.com/insight/content/doi/10.1108/IJHMA-02-2021-0023/full/html (2021)
7. M. Tomal, B. Marona, The impact of the covid-19 pandemic on the private rental housing market in poland: What do experts say and what do actual data show? *Metron,* **79**(1), 57 -91 (2021)
8. ECM. Hui, X. Zheng, The dynamic correlation and volatility of real estate price and rental: an application of MSV model. *Applied Economics*, **44**(23), 2985 - 2995 (2012)
9. J. Kuk, A. Schachter, JW. Faber, M. Besbris, The COVID-19 Pandemic and the Rental Market: Evidence From Craigslist. *American Behavioral Scientist*, **65**(12), 1623-1648 (2021)
10. M. Kashiwagi, A search-theoretic model of the rental and homeownership markets. *Journal Of Housing Economics,* **26**,33-4 (2014)
11. M. Viren, Is the housing allowance shifted to rental prices? *Empirical Economics*Volume, **44**(3), 1497–1518 (2013)
12. M. Vochozka, Formation of complex company evaluation method through neural networks based on the example of construction companies collection. *AD ALTA: Journal of Interdisciplinary Research*, **7**(2), 232-239. ISSN 2464-6733 (2017)
13. M. Vochozka, V. Machová, Enterprise value generators in the building industry. In Váchal, Jan, Vochozka, Marek, Horák, Jakub. *SHS Web of Conferences - Innovative Economic Symposium 2017: Strategic Partnership in International Trade*. 1. vyd. Les Ulis, France: EDP Sciences, 2017. s. nestránkováno. ISBN 978-2-7598-9028-6 (2017)
14. J. Vrbka, P. Šuleř, J. Horák, Analysis of companies operating in the construction industry in the Czech Republic based on Kohonen networks - identification of leaders in the field. In Ondřej Dvouletý, Martin Lukeš, Jan Mísař. *Proceedings of the 7th International Conference Innovation Management, Entrepreneurship and Sustainability (IMES 2019)*. 1. vyd. Prague: Oeconomica Publishing House, University Economics, 2019. 1017 - 1028. ISBN 978-80-245-2316-3 (2019)
15. R. S. Heralová, Life Cycle Costing as an Important Contribution to Feasibility Study in Construction Projects. *Procedia Engineering*, **196**, 565-570. ISSN 1877-7058 (2017)
16. P. Matějka, S. Vitásek, Utilization of BIM for automation of quantity takeoffs and cost estimation in transport infrastructure construction projects in the Czech Republic. *IOP Conference Series: Materials Science and Engineering*, **236**. ISSN 1757-8981 (2017)
17. J. Karásek, J. Pojar, L. Kaločai, R. S. Heralová, Cost optimum calculation of energy efficiency measures in the Czech Republic. *Energy Policy*, **123**, 155-166. ISSN 0301-4215 (2018)
18. T. Fráňa, Brněnský realitní trh v době koronaviru, [Brno real estate market at the time of the coronavirus], [online], *remaxpro.cz* [cit. 2021-04-29], Available at: www.remaxpro.cz/brnensky-realitni-trh-v-dobe-koronaviru/ (2020)
19. D. Klega, Vývoj a rizika realitního trhu v době koronavirové, [Development and risks of the real estate market in the coronavirus era] [online], *rfp.cz*, [cit. 2021-04-02], Available at: www.rfp.cz/post/v%C3%BDvoj-a-rizika-realitn%C3%ADho-trhu-v-dob%C4%9B-koronavirov%C3%A9 (2020)

20. J. Richter, Jak korona změní ceny bydlení v Česku, [How corona will change housing prices in the Czech Republic], [online], *aktuálně.cz* [cit. 2021-04-02], Available at: www.zpravy.aktualne.cz/ekonomika/ceny-realit-pod-natlakem-koronaviru/r~52d9127277e611eaa6f6ac1f6b220ee8/ (2020)
21. T. Fráňa, Vývoj cen nemovitostí v Brně a okolí 4. čtvrtletí 2019, [Development of real estate prices in Brno and its surroundings Q4 2019], [online], *remaxpro.cz,* [cit. 2021-04-02], Available at: www.remaxpro.cz/vyvoj-cen-nemovitosti-v-brne-a-okoli-4-ctvrtleti-2019/ (2020)
22. T. Fráňa, Analýza realitního trhu v ČR po krizi v roce 2020, [Analysis of the real estate market in the Czech Republic after the crisis 2020], [online], *remaxpro.cz,* [cit. 2021-04-02]. Available at: www.remaxeso.cz/clanek/analyza-realitniho-trhu-v-cr-po-krizi-v-roce-2020 (2020)

Platform Business in Korea: Advent and Growth of Kakao

Yoo-Jin Han[1,*]

[1]Department of Entrepreneurship, School of Global Service, Sookmyung Women's University, Cheongpa-ro 47-gil 100, Yongsan-gu, Seoul, Korea, 04310

Abstract. With the rapid development of ICT technology, a platform business is exerting a dominant influence in various fields. This type of a business has a great ripple effect in that it creates value by engaging and connecting various market actors such as producers, suppliers, and business partners to the platform ecosystem, rather than directly selling independent products or services. In this vein, the current study introduces the emergence and growth process of Kakao, which has a monopolistic dominance in the Korean mobile messenger market, as an example of a platform business. This company started to be distributed to Korean consumers in 2010 and is currently exercising its market influence in various fields such as commerce, entertainment, finance, and transportation, and more growth is expected in the future. Based on the case investigation, the background of Kakao's success can be interpreted with a combination of various reasons, such as Korea's developed ICT environment, a large game users' population, quick service launch, the strategy to utilize the founder's previous work experience, and service development geared towards Korean consumers.
Keywords: ICT, platform, Kakao, growth

1 Introduction

In recent years, platform business strategies have gained massive interests not only in the academic fields but in the practical aspects[1-5]. In this "Big Wave" era, Kakao Talk has become a national communication channel in Korea since its emergence in 2010[6]. Within this 11 year period, it became a representative company and expanded its business encompassing entertainment, finance, and transportation. In recent years, the number of Kakao's subsidiaries reached 118, following one of Korea's largest business groups, i.e., SK. In terms of market capitalization, it ranked 3rd, followed by Naver, which has been dominating the Korean Internet market since the late 1990s.

This company has positioned itself as a leader in Korea's mobile ecosystem by entering various areas, such as artificial intelligence, big data, internet of things, blockchain, and so on. It actively participates in technology development in the era of the 4th Industrial Revolution and maintains the leading position.

*Corresponding author: yjhan@sookmyung.ac.kr

As Kakao's business area has become deeply established in the daily lives of many people, related research to investigate its strategies and impacts has been conducted in various ways. The existing research with respect to Kakao shows the analysis of the growth process of Kakao Talk, motivation, and intention for continuous use, SNS mobile marketing services, the quality of mobile messenger interface design, and platform business and M&A strategies. In particular, it epitomizes effective platform business strategies in the era of the fastest-changing ICT envoronment.

Preliminary studies analyzed Kakao Talk, related services, platforms, M&A, and strategies from various perspectives. However, research on Kakao's overall entrepreneurial growth process remains scant. Thus, this study examines the entire process from the start-up to the present and takes three stages into consideration. In addition, the environmental characteristics of the Korean market that acted as the background of Kakao's success must be analyzed to derive implications.

2 Advent of Kakao

Kakao's chairman, Bum-soo Kim, joined Samsung SDS in 1992 after obtaining his aster's Degree in Industrial Engineering at the Seoul National University. The company was responsible for the development and operation of PC telecommunications service called " Unitel." In this process, he realized that the Internet era would change the business environment, which led him to quit a stable job in 1998. Then, he established an Internet-based computer game company called "Hangame" in early 1999 and attracted three million members in five months. The first-mover advantages, including technological leadership, preoccupation of scarce resources, and less consumer conversion costs were achieved (Lieberman and Montgomery, 1988). Soon, the number of subscribers reached 10 million owing to the Internet boom.

Fig. 1. Bum-soo Kim: Founder & Chairman of Kakao
Source: K.E. Lee, S.W. Jung, Bum-soo Kim, Kakao founder, handed out his shares to his family and relatives, *Chosun Ilbo*, Jan. 19 (2021)

The company has grown rapidly since. Meanwhile, Navercom, provided an Internet search engine with a limited number of users by spinning itself off from Samsung SDS. To overcome the limitations of each company, Chairman Bum-soo Kim and Chairman Hae-jin Lee merged Hangame and NaverCom in July 2000. Then the company changed its name to NHN in September 2001. NHN initially generated revenues from the sale of game items, and it started to gain popularity among Korean users through JiSikIn, which is a knowledge exchange service embedded in the Naver portal. Later, it became the No.1 portal company in Korea as it started to earn profits from keyword-based search advertisements.

As the major revenue source for Naver changed to these areas, the roles of Chairman Bum-soo Kim have been reduced, which compelled him to quit and start a new business. After moving to USA, he launched various social network services, such as "Buru.com" and "Wisia.com," which eventually failed. In this era, Apple successfully launched its first smartphone in 2008, which influenced him to switch his business to a 100% mobile-based one. Figure 1 shows the main log-in screen of Kakao Talk.

Fig. 2. Log-in screen of Kakao Talk
Source: www.malavia.com

Kakao initiated its business by providing a free platform, which is similar to Hangame. This strategy is highly similar to Google, which provided a search engine for free for two years after its inception to engage advertisers. Facebook also secured enough subscribers for two years through the same strategy. For the two years, 2010 and 2011, Kakao spent 15 million dollars (15 billion Korean won). However, it did not earn any revenues. To strengthen competitiveness in the domestic market and enter the global market, a domestic game company, Wemade, invested 92 million dollars (92 billion Korean won) in 2012. However, the revenue model remained unclear.

With the release of "Playing Kakao Games" in July 2012, the situation began to improve. By forming strategic alliances with seven partner companies and 10 games, including Anipang have been released. At that time, social games were considered a successful business model for overseas companies, such as Facebook and MySpace. In this game space, exiting users invited their friends on the basis of offline relationships. Anipang was a simple mobile game in which people were able to establish relationships with one another by sending invitations, which is represented through a "Heart" emoji. Through this, Kakao successfully influenced middle-aged consumer groups to be interested in using a "new" device called "smartphone." In 2012, Anipang was able to generate 10 million won (10 billion won) in sales. The revenue allocation principle of the App Store:KakaoTalk:developers = 30:21:49 resulted in a 2.1 billion revenue. To secure subscribers, it is often forced to use the Kakao platform, which brought market dominance to Kakao.

With the "free installation policy" and "relationship-based invitation", the number of users reached 45 million as of 2012, which is 90% of the entire population in Korea as shown in Figure 3. It was part of a platform envelopment strategy[7-8] that extends the scope of the usage to other areas on the basis of power[9-13], and it caused a lock-in phenomenon that cannot escape from Kakao Talk.

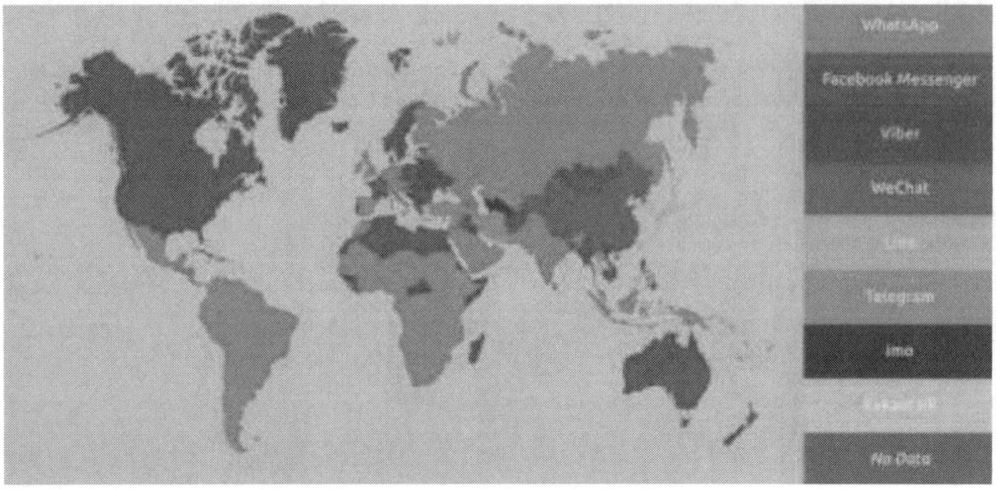

Fig. 3. Market dominance of mobile messenger platforms worldwide
Source: www.messengerpeople.com

3 Growth of Kakao

In May 2014, it was merged with a web portal, Daum Communications, and changed the company name to Daum Kakao. The merger is at a rate of 1(Kakao):1.556(Daum), given that the mobile market is expected to grow faster in the future. Then, the company began to provide global services in 15 languages in 230 countries by reinforcing itself.

Kakao's "fast-follower" strategy, which quickly pursued the Facebook's WhatsApp, was driven by Chairman Kim Bum-soo's quick decision making after the failure of Buru.com and Wisia.com. The background of Kakao's fast-follower strategy can also be found in Naver's case.

As a revenue source, first, Kakao Games was spun off, which was a big stepping stone for growth. Concurrently, with the establishment of Oh, K-Cube Ventures, which was established to invest in promising startups, was incorporated into an affiliate. Later, it was reorganized into a holding company (KCUBE Holdings) and a venture capital company (Kakao Ventures). Kakao Investment focuses on start-ups larger than Kakao Ventures investments. Similar to Kim, Kakao Investment will take over a company where businesses are based to a certain extent. It looks organized into The companies that Kakao Ventures invested in are the American augmented reality startups. Other than this effort, Kakao Pay, Kakao Commerce, and Kakao Mobility were established.

To promote the music industry, Loen and Entertainment, which used to run Melon, was incorporated into Kakao M. Recently, by acquiring numerous music-related software developers and entertainment agencies, the company is integrated. the company is expanding into an entertainment company.

Kakao started with "Seoul Bus," a bus arrival information app, after the merger with Daum Communications. It acquired (6.67 billion won), K-Cube Ventures (5.55 billion won), Cellit and a Global Positioning System (GPS) app in May 2015. It acquired Rock and All, which is known as an application called "Kim Gisa," for $62.6 million (62.6 billion won) and Path, Indonesia's top three social networking websites $26.3 million (26.3 billion won).

Kakao's M&A strategy peaked in 2016. On January 11, Kakao announced its acquisition of Loen Entertainment, which was famous for its music platform, Melon. The acquisition amounted to $1.8B (1.8743 trillion won), which was extremely exceptional. This was surprising considering that Daum Communications was only worth 100 billion won when it acquired Lycos. The acquisition of Loen, which Kakao announced, aimed to "secure growth engines for mobile content platforms and strengthen competitiveness." Many reasons exist for this merger. First, Kakao could enjoy many advantages, such as adding Kakao Pay as a payment method for Melon, selling music on Kakao Page, and providing exclusive contents of Loen Entertainment artists on Kakao TV. Second, Kakao attempted to lower the dependency on gaming by diversifying revenue sources to Loen, which constantly generated $60 million (60 billion Korean won) of sales yearly. Last, Loen utilized K-pop music to enter the global market, which was a good choice to leap to the platform. Figure 4 shows the snapshot of Kakao's diversified business areas.

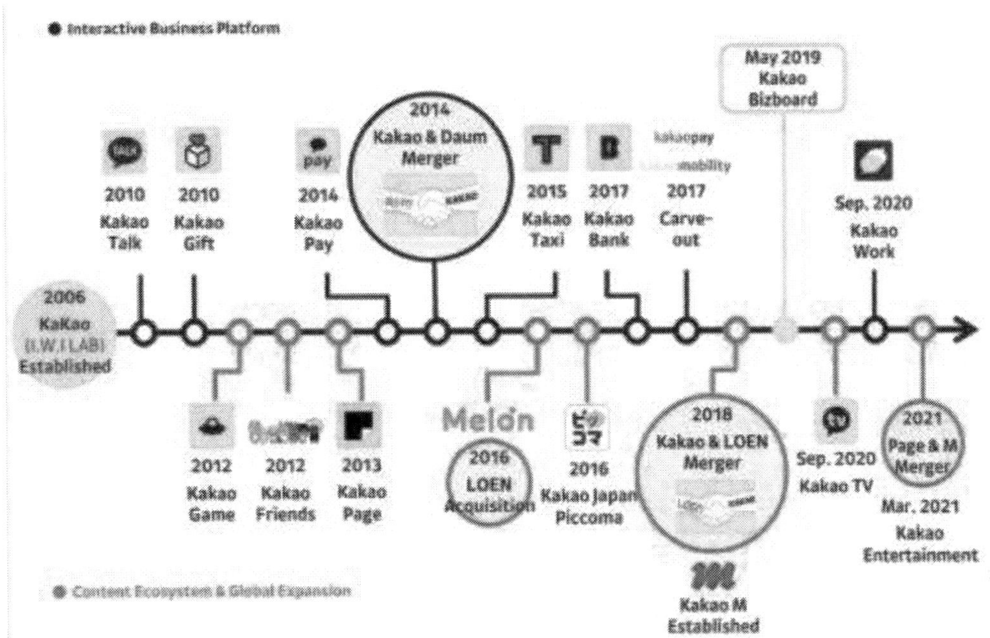

Fig. 4. Kakao's Business areas
Source: Kakao. *Kakao Investor Relations*. Feb. (2021)

Based on its brand power and popoularity, Kakao became the No. 1 company to work for, followed by many competent Korean companies.

Table 1. Best Korean companies to work for (2019/2020)

Rank	1	2	3	4	5
Company	kakao	SAMSUNG	NAVER	CJ ENM	KOREAN AIR
Rank	6	7	8	9	10
Company	HYUNDAI	AMOREPACIFIC	LG Household & Health Care	CJ CHEILJEDANG	SK innovation

Source: S.B. Lee. Kakao selected as the best companies to work for. TV Chosun, Jul. 21 (2021)

4 Future of Kakao

Kakao's stellar growth shows Korea's industry structure change from a manufacturing industry to an ICT service-based one. Few people expected that Kakao was able to be included in the top five list of the business conglomerates in terms of market capitalization. It successfully surpassed conventional global companies, such as LG, POSCO, Hyundai, and so on.

There existed several drives for this noticeable growth. First, its holding company, Kakao Corp's market value grew from \$28.9B (34 T Korean won) to \$53.6B (63 T Korean won). Second, the market value of the major subsidiaries, such as, Kakao Games, increased 228.12%, from \$1.4B (1.7T Korean won) to \$4.9B (5.8T Korean won). Third, KakaoBank was valued at \$15.8B (18.5 Korean won) on the basis of its IPO price, and the market capitalization increased to \$35.3B (41.5 T Korean won).

More affiliates, such as Kakao Pay Corp., a mobile-payment service, Kakao Mobility Corp., a transportation service, and Kakao Entertainment, a music and contents company are expected to grow as next-stage cash cow sources.

Kakao has strived to expand its service from a digital world to a physical service, which is dubbed as Online to Offline (O2O). A golf service could be a good example, which is implemented by Kakao VX, Kakao Games's subsidiary.

The next focus is the culture and entertainment, which could be propelled by one of its affiliates, Kakao Entertainment. This company was formed through the merger of a music streaming service platform, Kakao M, and a content providing company, Kakao Page. Kakao Entertainment acquired a 100% stake of Antenna Music, a pioneering indie music label in Korea, and through the negotiation process with Soo-Man Lee, who is one of trailblazers in the modernization of Korean music production industry.

Webtoons are another major pillar in Kakao's entertainment business. It recently acquired Japan's webtoon platform, Piccoma, with \$600M (708B Korean won). In Japan, Kakao is trying to possess additional intellectual property (IP) rights, which will be one of revenue sources for the entire business group. With these efforts, Kakao's global market size has been steadily increasing as shown in Figure 5.

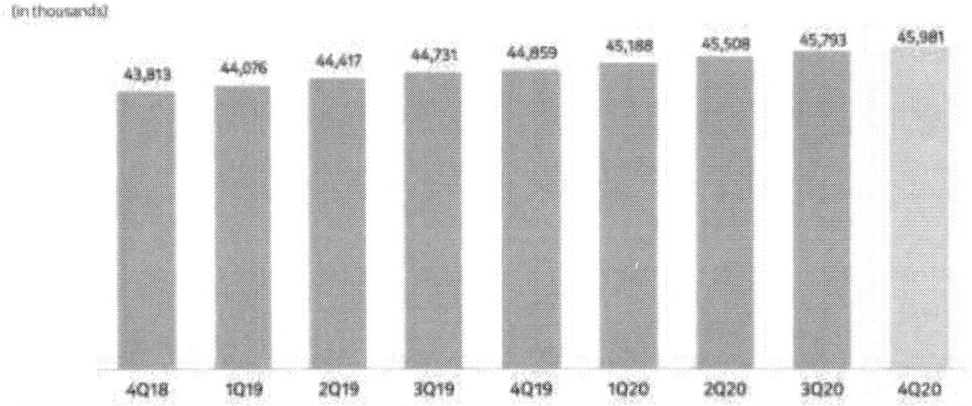

Korea	43,813	44,076	44,417	44,731	44,859	45,188	45,508	45,793	45,981
Global	50,220	50,548	50,882	51,374	51,495	51,768	52,133	52,299	52,223

Fig. 5. Kakao's Domestic/Global Market Size
Source: Kakao. *Kakao Investor Relations*. Feb. (2021)

5 Concluding Remarks

Many reasons exist for this success. First, the Korean market has a larger share of the national economy in ICT than in other industries. ICT accounts for roughly 10% of the total value added generation, ranking first among OECD countries[14]. In addition, 4.7% of all employment is in ICT, ranking first among OECD countries[15]. Such a well-established environment related to ICT played a major role in Kakao's development of mobile technology and its rapid commercialization. Given that there exist large and small ICT service providers, including smartphone developers, such as Samsung Electronics and LG Electronics, and games, such as Netmarble, NCsoft, and Nexon, they were able to quickly procure skilled ICT personnel and catch up with WhatsApp, which launched mobile messengers. In addition, when the Internet revolution broke out 20 years ago, a company called Naver survived fierce competition in the browser and portal industries. Given that most of Kakao's founders and early members are from Naver, the presence of people with a sense of and skills in existing ICT services has contributed to this. Although Google dominated the search engine market in most countries with its advent, Naver remains influential in Korea. Moreover, Koreans are the top users of mobile apps compared with major countries. With the increasing amount of mobile app users through platforms, such as Kakao Talk, the use of tools to communicate with one another has also increased.

Second, in terms of growth, "Playing Kakao Games" played a significant role, which utilized an advantageous position in Korea, the fourth largest game market in the world.

Third, many messenger developers, including Daum's My People and Naver's Line, delayed the release period and gave Kakao Talk time to promote. In addition, although mobile carriers, such as SKT or KT levied 10 to 20 Korean won per use, KakaoTalk offered the platform for free. In this period, WhatsApp maintained a pay-per-use policy until it was acquired by Facebook. In addition, although Line expanded to Japan and Southeast Asia, Kakao would have been able to compete in the Korean market. It launched Kakao Plain and Brucch services, which epitomized the "snack" culture. Its archrival Naver blocked advertising posts and provided UX, which is ideal for writing. However, it secured a quality blogger with a system that supports writers with good ideas and skills.

Fourth, the founder, Kim Bum-soo, made use of the experience he acquired in his previous workplace, i.e., Naver. After gaining a user base, Kakao provided expanded services, which is highly similar to NHN's growth strategy. Next, the company read the growth pattern of the global mobile messenger market and released it in the domestic market quickly. One "fast follower" strategy worked effectively. Subsequently, the company succeeded in attracting investments from outside and platforming them while risking deficits. On the basis of the platform, various services, such as banking, transportation, music, content, commerce, and so on are provided through active M&A [16-20].

Finally, Kakao is strictly targeting Korean consumers. The service has been launched although it has been criticized due to its classic 'copycat business." Although well-known conglomerates, such as, Samsung Electronics and Hyundai Motor, focus on overseas markets, most ICT companies in Korea are considered to have a "copy cat business." The company is bound to be underrated in terms of its performance. Nonetheless, Kakao quickly adopts new technologies, such as blockchain to the domestic market. Ultimately, Kakao is considered a leader owing to its ability to create a business model that has succeeded in gaining control of the market while ranking first in the global market.

References

1. K. Boudreau, Open platform strategies and innovation: Granting access vs devolving control. *Management Science*, **56**(10), 1849-1872 (2010)
2. K.J. Boudreau, L.B. Jappesen, Unpaid crowd complementors: the platform network

effect mirage. *Strategic Management Journal*, **36**(12), 1761-177 (2015)

3. A. Chakravarty, A. Kumar, R. Grewal, Customer orientation structure for Internet-based business-to-business platform firms. *Journal of Marketing*, **78**(5), 1-23 (2014)
4. M. de Reuver, C. Sørensen, R.C. Basole, The digital platform: A research agenda. *Journal of Information Technology*, **33**(2), 124– 135 (2018)
5. J.T. Eckhardt, M.P. Ciuchta, M. Carpenter, Open innovation, information, and entrepreneurship within platform ecosystems. *Strategic Entrepreneurship Journal*, **12**(3), 369-391(2018)
6. J. Han, O. Cho, Platform business eco-model evolution : Case study on Kakao Talk in Korea. *Journal of Open Innovation: Technology, Market, and Complexity*, **1**(6), 1-14 (2015)
7. T. Eisenmann, G. Parker, M.W. van Alstyne, Platform envelopment. *Strategic Management Journal*, **32**(12), 1270-1285 (2011)
8. J.A. Fehrer, H. Woratscheck, R.J. Brodie, A systemic logic for platform business model, *Journal of Service Management*, **29**(4), 546-568 (2018)
9. D.S. Evans, R. Schmalensee, Failure to launch: Critical mass in platform businesses, *Review of Network Economics*, **9**(4), 1-26 (2010)
10. G.G. Parker, M.W., van Alstyne, Innovation, openness, and platform control. *Managment Science*, **64**(7), 3015-3022 (2018)
11. R. Perren, R.V. Kozinets, Lateral exchange markets: How social platforms operate in a networked economy. *Journal of Marketing*, **82**(1), 20-36 (2018)
12. K. Rong, Y. Lin, Y. Shi, J. Yu. Linking business ecosystem lifecycle with platform strategy: A triple view of technology, application and organisation. *International Journal of Technology Management*, **62**(1), 75–94 (2013)
13. K. Täuscher, S.M. Laudien, Understanding platform business models: A mixed methods study of marketplaces. *European Management Journal*, **36**(3), 319– 329 (2018)
14. OECD. *Employment in the ICT Sector and Sub-sectors, 2015* (2017a)
15. OECD. *Value added of the ICT Sector and Sub-sectors, 2015* (2017b)
16. A. Hatiu, E.J. Altman, Finding the platform in your product: Four strategies that can reveal hidden value, *Harvard Business Review*, **95**(4), 94-100 (2017)
17. J. Oh, B. Koh, S. Raghunathan, Value appropriation between the platform provider and app developers in mobile platform mediated networks, *Journal of Information Technology*, **30**(3), 245–259 (2015)
18. Y. Zhao, S., Delft, A. Morgan-Thomas, T. Buck, The evolution of platform business models: Exploring competitive battles in the world of platforms. *Long Range Planning*, **53**(4), 1-24 (2020)
19. F. Zhu, M. Iansiti, Entry into platform-based markets. *Stratetgic Management Journal*, **33**(1), 88–106 (2012)
20. F. Zhu, M. Iansiti, Why some platforms thrive and others don't. *Havard Business Review*, **97**(1), 118-125 (2019)

AUTHOR INDEX